Mangifera indica

Mangifera indica

A BIOGRAPHY OF THE *Mango*

SOPAN JOSHI

ALEPH

ALEPH BOOK COMPANY
An independent publishing firm
promoted by *Rupa Publications India*

First published in India in 2024
by Aleph Book Company
7/16 Ansari Road, Daryaganj
New Delhi 110 002

Copyright © Sopan Joshi 2024
Illustrations copyright © Somesh Kumar 2024

The author has asserted his moral rights.

All rights reserved.

The views and opinions expressed in this book are those of the author and the facts are as reported by him, which have been verified to the extent possible, and the publisher is not in any way liable for the same.

The publisher has used its best endeavours to ensure that URLs for external websites referred to in this book are correct and active at the time of going to press. However, the publisher has no responsibility for the websites and can make no guarantee that a site will remain live or that the content is or will remain appropriate.

No part of this publication may be reproduced, transmitted, or stored in a retrieval system, in any form or by any means, without permission in writing from Aleph Book Company.

ISBN: 978-93-95853-77-4

3 5 7 9 10 8 6 4 2

Printed in India

This book is sold subject to the condition that it shall not, by way of trade or otherwise, be lent, resold, hired out, or otherwise circulated without the publisher's prior consent in any form of binding or cover other than that in which it is published.

To Vishnu Chinchalkar

Artist Wayfarer Maestro

Contents

Prologue: Desire and Risk ix

Book One
The Fruit of India

1. *Fruit:* India's Favourite Bribe 3
2. *Variety:* Better the Mango You Know 24
3. *Culture:* Escaping Our Refugee Camps 39
4. *Season:* A Festival in Search of Revellers 55
5. *Religion:* Bones of a Civilization 73
6. *History:* The Great Leveller 88
7. *Origin:* A Unit of Diversity 112

Book Two
The Fruit of Wilderness

8. *Angiosperms:* An Abominable Mystery 129
9. *Primates:* And Then Happened a Miracle 146
10. *Genetics:* Randomized Field Trials 164
11. *Ecology:* Biocasino 185
12. *Cultivation:* Who Grows Your Mango? 204
13. *Markets:* Caution to the Wind 225
14. *Exports:* Source: 'Exposed' 244

Book Three
The Fruit of the Senses

15. *South:* Tradition and Individual Talent 263
16. *West:* What Money Cannot Buy 287

17.	*North:* Memory and Loss	312
18.	*East:* Decadent and Digestive	336

Epilogue: Eat, Discuss, Repeat — 360
Acknowledgements — 364
Notes — 367
Index — 389

Prologue
Desire and Risk

*'Take a chance. Play your hand. If you win, go home a winner.
If you lose, you still have a story to tell.'*

—SUBHASH JOSHI

Throughout the year in Delhi, we children waited for the summer vacations and the Dussehra break. In the custom of migrant families pining for their 'native place', we went to the city of Indore. There, we met our beloved cousins. At night, we climbed on to the kuchha tiled roof of our ancestral house. We could chat there till late at night, escaping the disciplining eye of adults. Our bedding smelled of insect repellent because it was impossible to set up mosquito nets.

The nights were always pleasant in Indore, which lies on the Malwa Plateau, 50 kilometres south of the Tropic of Cancer. We gazed at the stars, free of all care, discussing pressing matters of the day. One night during the Dussehra holidays, the conversation drifted to which season was the best. Arguments began raining in. I weighed in on behalf of winter. The dispute became heated and loud, as such sensitive matters sometimes do. At any moment an adult was likely to wake up and reprimand us for the rumpus. 'But the mango comes only in the summer,' said a cousin. That was that—the matter settled. Everybody went quiet, retreating to positions antebellum.

When we returned to Delhi after the summer vacations, I envied those who talked of climbing mango trees in the orchards of their ancestral villages. Both sides of my family were from cities. During summer breaks, I cycled through crowded lanes with my cousins or ran around barefoot, chasing drifting kites, or played cricket and assorted games on dusty grounds. There was no ancestral village with orchards and fruit trees to climb. But there were mangoes. And in Indore there was Subhash Chacha, my oldest uncle. His temper was as fierce as his affection was intense. He indulged us kids in a number of ways, particularly with his culinary skills. He was an extraordinary cook, even by the standards of that food-obsessed city. His friends were as interesting as him—artists, photographers, poets.

Sometimes, I got on to his vintage Priya scooter and accompanied him to the mandi, the market for fresh fruits and vegetables. I tugged at his kurta or his callused hands while he negotiated the market. He had a discerning eye for fresh produce. He knew the regular vendors by name and enquired after their families while assessing their wares. I actually saw a vendor refuse to sell her produce to him; it wasn't good enough for him, she said flatly. One time, I felt a rush of delight after he bought an entire basket of small sweet and sour desi mangoes sold in the Itvariya mandi by farmers from neighbouring villages. I emptied the basket into our bag, touching each fruit.

On the way back, I stood in the pit of the scooter, holding the handle on either side of the headlight, where the worn grease from the gear-shifter smeared my hands. His right hand worked the accelerator, a cigarette tucked between forefinger and middle finger—tapering and crooked at the joints between the phalanges. I had the wind on my face, a sackful of joy by my knees.

Back home, the bag was emptied into buckets of water. We'd sneak a mango or two out of impatience. After a while, all the cousins sat around the buckets and drew mangoes one by one. Subhash Chacha sat the youngest child on his lap—or the one most in need of attention! He selected a sweet mango. His powerful sculptor's fingers massaged it with calibrated strength, just like they worked the clay and wood that he sculpted into art. Then, he knocked off the black tip and passed the fruit. The juice gushed into our mouths. The world felt joyful, secure.

One summer, when we reached home, Subhash Chacha hugged me and took me to the treat he had prepared for our welcome: a large amount of aam ras—mango pulp prepared with milk and sugar. In our nuclear family in Delhi, we were used to smaller quantities of fruit—and also to some English words. That pool of mango pulp must've excited me; I do not remember when I exclaimed in delight about so much 'mango juice'. In a clan steeped in desi language and culture, those two words in English stuck out, instantly becoming a joke on the silly cousin who could not recognize his aam ras.

Thereafter, each time we met, Subhash Chacha invoked the mango juice anecdote, accompanied with a vivid re-enactment of my delight. It embarrassed me a little but I knew that he did not wish to let go of my innocent excitement. When we met in the summer, he went to great pains to get good mangoes, preparing the aam ras with sugar, milk, and ginger. Then, he sat us down and fed us with ferocious insistence till we felt sick. And he regaled the gathering with the mango juice episode, leaving me abashed.

If there were any mangoes around, he kept the best bits for me. He watched my face intently for a sign of excitement while I had them; but I'd grown up and learned to mask my feelings. He did this right until his last days, when his large frame had shrivelled from senescence and a lifetime of excess, chasing his numerous passions, demanding elaborate meals the doctor had forbidden, and consuming alcohol in volumes that confounded medical science. (I did not know then that our desire and capacity for alcohol is deeply linked to our taste for fruit!) It had become difficult for him to take drags from his cigarette; yet nobody could get him to quit smoking.

The last time I saw him in Indore in 2017, he could not get out of bed on his own. When he recognized me, his creased face lit up; that old spark returned to his eyes momentarily. He drew up his shaking arm, the skin sagging around the bones and atrophied muscles. He curled his hand like he was holding a cricket ball. Only two words came out of his mouth: 'mango juice!'

The mango is a unit of measuring relationships in India. Just a mention of it triggers nostalgia. People go into raptures, retelling stories of emotional connections. An interviewer once asked the Bikaner-born Pakistani singer Reshma about her child's birthday. She said he was born in the season when mangoes and musk melons arrive in the market. It indicated rootedness and the fact that fame had not changed her. It also revealed how the fibres of our memories are tangled up in this fruit. One way to remember something is to link it with the mango. It is a memory disk, the involuntary Aadhaar Card of India's socio-emotional life! Everything's linked!

Many deities old and new are linked to the mango. None more so than Kamadeva, the ancient god of desire and love, of sex and procreation. He is a trouble maker in ancient stories. Sporting a bow, he shoots arrows of love. His targets drop their better judgement, their fears and insecurities. He distracts them from the straight and narrow, from the safety of the familiar. He drives them beyond the edge into the unknown, towards temptation, towards adventure. Kamadeva's fruit is the vector of desire!

The mango is a luxury, not a necessity. India has numerous other fruits, sweets, decadent foods. But no other comestible fuels such an elaborate culture, business, and hysteria. Our passion for mangoes has turned an indulgence into a requirement. But the craving is actually atavistic. Watch a person eat a good mango, even in a refined and cultured setting. The mango has a way of drawing out the child inside. The politesse slips away like the skin off the flesh! There is no dainty way to truly enjoy a mango; it's been called the bathroom fruit. If you look clean after eating one, most likely you did *not* enjoy it.

It is a result of the risk we take for desire. Who takes it? Everybody! You could be a consumer, buying fruits in the market, weighing the odds. Is it ripe enough or overripe? Is the price right? Is this the best time and place to buy? Is this the best variety? Will it bring smiles to the family or to those who will receive this gift? You do not have the information to make a thoughtful choice. Yet you will set aside the calculations and give in; you will buy the mangoes. No other fruit is bought in quite the same way in India.

Or you could be a trader. Mangoes arrive in the market via a vast, invisible trade network. Whether you are a retailer or a wholesaler, each deal is founded on speculation and little else. What price will secure the deal while keeping it profitable? Again, there's no benchmark, no data to accurately judge the price that will maximize profit or minimize loss. Each decision is a gamble. There are no guarantees.

Or you could be a mango producer, the most risk-prone part of the mango business. To grow mangoes commercially, you have to tackle a wide range of variables. Capricious weather, pests and diseases, labour shortages...so much is up to chance! The markets for agricultural commodities are volatile. Poor production results in a waste of labour, time, and capital. A bumper harvest is even worse; the glut drives down prices. There's so much risk at every stage of the mango sector, it is surprising that growers grow the fruit, traders trade in it, consumers consume it!

It really does go all the way. All the way up to the mango tree, which also takes numerous risks to produce the fruit—with the soil, weather, pests, and plant diseases. What does the tree desire? To spread its seed, contained inside the fruit. Plants are rooted to the ground. The mango tree cannot disperse its seeds without animals carrying them to new grounds; so it attracts them with fruit. Yes, the tree takes all those risks to give you pleasure, seducing you to do its bidding. It is driven by a compulsion going back millions of years!

Two words appear repeatedly in conversations about the mango: risk and shauq. Over and over, from India's north to south, in one form or another! In orchards, markets, research labs, and drawing rooms. Shauq means several things in Arabic: desire, craving, longing, fancy, pleasure, curiosity, nostalgia and, above all, passion. If you are kam-shauq, the Urdu dictionary says you are indifferent.

Biology abhors indifference. The story of life is propelled by seduction, by taking a chance, by the kind of give-and-take that makes it difficult to tell a winner from a loser. Did we humans domesticate the mango? Or, as the author Michael Pollan might ask, did the mango domesticate us? Your answer is determined by who you are. If trees had their own eighteen-minute TED Talks, they might hold up the mango as a case study. Title: 'How to domesticate a primate and compel it to propagate your seed'!

Visiting one's old work can be stimulating—and humbling. The first time I wrote on the mango was in April 2002. The editor of the monthly magazine where I worked asked me for a piece on the coming mango season. In the time-honoured manner of hard-working investigative reporters, I made two phone calls to the wholesale fruit market, sampled the internet, and filed a one-page article titled 'Pleasures of the Flesh' that very afternoon. I recently dug up the old issue and read it. It has embarrassing puns on the pleasures/pains of the flesh, not to mention one or two factual errors. Whoever called journalism the first rough draft of history—or was it literature in a hurry?—did not foresee the fallout.

Many years later, I asked that editor about this compulsion to write on mangoes each year. He said the fruit is such a big deal because it is associated with memories of seasons. 'It is a marker of time and place. People get nostalgic and passionate about their favourite mangoes,' he said. I have not been approached by editors to write on any subject as much as the mango. A former colleague, now a newspaper reporter, called me one time for some phone numbers of mango growers and said: 'People always read a story on mangoes.'

Even if editors have not asked for it, they will say yes to any half-decent idea on the mango. If you bring in something fresh, you are guaranteed column inches or airtime. A 'human interest story' is a much abused term in journalism. If you make a pyramid of what comprises 'human interest' in India, the mango is right on top.

Spare a thought for *Sulabh Swachh Bharat*, a publication of Sulabh International, an NGO that makes public toilets. Its May 2018 cover story was 'the special story of the mango'.[1] An all-time favourite is a 2015 report from Jaipur about three boys in their early teens.[2] They were talking about mangoes when suddenly one of them proposed a trip to Allahabad to visit a relative's mango farm. They left right away by train without telling anybody. The families panicked and informed the police. When the boys returned, the families made it a point to feed them mangoes to their heart's content. Another chestnut is that mangoes elevate a marijuana high. I've asked herbal connoisseurs; they swear by it. The chemical details are readily available on the internet, although their reliability is questionable.

Mango trees feature frequently in reports of crime and suicides. Orchards are quiet, deserted places not far from habitation. It's possible to covertly dump a body there, or to kill somebody unobserved or to hang the body from a tree to make a statement. Hanging oneself is the most common mode of suicide across the world.[3] In India, mango trees are commonly found near settlements. Each year, there are reports of assaults, even shootings, with orchard owners confronting mango thieves. Some result in murder; search the internet for 'murder mango orchard' and set aside an afternoon. Even if the mango orchard is irrelevant to the crime, the headlines emphasize the fruit. Why? Because just mentioning the mango draws eyeballs. In 2010, when archer Deepika Kumar won two gold medals in the Commonwealth Games, the headlines of her profiles stressed that she grew up perfecting her aim by throwing stones at mangoes. Winning gold medals for the country is not catchy enough. For traction, you need the mango! An activist near Nashik, Maharashtra, claimed in 2018 that 150 childless couples had had a child after eating mangoes from his trees. He went a step ahead, claiming that couples who wished to have only male babies should consume his mangoes. After a case was registered against him, reports said he was bailed out for ₹15,000.[4]

Then there is the sub-genre of the mango feature, a must for all media outlets in the summer. I have a check-list of what comprises such an article: the delicacy and taste of one (or a few) prize varieties; the history of those varieties, preferably with stories of kings and empires and geopolitics; nostalgia; India's 1,000-odd mango varieties; the name of an exclusive variety nobody has heard of; poetic descriptions of the mango through history; recipes of pickles and chutneys; the threats to mango quality; reactions from the markets; mango connoisseurs; and a feeling of something special, something hat ke, something...worth a conversation. The formula works over and over, season after season, writer after writer. This book results from just such a piece that I wrote a few years ago. (Journalism tip: Writing on mangoes gets you noticed in India, even if you write on dull environmental topics nobody wishes to read!)

We say and write so much about the mango because so much is heard and read about it, because so much is craved and eaten. No fruit is awaited with such anticipation. The most dramatic outpouring of this is the annual mango festival. Almost all major cities hold such festivals; often, they are elaborate fairs. Growers from across the neighbouring region bring their best produce. For citizens, it is a thing to do; families turn up in their Sunday best. The venue is decorated with festive colours and bright lights. Mango eating contests are common, as are stalls of fancy clothing and food, toys and games, and charity promotions.

Each city showcases itself with razzmatazz; dance and music programmes, fashion shows, other performances. It is Diwali and Eid and Christmas rolled into one. Amidst pageantry, a range of mango varieties are displayed; there are competitions for size, shape, sweetness, and suchlike. Victorious growers proudly display photographs of the prizes they or their forebears had won for their outstanding fruit. They are framed on the walls next to the trophies in their living rooms.

Why do growers and buyers flock excitedly to such festivals? I've asked many of them. Everybody finds the question stupid. Some typical replies: 'Obviously! It is the mango!' 'What else is worth a festival!' 'It is beautiful, just look at it,' 'Why don't you taste and figure it out yourself'. Prakash Bhatambrekar, a Marathi writer in Mumbai, knows the city's culture. I asked him why people go bonkers over the mango. 'Why do you say hello when you pick up the phone? Why is vegetable oil called Dalda? Why is toothpaste called Colgate? Why is water wet?' he asked. 'The mango is like that. It is so obvious you cannot question it.'

Mango festivals do not need promotions; journalists and photographers appear readily because it is the most convenient way to cover the mango. Journalism is an urban occupation; its economics is tilted towards cities. Journalists live in cities and generate content consumed in the cities. But the mango comes from the rolling countryside. Farmers are not media-savvy; they are doers, not talkers. When they talk, they have their own codes, their own lingo. Agriculture scientists are no good with sound bites either. Mango agronomy is so complex that even growers and scientists are not very clear about it. Horticulture does not lend itself into saleable stories for urban readers and viewers.

This is one reason the media does not cover agriculture, even though it sustains more people than any other sector. It is the definition of what journalists call a 'slow beat'; everything requires a lot of non-dramatic waiting. A crop grows over months; finances are long drawn; all action is tedious. A lot depends on the weather, which is as variable as it is complex. Writing about such matters requires a scientific framework and training; journalists tend to come from the humanities. Even when reporters look beyond the mango festivals, the breathless pace of journalism stymies them.

'It is really irritating to deal with reporters during mango season,' a mango grower told me. 'They want nothing more than a few quotes to go with a prefabricated storyline. They do not want to get to know the farmer at all. Are we not human? Are we not worth knowing?' Many growers have ranted at me in a similar vein. In fact, they do not know how to react if a reporter shows interest in them, in their work, in their life. Scientists, too!

Editors, though, need mango stories each season. They go to the features desk, where the smart writers with degrees in literature and history can produce attractive text without getting any reporting dirt on their shoes. Their literary bent allows them to draw from poets like Kalidasa and Mirza Ghalib. Their familiarity with history brings forth grand stories of nawabs and kings. Talented food writers can whip up a hyper-detailed write-up on the just-so taste of an exotic variety in a trice. This means each season produces several readable feature articles on the mango, on festivals and the choicest varieties. There is little reportage on the actual fruit or those who grow it or those who research it. Many Indians consider themselves mango experts; almost all are mistaken and poorly informed—from consumers to traders to growers.

Scholars and scientists also get things wrong. More often than not, historical details and backstories of mango varieties are twisted, embellished with apocrypha for special effect. Such material pops up in research papers and gets published in academic journals! This explains why a mango-obsessed country actually knows so little about its obsession! Not many know that trees of almost all prize varieties produce fruit only once in two years. That only two dozen varieties out of India's 1,000-odd have any commercial value. That you cannot rely on most names of mango varieties because they change from region to region, sometimes within a region.

Even so, each season does produce a handful of nuanced stories, rigorously reported by enterprising journalists, who get up close with varied sources, including growers and scientists. Because it is about the mango, the sub-editors do not cut out the tedious details that otherwise would not pass muster. No crop gets similar editorial leeway! A subject so tasty can carry even the tiring accounts of farming and agricultural research that are anathema to the media's short attention span.

Researching, reading, and writing about it are just more ways to enjoy the mango. Sometimes a greater pleasure than sinking your teeth into a fruit! Yet I have met a few who do not care about mangoes. Perhaps they have allergies. Perhaps they did not have their Subhash Chacha. I owe to him my interest in flowers and fruits and vegetable markets and plants in general. A keen gardener, he often summoned me to visit nurseries and get him bulbs or cuttings or seeds. His poems teem with the metaphors of flowers and seasons. He died in December, when Malwa's mango trees have no flowers or fruits. But knowing India's diversity, I'm sure trees along the Malabar coast or the Andaman and Nicobar Islands would have been in flower. Some all-season varieties must have been in fruit.

I imagine how Subhash Chacha, ever the sporting man, might have reacted to someone who did not like mangoes. He would say not eating or talking about mangoes carries a bigger risk: the risk of missing out! Do not miss out, he often exhorted us: 'Take a chance. Play your hand. If you win, go home a winner. If you lose, you still have a story to tell.'

Storytelling is what we do, it is how we get by. Evolutionary biologist Stephen Jay Gould once said humans are primates who tell stories. Primates love fruits, above all else. A story about a fruit? Have at it!

BOOK ONE

The Fruit of India

Chapter 1

Fruit
India's Favourite Bribe

'Nobody says no to a case of mangoes.'
—MANISH SHUKLA

'I feel like Santa Claus during the mango season.'
—YOGITA MEHRA

It was the third time in her month-long hospitalization. Meenakshi Devi, sixty-four, had been admitted in late April 2014 with a severe lung infection, complicated by diabetes; her condition was deteriorating. On 31 May, she asked her son Mandar for mango pulp. He asked her to wait till the evening, trying to put it off. But just like the previous two occasions, he could not say no. He fed her the pulp that she so loved. Ten minutes later, she died. On her breath was the fragrance of the choicest Alphonso mangoes, Hapus in Marathi, from their family orchards.

Eleven days later, the family was conducting the death rituals. Rice cakes were placed out in the open to attract crows, in whose bodies ancestors are believed to return. The shraadh ceremony is incomplete without crows accepting the offering. Mandar watched as the crows stayed away. Then, his father reminded him her love for mangoes. Mandar fetched mango pulp and put spoonfuls on the rice cakes. The crows obliged instantly. The family could let go.

Not that Mandar needed his mother's shraadh to learn the

supernatural powers of the mango! His great-grandfather had planted Alphonso trees in Ratnagiri. Four generations on, the family name is perhaps the most valued mango brand in India and the world. Desai Bandhu Ambewale have extensive orchards in Ratnagiri, Maharashtra, and a thriving retail and wholesale business in Pune, where Mandar is based. The fruit has achieved the impossible for the family several times. For instance, in 2013, the automaker Mahindra was advertising an upcoming sport utility model. Mandar's little son Atharva became fixated on it, and kept asking his father to buy it. Mandar went to a dealership and paid the booking advance. He was told it would take months to deliver the vehicle; due to a tremendous demand, there was a backlog.

The boy kept hounding his father. Mandar did not want to disappoint his son. He got some influential people, like politicians and the local MLA, to call the dealership. Still, no luck. Each day, the kid saw adverts on TV and turned questioning eyes towards his father. Then, Mandar had an idea. It was the first week of March and the first crop from the family's mango orchards had started coming in. 'I had two boxes of our top quality fruit delivered to the manager of the dealership, with my visiting card,' he told me. Within an hour, the manager called. He was beside himself with joy, repeating over and over that his family was going to be thrilled to get such good mangoes so early in the season. He asked if he could do anything for the Desai family, anything at all! Mandar mentioned the automobile. 'On the third day, I got a call from the dealership. They said my vehicle was ready for delivery. And I did not have to pay one extra rupee,' he chuckled, sitting in his small office on the ground floor of a multi-storey mango showroom on Karve Road in Pune.

Mangoes open doors. The fruit has brought fame, literally and figuratively. The family has letters from prime ministers. 'Indira Gandhi used to call people coming from Pune to Delhi to bring her our mangoes. Atal Bihari Vajpayee's friends knew the best way to catch his attention was to meet him with a box of our best produce. If they were coming from Pune and went empty-handed to see him, he used to tick them off,' said Mandar. Bhimsen Joshi, a colossus of Hindustani classical music, was their client for forty years. He needed their mangoes each day of the season. Till a year before his death in 2011, he called Mandar's father before each mango season. 'He became a family friend, our promoter and brand ambassador even.'

Mandar was a big fan of Laxmikant Berde, a superstar of Marathi cinema and a character actor in Hindi films. He was desperate to meet his idol. 'He was a star, I a mere fan. But I had mangoes,' he said. 'He met me just because I went with a crate of our best Hapus. He soon became a friend. When I bought this extensive mango showroom in 2003, he insisted on inaugurating it himself, which meant free publicity.' Berde came and 'stayed for two full days' when Mandar got married. 'Once, he called from the gym because his trainer was asking him to keep away from mangoes to watch his weight. In his last days (he died in 2014), he begged the doctor to let him have a mango daily.'

A gift of mangoes carries much more value than the dry fruit boxes that have become the Diwali norm to grease relationships, Mandar said. In Mumbai, India's financial capital, mangoes are an essential part of business. Each year from April to May the city goes mango mad. The streets are overwhelmed with vendors selling Alphonso mangoes. From small-time traders to the largest corporations, from petty contractors to CEOs, a gift of expensive mangoes is the best way to impress people. 'I spend a minimum of ₹1 lakh each year on gifting mangoes, especially to the government officials who handle my business,' said a contractor from Mumbai who sells large machinery across Maharashtra. 'It ensures deals happen smoothly, paperwork moves faster, complicated financial matters get simplified. The mango is an elemental part of business investments here.'

At Mumbai's Crawford Market, I met Gurmeet Singh, an exporter-importer of automobile parts. As he dug into a mango to check for taste, the burly middle-aged trader turned into a child. Right there, he paid ₹5,835, including ₹2,300 just for delivering the five gift boxes at addresses that he provided, including one in Delhi. 'One does not count money with Hapus,' he said, walking away, sated. The shopkeeper showed me orders received over the phone: 'A rich man will transfer the money into my account and email me a list of more than 100 people who are to be delivered a box of my choicest Alphonso mangoes. Some people order 500 boxes or more. This city does not want money. It does want quality mangoes, though.'

What's so special about a gift of mangoes? I've asked several mango traders and businessmen in Mumbai and Pune. The answer was the same: The best way to please a person is to please his family. 'Nothing delights a family in this part of the world like a large number of quality mangoes, especially if you get the fruit before it has hit the market,' said one of the largest mango wholesalers in Mumbai. 'If a family lays its hands on the early produce, they invite guests for dinner, just to flaunt their exclusive access to premier fruits. It marks your status.' I've seen Alphonso mangoes kept in ornate baskets in Mumbai drawing rooms, displayed for guests; other fruits belong in the kitchen or the dining area.

Good mangoes are not just to be eaten; they are adornments like jewellery. The rules of the market push the best produce towards those willing to pay the highest price. The biggest patrons of Alphonso mangoes are the city's rich Gujaratis. 'Variety and refinement of food results from prosperity after basic needs are satisfied. The Gujaratis in Mumbai are generally into business, many of them quite rich,' said Rushina Munshaw Ghildiyal, a food and beverage consultant born to a well-to-do Gujarati family in south Mumbai.

The mango is not a physiological need, she said; it features higher up in Maslow's pyramid, between esteem and self-actualization.[1]

We met at her food studio in the city's western suburbs. She had prepared an evocative meal: it included kairi-nu shak, a preparation using the fruit. She described mango recipes handed down over generations with child-like excitement. And then related stories of wealthy families habituated to storing mango pulp in large refrigerators. Their children do not eat their meals without it, she said. I've heard others mention Mumbai households that maintain special deep freezers with airtight containers of mango pulp that, sometimes, last the whole year.

Mumbai's rich and powerful are hooked on the Hapus. A journalist had told me that the fugitive criminal and terrorist Dawood Ibrahim craves the Alphonso every summer; special mango boxes are flown to him via intermediaries. Born in Ratnagiri's Khed town, he made it big in Mumbai's high-rolling world of smuggling in the 1980s. He fled the city much before investigations revealed he had masterminded the 1993 bombing that terrorized Mumbai and killed about 300, and injured hundreds.[2] Police claim he is holed up in Karachi, Pakistan. I asked a crime reporter in Mumbai to confirm Dawood Ibrahim's obsession with mangoes with his underworld sources. They told him 'Bhai' has a big craving for the Hapusand gets his boxes delivered via Dubai.

In Pune, law-abiding, freezer-eschewing seekers of the Alphonso book their produce in advance. In the basement of the Desai Bandhu building is a godown where I saw people queued up to the street, clutching pieces of paper. They had bought coupons of ₹4,000 each to book their mangoes at advance rates. 'It is a hedge against price fluctuations,' said Vinayak Hirlekar, who had taken early retirement from a career in the IT sector. He did not seem like somebody bothered by prices; yet this was his sixth year of booking his produce in advance. 'It ensures quality,' he said. He remembered fighting over mangoes with his cousin when he was a child. 'One loses interest after growing up, you see.' But he still wants to get the best mangoes for his wife and son. He spent six years in Chicago, USA, and never bought a mango there: 'The mangoes there are tasteless.'

On the ground floor, in his small office tucked away beyond the display area, Mandar Desai was fielding phone calls from all over. With gold rings on three of his fingers, he pointed towards a framed photo of a Ganesh idol. Desai Bandhu has been decorating the famous temple of Dagadusheth Halwai Ganapati in Pune with 11,000 mangoes each year; the fruits are gifted to orphanages and hospitals after the puja. His son Atharva walked into the office and sat on a chair next to me. After the call ended, Mandar introduced us, complaining about not getting to spend enough time with his son. The opening of the mango season in March, he said, coincides with annual exams in schools. The results come out in April–May. Parents with modest means often walk into the showroom with their children, looking to reward them for doing well in the exams. Several times, they enquire about the rate of Alphonso mangoes and turn around, telling the child it is too costly. 'We never let them leave empty-handed. Our staff accepts whatever they can afford to pay,' said Mandar. Atharva sat still, listening.

The phone rang again. It was a journalist calling for a quote on the upcoming mango crop, for estimates of produce and prices. He answered the questions in Marathi. As soon as he disconnected the call, the phone rang again. He chuckled: 'This is how it is during the season.' The calls that stand out, however, come before or after the season from those enquiring if there are some good mangoes or even quality pulp. Usually, it is for a dying relative whom they wish to feed some quality mango as a last meal. 'Each year, I field more than 100 such calls. We try to do our best.'

You do not miss the mango in off-season if you are Mukesh Ambani, chairman of Reliance Industries Ltd. (RIL), India's largest public corporation. Reliance has set up an orchard that produces mangoes almost through the year. That, too, in a hot and arid desert with saline soils in Saurashtra, Gujarat. Ambani's father, the late Dhirubhai Ambani, was immensely fond of mangoes. So Reliance hired a range of horticultural scientists to turn the saline wasteland into what is perhaps India's—some say Asia's—largest fruit orchard.

The mangoes that the corporation gifts to people in Mumbai are grown here. They are often a variation of Saurashtra's famous Kesar variety; this one is developed by the company's horticulturists, so it's called Reliance Kesar. It's a large and attractive fruit. At the orchard, horticulturists have tinkered around with two varieties, Bajrang and Niranjan, which flower when the Kesar is producing fruit. They begin to bear fruit after the mango season has ended— and other varieties begin flowering then. The mango season in the household of the Reliance boss probably lasts throughout the year.

The culture wars over the mango are a very new thing. For most of our history, Indians consumed mangoes grown only in their own region, unaware of varieties grown elsewhere; the exceptions were kings and rich merchants and travellers. Elaborate mango markets have developed only over the past half-century or so. This period has witnessed a rapid expansion of infrastructure in trade, transport, and travel. Mango varieties are traded across faraway regions today; like people, they get to travel a lot more. Markets of most urban centres sell at least some non-local varieties. High-quality fruit from a long distance away is too expensive. And the best produce of one region seldom goes outside. Consumers in Uttar Pradesh have not seen or tasted the best fruit from Andhra Pradesh; in fact, it is very likely they have not tasted the best mangoes from neighbouring Bihar! Consumers in Mumbai will not get to taste the best mangoes from West Bengal or even Goa, no matter how high a price they are willing to pay.

Your understanding and definition of the mango is shaped by what you have experienced. 'You grow up with a taste and that does not change with age. It marks your past, your identity,' said Faiyyumbhai Karimbhai, a third-generation mango trader in the

Talala mandi near Junagadh in Gujarat. 'We've eaten Kesar mangoes from childhood. No mango compares for us. Even if I travel the world to taste the finest things, the memory of that taste will remain untouched!' It's how I feel about the seed-grown juicy mangoes of Malwa that my uncle used to buy.

People are quick to reject unfamiliar varieties in order to celebrate their own mangoes. The Alphonso fan club gets carried away, routinely declaring the Hapus as the best and the greatest variety in the world, and as the undisputed king of the king of fruits. No other variety has so many claims of peerless excellence made on its behalf. 'Nobody who can afford the Hapus eats anything else during its season here,' said Anand Desai, Mandar's brother who manages operations in Pawas village of Ratnagiri. In Mumbai, a kindly old man from a rich Gujarati family once asked me, his face overcome with pity: 'What do you eat in the north? You do not even get to see the Alphonso there!'

A backlash is inevitable. It gets tangled up in the familiar Delhi-Mumbai rivalry, with the Alphonso of Maharashtra, and the Dashehri/Langda of Uttar Pradesh sent into the ring. A fitting respondent to Hapus chauvinism is Sohail Hashmi, a documentary film maker-cum-social activist-cum-history buff-cum-bon vivant. He is quite knowledgeable about north Indian mangoes. I met him in the Delhi Press Club one evening. 'You never have to apologize for a Dashehri mango. If it's ripe, it will be good. But I've never tasted an Alphonso for which the host does not have to apologize,' he said. 'They will inevitably say this one is not as good as the best the Hapus can be.' He said the Alphonso tastes like aam papad or leathery mango bars sold in sweet and salty variations.

Hashmi organizes walks through Delhi's historical sites and food trails and is much in demand during the mango season. His range extends to an annual mango eating picnic near Delhi. Hashmi, as the raconteur, is as big an attraction as the setting and the mangoes. He recounted an altercation between two famous and friendly personalities many years ago.

The first was Mumbai-born screenwriter Saeed Akhtar Mirza, creator of the popular 1980s tele-series *Nukkad*. The second was Rajan Prasad, a socio-political activist with SAHMAT, the Delhi-based collective formed after Hashmi's dramatist-activist brother Safdar was murdered in 1989. Mirza had served Rajan Alphonso mangoes in Mumbai. The meeting soon descended into the familiar Hapus *vs* Dashehri script, becoming a sub-text to the Delhi *vs* Mumbai wrangle. Rajan claimed it was not even a contest, that Mumbai cannot be called a city because it lacks the two basic qualifications for a serious claim to urban culture: a poet like Mirza Ghalib and a mango variety like the Dashehri.

Just as the honour of families is popularly linked to the bodies of women—consider common swear words—the honour of cities/regions rests in their signature mango varieties. To disrespect a city or a region, all you have to do is insult its mango. A Delhi-born friend of mine, now living in Mumbai, never misses an opportunity to run down Mumbai's obsession with the Hapus. For the longest time, I did the same and wrote about it, too.

I first heard of the Hapus in the mid-eighties. My father had gone to what was Bombay at that time; his colleagues dragged him to the markets to buy presents. He returned to Delhi bearing two gifts: a box of Alphonso mangoes and colourful shirts of the Charagh Din brand. We had heard that both the items were expensive and much sought after. Accustomed to cut-price khadi and other cheap fabric in varying shades of dour, we kids were impressed. The canary yellow of my shirt was brighter than the mangoes. It soon became an embarrassment because it drew ridicule from the other boys. It remained hidden in my closet till my mother gave it away in exchange for metal utensils.

We got an Alphonso mango or two each. I have no memory of its taste, just of its link with the Charagh Din shirts—expensive signs of a city that we as kids knew only from the names on our desi variant of the board game Monopoly called Vyapaar; I loved the sound of Crawford Market and bought it the first chance I got. Alphonso mangoes came to mean the same: hype and extravagance, to be enjoyed with one eye on the price tag. What's the point of a box of good Hapus if its cost does not compete with the monthly earnings of a poor family! Later, I lived in Mumbai for a bit and was even more impressed by the real Crawford Market. I did get to taste some nice Alphonso mangoes. But they reminded me of packaged mango drinks like 'Maaza' and 'Frooti'; they all use a small amount of Alphonso pulp to give it that taste. Conditioning is a very big part of taste.

Varieties of the east and the north like the Langda, Dashehri, Chausa, and Himsagar are often very sweet. The more famous varieties of the south are more flavourful, with a subtle sweet-sour taste. The Alphonso's marquee taste owes to a particular blend of sugar and acid. The most widespread southern variety is probably Andhra Pradesh's Banganapalle; it is called Benishan/Baneshan in Telangana, Safeda in North India, Badam in Maharashtra, and Badami in Karnataka. This, too, has a characteristic sugar-acid blend.

'Flavour in mango results from heat,' said Bharat R. Salvi, one of India's best known mango scientists based in Maharashtra. Watery conditions decrease both the sugar and acid content of the fruit. Tanveer Hussain of Bengaluru, India's largest mango pulp exporter, described another difference: 'The taste and flavour depend on the soil. In coastal areas, the skin of the mango tends to be much thinner; it gets thicker as you move inland. Closer the sea, sweeter the mango!' However, no one rule can explain everything. The signature taste of most varieties owes to the soil and weather of a particular region. When grown outside that region, the same variety tastes and looks slightly different.

Southern varieties with a sugar-acid blend travel better. The excess sugar in northern mangoes makes them ripen quickly, reducing their shelf life. As a consequence, North Indians end up eating more southern mangoes than vice versa. There are other reasons, too. Horticultural operations in southern India are better organized than in the north, where rough handling cuts down their visual appeal. The biggest reason, though, is the seasonal variation: the summer begins much earlier in southern India and the monsoon arrives earlier. The mango flowers and fruits much earlier there, about two months ahead of north India. Mangoes from the south begin to flood north Indian markets in April-May; but the markets in the South do not see as much of the late produce from the North in July-August.

This influences consumption patterns. 'In Mumbai, the Jains stop eating the mango with the onset of the monsoon out of a religious fear of ingesting insects,' said Bhatambrekar. 'Right after the first shower, you cannot find one customer for mangoes in this city,' said a trader in the Crawford Market. When the monsoon hits Mumbai in early June, the mango market ebbs away.

Mid-June is when the major varieties of North India begin to hit the markets. In the mango belts here, people say the mango is to be enjoyed only after the rains set in. 'We've heard from our childhood that only after the rains does the mango acquire shree (grace, richness, and plenitude),' said Shyam Sunder Jaiswal, a food connoisseur of Varanasi. These differences get expressed in the greatest exhibition of the fruit, the mango festival. Each region holds its festival at the peak of its season, when the best fruit from other regions is not available. South Indians have not consumed the best mangoes from the north, and vice versa. I've eaten many a humbling mango over the past few years, regretting my ignorance.

A few years ago, a friend from Goa invited me over during mango season. The small state has a mango culture all its own, without any major commercial plantations. Many people own a small number of trees and the fruits are consumed locally. I met Vivek Menezes, a writer and photographer who has evangelized

Goan mangoes in the English press over the past few years. He has argued that Goa has the best mangoes in the world; one article was headlined 'My mango is better than yours'. Speaking to me at his office in Panaji, he was more forthright. He said he had tasted very good mangoes from other parts of India. His stand on Goa's mangoes was in the spirit of gamesmanship, a habit he had picked up from playing basketball. He has brought attention to several remarkable varieties of Goa.

Vikram Doctor, a journalist and food writer in Mumbai, has produced several fine articles and podcasts about the mango. In 2014, he wrote one with the headline: 'Mumbai's Mad Love is Killing the Alphonso Mango'. When I met him in Mumbai, he told me his favourite variety is the Imam Pasand of Andhra Pradesh and Tamil Nadu. The disproportionate interest in the mango irritates him: 'Each turn of season you have to produce stories on the mango. Yet in April, we get great sapota (chiku) from Dahanu in Konkan. They are the size of cricket balls.' But it is not considered worth a conversation; nobody has arguments on the microblogging platform formerly called Twitter about which sapota is the best. You cannot insult a person or a city or a region by insulting their bananas or oranges or grapes.

For all its entertainment value, the North-South mango debate is the noise of two players arguing loudly when the match is taking place elsewhere. Bihar and West Bengal are even more taken with their mangoes. As Mumbai and Delhi have gained importance, becoming centres of political and financial might, Kolkata and Patna have lost out. Their elites do not have a comparable all-India voice and presence. But they have numerous fine mango varieties, even if the rest of India is clueless about them.

The only rival in fame to the Langda of Varanasi is the Langda of Digha, a Patna suburb. Here, they do not call it the Langda but the

Doodhiya/Dhulia Malda. (Lesson in variety names: in West Bengal's Malda, the same variety is called Laingda; in eastern Uttar Pradesh, it is called Kapooriya, a name also used for other varieties in other regions.) Like so many fine varieties, the Doodhiya Malda does not go to the market; connoisseurs go to the orchard and buy it right there at any price the grower quotes: I've heard ₹1,300 per kilogram. east of Patna is Bhagalpur, where you get the flavourful Zardalu, a slender and delicate variety. If you keep travelling along the Ganga's south bank, you go past several centres of horticulture. There's Pakur, a Jharkhand town with numerous old orchards; the mangoes here include fine mangoes developed from courtly patronage as well as heirloom varieties of forest-dwelling tribes. Northwest of here is Malda, a major centre of mango cultivation. Much of the rich mango-growing area beyond this point is now in Bangladesh.

About 60 kilometres downstream of the Bhagirathi/Hooghly, a distributary of the Ganga, is Murshidabad. It is perhaps the greatest centre in the history of fine mangoes. It's worth a quick detour into history here. In the early eighteenth century, the richest province of the Mughal Empire was Bengal; it included the states of Bihar and Odisha, along with what is now Bangladesh. In 1717, the capital of the province moved from Dhaka to Murshidabad. The city soon became the seat of the opulent nawabs of Bengal. Jagat Seth is a prominent title of the time, meaning 'banker to the whole world'! The Mughal court gave this title to Seth Manikchand, an Oswal Jain merchant who used his banking network to deliver the Bengal nawab's tribute to Delhi; thereafter, the title became hereditary.

The Jagat Seth family bank, the world's wealthiest at its peak, became central to Bengal. At one point, the family ran the state treasury and had its own mint, controlling the financial markets and the thriving trade. Gradually, other Jain clans from Rajasthan joined them. Within a few decades, Murshidabad became the centre of all things wealthy and refined, from banking to architecture, from silken weaves to craft mangoes. However, the 1757 Battle of

Plassey made the nawab irrelevant; the patronage of Murshidabad's fine mangoes fell to the wealthy Rajasthani merchant families. The power centre shifted to the upcoming British city of Calcutta.

It was in Kolkata that I met a suitable person to evaluate Murshidabad's mangoes. Lata Bajoria was born in Mumbai in 1950 into a Rajasthani family. She was very keen on the outdoors; as a young girl, she climbed trees in the family's mango orchard in Thane. They had the best of mangoes available in Mumbai. Then, she was married into a prominent business family in Kolkata; among other things, it owned India's largest jute business that she now runs. During the summer, relatives in Murshidabad sent the Bajorias baskets of mangoes. 'Each basket had a label with the variety's name on it. I remember Bimli, Champa, Rani Pasand, and so many others. Each of those mangoes had a fascinating taste. They were like magic to me, I used to transform into a child, excited to see those mangoes, to eat them,' she told me in the remarkable garden she has set up around her house on Garden Reach Road in Kolkata's Khiddirpore area. I asked her to compare the mangoes she'd had in Mumbai with those in Kolkata. She said the Bengal mangoes were far superior in taste, variety, fragrance, and appearance to the ones she had had in Mumbai.

Over time, almost all the Rajasthani merchant families of Murshidabad moved to Kolkata. But their luxurious mansions and ornately-decorated houses still stand in the twin towns of Azimganj and Jiaganj on either side of the Hooghly, north of Murshidabad. Their relatives in Rajasthan called them 'sheher wale' because Murshidabad was the big city of the time, the centre of refinement. Subsequent migrants from Rajasthan to Kolkata also addressed the older families of Murshidabad as 'Sheherwali'.

Some families retained their mansions and orchards in Murshidabad; even though they live in Kolkata now, they've joined hands to revive the city's cultural heritage, forming the Murshidabad Heritage Development Society. One of its activities is an annual mango festival in Kolkata to showcase Murshidabad's fine mangoes and their attendant cultural refinement, like the ritualized cutting of mangoes by the Sheherwali women. Several members of the community say they cannot eat mangoes outside their house because they are not cut properly.

One of the prized varieties here is the Kohitoor. It is so delicate, legend says it was kept in cotton wool and turned every few hours so as to not put too much pressure on any one side and also to ripen it evenly. It was preserved in honey, its flesh cut with bamboo blades because metal was considered too coarse! Legends apart, I've met growers who still sell one Kohitoor mango for up to ₹800! Such mangoes are, however, limited to Murshidabad and the Sheherwali Jains. Ordinary folk have not heard of or tasted them, which is why these mangoes are not a part of Bengal's mango culture.

Bengal, though, is as passionate about the mango as any other region. 'The mango tree is like a family member to us, like a friend or a neighbour,' said Swati Bhattacharjee, senior assistant editor at the daily *Anandabazar Patrika* in Kolkata. 'We count our years and seasons by the cycle of the mango tree. We do not say "how many springs", we say, "how many summers of the mango" here,' she said. If a Bengali family has land, it is a given that they will have at least one mango tree; children develop their sense of adventure and nature by climbing mango trees to pluck the fruits: 'Girls are especially fond of raw mangoes with salt and chilli powder.' Her neighbours have a dog named Mango because when he was a puppy 'he was so sweet!'

She mentioned some of the numerous ways in which the mango dominates conversations. 'Bengalis love to discuss if this is a good year for yields or not. The price and quality of mangoes in the market is discussed obsessively. They describe in great detail the price they got, comparing it with what others paid,' she said. The fruit of choice in most of West Bengal is the Himsagar, a very sweet variety with firm flesh, little fibre, and a hint of turpentine. Then, there are the festivals! The sixth day of the month of Jyeshtha is called Jamai Shashthi, when Bengali families gift mangoes, jackfruit, and other fruits to the son-in-law. 'A lot of mango gifting happens on that day, especially among those who own orchards,' she said. During Saraswati Puja, the panchamrut mixture is slurped from a stone bowl, accompanied by the chewing of mango flowers.

'It is the Michael Jackson of fruits, the mango,' said her husband, Biprodas, who is a retired school teacher. 'Do you remember the effect MJ had on his fans when he appeared on stage with the blitz of sound and light? That's how Bengalis react to the mango! People identify themselves by their favourite mangoes, and it becomes a part of their personality.' Their thirty-three-year-old son Bornil lives in Texas, USA. 'He says the only thing he misses there are Indian mangoes,' said Biprodas. After Biprodas's father had died, the family went to Gaya to conduct the shraadh ceremony. As part of this, the spouse takes a vow to stop eating one beloved food item; his mother made the toughest choice, vowing to not eat mangoes. I've met other people who have vowed at religious ceremonies to stop eating mangoes for a particular duration, depending on the requirement; it is the greatest sacrifice, the most rigorous test of self-control!

Then there was the eminent Hindustani classical vocalist Samaresh Chawdhury. 'He could eat up to twenty-five kilograms of Himsagar mangoes at one go. He'd say he cannot stop himself, that he was going to live only once and he wanted to enjoy his mangoes,' said Biprodas, his friend of thirty years, from the time when they were young teachers in the same staff room. 'He was admitted to the hospital because of poorly controlled diabetes. He died in 2019 because he could not control his love for food, especially mangoes.' There's also his friend Binayak Deb, a reputed cardiologist who left India in 1986 to practise in Europe, then Singapore. 'When he returned to India for good in 2005, I asked him why he had come back. He said he missed Indian mangoes,' said Biprodas.

I took Dr Deb's number and called him. He verified that he had said that; it was symbolic of his desire to return to his roots. The mango stands for all that he loves about his land and his people. 'I missed India's seasonal fruits. The fruits one gets there look great but are poor in taste,' he said. He was a mischievous child who often climbed the neighbours' trees to steal mangoes; when he got caught, he got a beating from his father. 'Even my patients know of my love for the fruit and often gift me mangoes. One patient returned yesterday from Mumbai and brought me a box of the Alphonso. During mango season, I have at least four mangoes a day. My favourites are Himsagar and Langda,' he said.

Some of the greatest untold mango stories are in India's east. Not in Bengal, however, but in north Bihar. Not the western part of north Bihar, the Champaran region, which also has rich orchards and several characteristic varieties, the flagship being a small, round one called the Zarda. No, this story belongs to the city of Darbhanga, the erstwhile capital of the state of Tirhut, later called Raj Darbhanga; it stretched all the way to Nepal in the north. In the summer of 2022, farmers from a village here called Sarisab Pahi got together to informally organize a mango festival. They exhibited 139 varieties! Taking a cue from them, a nearby village called Deep held its own festival; they had 155 varieties. In Bihar, it is common to find grafted varieties of even sucking-type juicy mangoes, not just the table varieties that yield firm pulp that's cut with a knife. Perhaps the only region that can stand up to north Bihar on high-quality grafted varieties of sucking-type mangoes is coastal Andhra Pradesh. North Bihar's mango wealth, its remarkable history, or its ubiquitous presence in the culture and life of the region has not been studied in depth in recent years.

Outside of Bihar and West Bengal, Darbhanga and Murshidabad hardly feature in the accounts of the mango. That says less about the tapering influence of these historical cities and more about how the mango is discussed in modern India. Yes, Indians love to talk about their mangoes with passion. No, they do not know their mangoes as well as they'd like to believe.

Malihabad offered a few hints on why we go crazy over the mango. The village is about 30 kilometres northwest of Lucknow, Uttar Pradesh, surrounded by dense mango orchards. The region has many fine mangoes but its lore is built around one of North India's signature varieties, the Dashehri. The name comes from the village of its origin in Kakori. All stories of mangoes here are replete with anecdotes of nawabs and their passion for fine fruit. But the theme can be seen all over the country. 'Mango bole to royal mein aata hai (The mango is in the royal category),' said a trader in Mumbai, raising his hand and bowing his head in a courtly gesture, contrapuntal to his street slang.

It's not merely fame, money, and power; it's not just the taste and look of the fruit; the mango has a unique emotional power in India. It makes people feel special. Naturally, those who deal with it, even for non-commercial reasons, become special. 'People are so grateful for what we do,' said Yogita Mehra, who runs a training programme on kitchen gardening in Porvorim, Goa. Her operations extend to coordinating the distribution and sale of some of Goa's finest mangoes, grown on the island of Chorao. She described a call she had received at 11.30 the night before I met her. It was a seventy-year-old man, overcome with emotion. 'He had just come home to find the mangoes we had delivered. He could not wait till the morning to thank us, because he had not tasted such fine mangoes since his grandmother died.' Such moments occur often. 'I feel like Santa Claus during the mango season,' said Mehra.

'I could not understand Goa's craze for the Mankurad mango variety when I got posted here in 1996,' said Adavi Rao Desai, senior scientist in charge of horticulture and fruit science at the Central Coastal Agricultural Research Institute in Old Goa. 'The earliest fruit is available by end-January. Some buy it for ₹500 per fruit and then boast about it.' It's not about making money, either. Orchard owners in Chorao can earn a lot more if they cut down mango trees and plant cashews instead. Yet they choose to suffer the financial costs to keep the old mango trees that make them stand out!

In Maharashtra's Vidarbha region, orange cultivation has led to widespread destruction of traditional mango groves called amrais. 'If we plant oranges on the land we have under mangoes, the profits will be in lakhs of rupees,' said Karuna Futane of Rawala village in Amravati district of Vidarbha. 'We keep the mango trees because not all value is monetary. The mango is the basis of our relationships with all kinds of people.' Then, she pointed at me: 'Look at yourself. You came here for our mangoes.' I saw a man from a city come there to pick up a mango consignment; he patiently waited in his car for more than an hour for the fruits to be harvested. In rural India, access to hard-to-reach government officers means a lot. 'All kinds of important officials come to our door for our mangoes,' she said.

'When an official cannot be bribed with money, he is gifted mangoes. Nobody says no to a case of mangoes,' I heard the poet Manish Shukla say at a Lucknow mango festival. Many people from across the country have told me the same. Even the most honest and circumspect government officer will not say no to a box of mangoes. It is India's favourite bribe, so much so that it is not considered a bribe, merely a social courtesy.

Mehra said what adds to the mango's importance is all the eager waiting over months. The Indian year can actually be divided into parcels of mango time. Most Indian calendars start the new year in spring, a season signified by mango flowering. The early summer is punctuated with food items made of raw mangoes, especially aam panna, a drink made of raw mangoes; the name likely comes from the Sanskrit prapanak, meaning a beverage made of fruit pulp. The variant I crave is garnished with cumin, salt, and asafoetida.

In summer, I frequently visit aunts who make it well. They always tell me stories of how it helps ward off heatstroke. If there is no book on the numerous panna recipes, here's an idea waiting for a food writer.

The form in which the fruit marks itself throughout the year in the Indian diet is the mango pickle—a topic very well covered in recipe books. There are households that store mango pickle by their year, like wine. Much more than an item of refinement, the pickle has always been the condiment that salvages the blandest of meals. Several industries rely on mango pickles, preserves, and pulp.

The mango has a major impact on public health. Vijay Singh, a prominent mango grower in Mall village near Malihabad, puts on 10-15 kilograms each mango season. Doctors hand out special warnings to diabetic patients during mango season. For traders, the season can be very stressful. 'By the time the mango season ends I need to run away from it all. I'm drained, restless, and tense,' said Tanveer Hussain. 'I sweat it out on the treadmill. I just run and run and run for a week to decompress.'

Not everybody has such athletic remedies to recover from the season. I know several mango lovers who get depressed at the end of the season. It's like the end of the Dussehra–Diwali season or an Indian variation of the seasonal affective disorder (SAD), which afflicts the northern reaches of Europe and North America as winter approaches. At the Talala mandi in Saurashtra, I was interviewing a trader when a man came in to speak to him. He overheard our conversation and became quite animated. 'At the end of the mango season, my twenty-one-year-old daughter goes into such a depression that she needs medical help,' he said. I could not tell if he was happy or sad to tell me this. There exist many tricks to make the mango last a little longer. Sohail Hashmi recalled that his mother made aam balaai or maana ka salva, a simple preparation of salt, chillies, and mint with mango remains. (Maana is the biblical 'manna' that dropped from heaven.)

In Pune, Mandar Desai told me about a particular kind of phone call they receive ahead of the mango season: parents who have a baby ready to wean. In several parts of India, the ceremony is called 'anna prashan', in which near and dear ones gather for a meal and celebration as the baby is fed its first non-milk food. There are peals of joy as the family watches the baby's reaction to concentrated sugar. The food for this occasion in many parts is honey or rice kheer. But the parents who call Mandar make their baby wait for the mango season to taste its first worldly food.

In some other parts, too, the mango replaces honey that is much sweeter but can still be cheaper than prize mangoes. The poet Ghalib called the mango a sealed glass of honey. One Sanskrit name of the mango tree is madhu awas, the abode of honey.

Chapter 2

Variety
Better the Mango You Know

'They ought to be cut with a very sharp knife, that the slice may not be injured, and I want to taste them first, for allowance must be made on account of the season.'

—GARCIA DA ORTA

A legendary officer in the civil services, M. S. Randhawa was the founding vice-president of the Indian Council for Agricultural Research. He wrote a comprehensive history of India's agriculture in four volumes. 'Basically, mangoes can be divided into two types: the sucking and the table types,' he wrote.[1] 'The sucking types have thin juice and have more of fructose. Hence even if the juice of a large number of sucking mangoes is consumed, it causes no discomfort…. Commercial table types have thick pulp and are cut with a knife, and the flesh is eaten with a spoon, leaving the skin. These varieties have more of sucrose, and not more than two can be eaten without discomfort.'

This difference is quite old. 'It is eaten in two ways: one is to squeeze it to a pulp, make a hole in it, and suck out the juice—the other, to peel and eat it like the kardi peach,' wrote Babur, the sixteenth-century founder of the Mughal dynasty. This difference between the juicy and the pulpy mangoes goes back to ancient times, although the details are not easily deciphered. Among the terms used for the mango in ancient texts, two have drawn much interest: rasala/rasalu and sahakara. Some researchers hold that the rasala is the sucking type of juicy mango and the sahakara is the table type with firm pulp. There is a lack of reliable evidence, so we cannot be sure.

Just as there are two ways to eat the mango, there are two ways to grow it. One is to plant the seed, beej in Hindi and tukhm in Farsi. A tree grown from seed is called beeju or tukhmi; several Indian languages have their own terms for it like naati in Telugu and gaothi in Marathi. Fruits from seed-grown trees are highly variable due to the mango's erratic reproductive character (its details are distracting so wait to read about it later!). For example, if you really like a fruit, you might plant its seed in the hope that after the tree grows it will produce the exact same fruit. That does not happen; it is quite likely that the fruit from the resulting tree will not taste the same. It might be slightly different, or it could be a new taste altogether.

Horticulturists work around this uncertainty by cloning the parent tree. They cut out a branch from the parent tree and graft it on to a sapling grown from seed. Below the line of joining, the tree is the one that came from the seed; above, it's a clone of the parent. After it has grown, it will produce exactly the same fruit that you initially liked. Such is the craft of grafting, the other way to grow mangoes. Almost all mangoes sold in markets today come from trees grown from a graft or a kalam, from the Arabic for a pen or a vegetative shoot or even sideburns! Such a tree is called kalami. An orchard of grafted trees is simply called 'kalam' in the Mithila region of northern Bihar.

Most Indians have always eaten fruits of trees grown from seeds. When did we begin to graft mangoes? That's a question for historians. Goa-born polymath and historian Damodar D Kosambi wrote in a 1956 book that Jesuit priests initiated mango grafting in Goa in the sixteenth century: 'At about the same time (AD 1575), the same active Society of Jesus introduced systematic mango grafting, which improved the Indian fruit out of all recognition, and created a further source of income for the horticulturist in the whole of India.'[2] In another book in 1962, he wrote: '...the Jesuits, through their early activities introduced the cashew tree, the pineapple, potatoes...and best of all graft mangoes developed from local varieties.'[3]

Among the first to address the grafting question in a modern framework in 1946 was Parashuram Krishna Gode, a Sanskrit and Prakrit scholar at the Bhandarkar Oriental Research Institute in Pune; the article was republished in a book in 1961, the year he died.[4] In a follow-up article on grafted mangoes, he cited nine references that range from the sixth century scholar Varahamihira's *Brihatsamhita* to *Hobson-Jobson*, an 1886 glossary of Anglo-Indian terms.[5] He concluded: 'It will be seen from all the data recorded above that the art of grafting was introduced into Indian horticulture only after about AD 1550 but its operation was confined to Goa say between AD 1550 and 1790. It was absent in Kashmir in the 17th century... It appears at Madras about AD 1798, when it was introduced there by Clive, the Governor of Madras.'

Among those who helped spread the grafted mangoes of the west coast in the nineteenth century was the British officer Proby Cautley, an engineer and fossil hunter who initiated the work on the Upper Ganga Canal in the 1830s; his work led to the creation of the India's first civil engineering college at Roorkee in Uttar Pradesh. The banks of the canals were planted with grafts of 'Bombay mango'.[6] In the late nineteenth century, English botanist Charles Maries set up the orchards around the palace of the maharaja of Darbhanga in Bihar; Maries is credited for some famous mango varieties in Bihar. He did use mangoes from the west coast. Other names also pop up. But the details are not well understood.

Since the 1960s, a few scholars have addressed this question and some new material has emerged. 'Though the technique of grafting was known to Indians as early as the 6th century AD, it was apparently not practised in (the) mango,' wrote agriculture scientist and historian Yeshwant L. Nene in a twenty-eight-page review of literature in 2001.[7] Nene founded the Asian Agri-History Foundation (AAHF) in Secunderabad, Andhra Pradesh, in 1994; it moved to Pantnagar in Uttarakhand after his death in 2018.

As with so many ancient texts, it is not possible to know the exact details of Varahmihira's *Brihatsamhita*, which has a separate section called 'Vrikshayurveda'; it lists fruits propagated by grafting the root or the stem, also describing grafting techniques. In his authoritative 1994 book *Indian Food: A Historical Companion*, food historian K. T. Achaya wrote about the *Brihatsamhita*: 'A number of trees amenable to grafting are also mentioned. These include the jack, plantain, lemon, pomegranate, grape, citron, jasmine and others, but not the mango.'[8] He wrote that grafting 'had been described as early as the 4th century AD in the *Kamasutra* as one of the 64 arts.'[9]

Grafting finds mention in a completely different text also titled *Vrikshayurveda*; it is attributed to one Surpala, about whom nothing is known as of now.[10] Its 1996 translation by AAHF was carried out after the original manuscript was obtained from a library in Oxford in the UK. Its date is uncertain, but the translator Nalini Sadhale estimates it is from the tenth century. Paul Craddock, a cultural historian in London, scoured texts and references from Europe and Asia to write *Spare Parts*, his 2021 book on transplant surgery from the sixteenth century to the present day. He told a reporter that the earliest skin grafts as a medical procedure in Europe were inspired by agricultural grafting techniques that had their origins in ancient India.[11] 'The Chinese and Indians were grafting plants at least as far back as 2000 BCE,' he wrote in his book.[12]

Written accounts of the Mughal administration inviting horticulturalists from Central Asia and Iran date back to medieval times; grafting was employed for the improvement of several fruits except the mango. Talking about the early Mughal rulers, Achaya noted: 'Grafting began to be widely practised in this period.... The Portuguese in Goa had employed grafting to produce excellent varieties of mango. Grafting had only been permitted at first in the royal gardens, but Shah Jahan lifted this ban, and the technique was applied to cherries and apricots in Kashmir, and oranges and mangoes in Bengal. Figs were grafted on mulberry trees, peaches on plum trees, apricots on almond trees, and vines on the apple.'[13]

Besides, grafting was not the only mode of improving plants. Several old texts describe methods to treat and modify seeds and soil and of particular ways to irrigate and nurture the tree to obtain desirable traits in fruit. AAHF has translated and published several articles and books on such texts. They include ancient ones by Parashar, Kashyapa, and Surpala, as also medieval ones by Mishra Chakrapani and Mahabat Khan. A few other scholars have reviewed and listed such texts. A prominent summary from 1987 is by Lallanji Gopal of the Banaras Hindu University's departments of ancient Indian history, philosophy, religion, and archaeology. His paper on agricultural techniques in medieval India and their Central Asian connections contains 116 references.[14] He ends by saying that even in Kashmir, 'grafting practices could not take firm root'.

As far as the available evidence goes, the modern practices of mango grafting began with Christian priests in sixteenth-century Goa controlled by Portugal. It was in Kerala that the Europeans first came across mangai or manga, the Tamil and Malayalam terms for the mango that make its English name. The botanical *Mangifera indica* means in Latin 'an Indian mango-bearing plant'. But why were horticulture and the mango of such interest to priests? The answer lies in the European 'Age of Discovery' and a monastic order sometimes called God's marines.

The Columbian Exchange is a 1972 book by American historian Alfred Crosby. Its title draws from the 1492 voyage of Genoese explorer Christopher Columbus to discover the naval route to the Orient, instead finding the Americas. Western Europe was desperate for new trade routes after the Ottomans captured Constantinople in 1453, closing the eastward routes for the merchant powerhouses of Venice, Genoa, and Florence.

The Portuguese explorer Vasco da Gama took a different naval route, reaching Kozhikode/Calicut in 1498 after rounding the southern tip of Africa. This led to the establishment of the Portuguese empire in India, dotting the western coast; by the 1530s, it stretched from Kerala to Gujarat.

On one side of the globe, traders, priests, and states from Europe were exploring the 'New World' of the Americas; on the other side, they found a direct connection with the 'Old World' of Africa and Asia. This set off an unprecedented planetary movement of people, animals, plants, and diseases. This wider change also came to be called the Columbian Exchange. This brought to Europe plants never seen before: potato, tomato, corn, peanuts, chilli peppers, several beans like rajma, pumpkin, cashew, papaya, guava, custard apple, pineapple, sapota, cassava, tobacco.... The list is long because the tropical Americas constitute one of the world's great reservoirs of biodiversity. Slowly and through diverse routes these reached Africa, Asia and, later, Australia. In return, the Americas got Afro-Eurasian crops like wheat and rice and lentils. From India arrived sugarcane, cotton and the mango.

With the colonies established, Europeans began propagating Christianity. The torturous voyages, however, were too much for 'secular priests', the ones who wear white robes, mind their flock, and attend to routine tasks of the church. They were not cut out to undertake unimaginably difficult travels, go to strange places, live in hostile conditions and attempt seemingly impossible tasks. That required a new, committed cadre with military training and resolve. Such traits come naturally to the 'monastic orders' that wear black robes. Their history goes back to the Crusades, giving them a military edge. Indeed, it was a soldier who established the Jesuit order that influenced world history, not to mention the mango.

Born in 1491 in the Basque region, Inigo of Loyola joined the Spanish army, impressed by tales of knightly chivalry. He was injured in a 1521 battle with the French. While convalescing, the only books he could access were religious. Their influence brought about a spiritual transformation. Eventually, in 1534, he took the monastic vows of poverty, chastity, and obedience along with a handful of companions, including one Francis Xavier. Six years later, this became the 'Society of Jesus', headed by Ignacio née Inigo as its Superior General. Its members came to be called Jesuits. Its founder died in 1556 and was later canonized as Saint Ignatius of Loyola.

In 1540, the Portuguese government requested the Pope to assign missionaries to spread Christianity in their Indian territories. Several Catholic orders arrived in Goa, setting up churches and monasteries. Francis Xavier was dispatched to Goa in 1542; the Jesuits soon became the most powerful Catholic order here. The power of the priests had a lot to do with their continuity. While Portuguese viceroys served for three years, an archbishop could continue for decades.[15]

Unlike other orders, Jesuit training was more open-ended. It included worldly knowledge, along with learning the languages of the people they were looking to convert. In Goa, Jesuits had to take two classes a day of Konkani as part of a six-month intensive course that included practising with native speakers and using only Konkani with fellow students! The society has produced a long list of scientists who have contributed important discoveries, especially in physics and astronomy. Jesuit priests introduced umbrellas from China to Europe; influenced the modern calendar; created the trapdoor used in theatre; and one man named Roberto Busa pioneered computational linguistics, making possible searchable text on computers and the internet.[16] They were at the cutting edge of science in the sixteenth and seventeenth centuries, as documented by several historians.

Botany came naturally. Jesuits brought quinine from the bark of the cinchona tree in Peru to cure malaria worldwide; it was called the Jesuits' Bark. The cashew was the most valuable crop they brought from South America; they also brought pineapple, potato, guava, and custard apple.[17] These brought them business, for they needed money to run the missions. Sale of their agricultural produce, including exports, was quite profitable.

I caught a glimpse of the Jesuit approach at the Xavier Centre of Historical Research (XCHR) in Alto Porvorim in Goa. 'St Ignatius wanted his folks to be learned,' said Anderson Fernandes SJ, a young entrant. Having just completed his education when I met him, he said a Jesuit's training is much more rigorous than that of a secular priest. 'In Jesuit training, "discern" or "discernment" are important terms,' he added. Father Anthony da Silva SJ, the centre's director, said, 'Jesuits are exceptional because their training is multifaceted.' How does that extend to horticulture and the mango?

'You have to imagine the life of a Jesuit five centuries ago. They travelled long distances on tough ship voyages. A major part of the knowledge exchange happened in transit. They often waited at, say, Mombasa in East Africa for the winds to turn before sailing to India. Such transit points harboured priests from all over the world. While they waited, there was nothing to do but talk and learn from each other,' the director said. From Mombasa, Goa, Macao, the Philippines to Beijing, each Jesuit centre reported to Rome. Each priest wanted to impress his superiors with detailed research, with the knowledge he had acquired. This created a unique system of scholarship. 'Jesuit letters were the Google search engine of that age,' he said.

Back to the mango. Kosambi did write that Jesuit priests first began grafting the mango in Goa, but he did not give a source for this information. We do not know exactly when and where the Jesuits began grafting the mango. Writer Vivek Menezes of Goa said this gap is a challenge to a historian willing to travel to

Rome and Portugal to glean through the enormous Jesuit archives in Portuguese and Latin.

Possibly, the location of this refinement was the island of Chodan/Chorao in the midst of the Mandovi River, north of Old Goa. The island was among the earliest conquests of the Portuguese and had a number of Christian converts by 1560, when the first Jesuit priest moved there. The Novitiate of Chorao was established in 1584; here, newly initiated priests learned Konkani, among other things. It closed in 1759 when the Jesuits left Goa in disgrace after the Pope ordered their 'suppression'. Their international political and financial power was so resented that European and colonial governments began to expel Jesuits. By 1773, their suppression was complete, revoked only in 1814, in the aftermath of the Napoleonic wars. But they left lasting impressions.

A more likely location for the Jesuit horticultural experimentation, said Menezes, was further north of Goa in the town of Bassein, or Vasai. It had hosted Xavier in 1548. Two decades later, the Jesuits had built a college there and commanded sprawling, wealthy estates with orchards.[18] Not only did the Jesuits have a centrally-run international network, but their priests came from across Europe, bringing the best knowledge of their countries. This could have been the site for trying out the latest grafting techniques on both Indian plants and those coming from the Americas.

The Jesuit story is not all so sweet. In 1546, St Xavier decided conversion was not proceeding according to expectations. He requested the Pope to set up the dreaded inquisition. In 1559-1561, the Goa Inquisition began, ushering a reign of terror, death, and repression that lasted 250 years; thousands were persecuted. The exact numbers are not known because the records were burnt in 1820 when the Inquisition ended in Goa. The state of fear among the laity in its duration can be sampled from a mango-related story: when the lord inquisitor sent a messenger to a Goan household for mangoes from his famous tree, the family sent a basket of fruit and then had the tree uprooted to avoid future attention.[19]

Seeds and grafts of select mangoes must have travelled along the coast, acquiring names like Alphonso, supposedly named after the admiral Afonso de Albuquerque, the second Portuguese governor in India. At some point they reached the island of Bombaim, the long-term lease for which was gifted in 1554 by the Portuguese governor to his friend Garcia da Orta, a physician and naturalist, who in 1563 wrote a remarkable book on the medicinal plants of India: *Colloquies on The Simples and Drugs of India*. It has an eleven-page section on the mango, including a description of the fruit from two trees in Bombaim that his tenant had brought along: 'They ought to be cut with a very sharp knife, that the slice may not be injured, and I want to taste them first, for allowance must be made on account of the season.'[20] He was waiting in Goa for mangoes from Bombaim!

Da Orta's Sephardic Jewish parents were forced to flee the Spanish Inquisition in 1497. By the time da Orta was born in 1500, they had converted to Christianity and settled in Portugal. There, da Orta grew up to become a doctor of medicine at the University of Lisbon.[21] When the Inquisition came knocking in Portugal in 1534, he fled to Goa. He secretly practised Judaism till his death in 1568; his Jewish name was Avraham ben Yitzhak. He escaped the Goa Inquisition only because he was a physician and friend to several powerful people. But the Inquisition got to him a year after his death. His sister Catarina was tried and held guilty of secretly practising Judaism. She was burnt at the stake along with her brother's remains that were disinterred from his grave.

A century later, his beloved island of Bombaim underwent some restoration drama. It was part of the Portuguese princess Catherine's dowry in 1661 when she married Charles II, recently restored to the throne of England, eleven years after his father Charles I had been executed in the English Civil War. Bombaim became an English possession called Bombay.

About 150 years later, one Maria Graham, wife of a British naval officer, wrote in 1809 an account of 'Mazagong, a dirty Portuguese village'; today it is a south Mumbai locality called Mazagaon. In her book published in 1812, she wrote: '(Mazagong's) celebrity in the East is owing to its mangoes, which are certainly the best fruit I ever tasted. The parent tree, from which all those of this species have been grafted, is honoured during the fruit season by a guard of sepoys; and in the reign of Shah Jehan, couriers were stationed between Dehli and the Mahratta coast, to secure an abundant and fresh supply of mangoes for the royal table.'[22] Such trees did not disappear quickly. Old residents of Mumbai have told me that Alphonso-bearing mango trees were a common sight in the city and its suburbs till a few decades ago.

We do not know how and when the fine mangoes travelled along the coast. There are references to state Jesuits in Goa handing out grafting lessons to farmers of Chorao Island to gain conversions.[23] Mango growers in Chorao Island can still recall families of trained graftsmen and horticulturists. The island produces some highly rated mango varieties, particularly the Mankurad. It's very easy to find grand mango trees in Goa that are 60-100 years old. Many people believe these old trees are grown from seedlings, and some of them are trained horticulturists and Goa-born mango cultivators. But a seasoned agriculture scientist born in Goa said they appear seed-grown because of their vigour and height. The grafting techniques used earlier did not leave apparent signs of cutting and grafting.

How did these techniques spread across India? Menezes puts it down to the three Jesuit missions sent in the late sixteenth century to Emperor Akbar's court in Agra. Akbar had run into the Portuguese during his campaigns in Gujarat; he invited Jesuits to his court. They reached Agra with gifts including fruits of the Americas, which had never been seen before in India. These became so popular that a single pineapple in Delhi cost as much as ten mangoes. The priests were looking to convert Akbar, paving the way for converting large parts of India to Christianity. A descendant of a fruit-obsessed family, Akbar let the Jesuits build a church in Agra in 1600. He gave them the impression that he was genuinely interested in conversion. It took the priests a long time to figure out that he was stringing them along. 'For the Jesuits, Akbar was a big disappointment. He turned out to be neither the visionary, or reformer or ecumenist he pretended to be, but a shrewd schemer using religion to further his political base,' wrote Jesuit historian Charles Borges SJ, former associate director of XCHR.[24] The Portuguese and the Mughals did not have diplomatic relations. This left the Jesuits as the only go-betweens. Agra went on to become the location of a colonial rivalry. The Jesuits were pitted against the British East India Company, which went on to replace Mughal power in South Asia. The Portuguese Estado de India shrank to the west coast, tormented by the growing power of Maratha rulers.[25]

In 1739, the Portuguese lost Bassein with its valuable estates to the Marathas, and were restricted to Goa, Daman, and Diu. Maratha rulers most likely inherited the Portuguese horticultural resources, including grafted mangoes and the crops from their American colonies, like chilli peppers, peanuts, potatoes, and tomatoes.[26] When the Maratha armies moved north, they brought along these crops. They are also said to have planted mango trees on their marching routes and halts.

The important question here is: did the Jesuits give their grafting techniques of growing true to type mango trees to Mughal gardeners? We do not know. Historians estimate that grafted mangoes did not catch on in North India till late into the eighteenth century; even then they were limited to certain orchards patronized by rulers and merchants.[27] The famed orchards of Awadh in Uttar Pradesh were mostly seed-grown. The planting material is said to have come from Murshidabad in Bengal, along with trained graftsmen and their techniques. Till India's independence, most of the trees in orchards were grown from seeds. In the late 1970s, government scientists conducted a pilot sample survey in Uttar Pradesh's orchards. They found up to 76 per cent of the mango trees in Varanasi and 57 per cent in Saharanpur were grown from seeds.[28]

The difference between seed-grown mangoes and clonal grafts shows, in a nutshell, a fundamental distinction within India's mango culture. Only the rich created orchards, in the form of a vatika or a bagh or a bageecha. In these pleasure gardens they employed specialized gardeners who could carry out experiments over years to grow refined fruits. A majority of Indians did not have such resources. But they did have desirable mangoes from amrais or groves. What's the difference? Amrais had mango trees in such density that light did not reach the floor. They were much, much more extensive. They stood on common lands and anybody could have their fruit. Their trees were grown from seeds. In the Mithila region, it has been a custom for families to donate their mango groves for public use in the memory of dead ancestors.

I wish I had paid attention when old people talked of the glory of seed-grown mangoes. Relatives, now dead, who spoke of beeju mangoes with pride! They made it a point to remind us kids that the wonderful mangoes of their childhood were all beeju, not kalami. The mangoes they loved were juicy fruits meant to be sucked by the bucketful. Each juicy mango, they said, had the potential to surprise you. One or two of them are still alive; they say the only mangoes that excite them are the small seed-grown ones purchased with abandon by the basket, not the grafted table mangoes weighed in cautious grams and cut with knives; they saw those as the marginal refinement of the rich. And yet those varieties monopolize today's mango conversations and '#mangowars'.

An old family friend had a special interest in fruits. He once told me about the seed-grown mangoes around Nagpur, recounting the extraordinary taste of small and juicy fruits sold by women from wide baskets near small-town railway stations. It sounded like standard old people nostalgia. Several years later, I happened to travel through Vidarbha during mango season. I did find such mangoes and when I tasted them, they were delightful. They reminded me of the Malwi mangoes of Indore.

In Ratnagiri, I heard from an Alphonso grower that leaves of grafted trees are not used in puja or other religious customs or in ceremonial decorations. Flowers and leaves offered on Shivaratri cannot be from a grafted tree. Only gaothi or desi seed-grown trees of a variety called Raival are favoured for all rituals. When I asked for the reason, people shrugged and said that is just how it is. I've heard similar stories from the Mithila region; it was customary to cut down a seed-grown tree to cremate a body. Never was a grafted tree or its products used in any ceremony.

It seems a grafted tree is not considered wholesome, so its parts are not fit for offerings to the divine. Could it be, I wondered, that the mango was not grafted in ancient India for the same reason? Seed-grown mangoes have been so deeply embedded in life and belief that it seems possible. Were people willing to take their chances with the mango's random character for a notion of wholesomeness? Was this embrace of wilderness and uncertainty over control and predictability behind the mango not being grafted, despite techniques being known from ancient times?

A researcher has claimed that seed-grown mangoes were preferred because grafted trees were taboo. He quoted Yajnavalkya, a Vedic rishi who influenced several schools of ancient philosophy. He is said to have lived in northern Bihar about 2,700 years ago. A verse appears in his Yajnavalkya Samhita describing items to be avoided in offerings to the gods: '[One should avoid]...the juice of a pot-herb, that of a red-coloured tree, that of trees born of cuttings, meat of animals not offered in sacrifices...'[29] I got a Sanskrit professor to find the original and scrutinize the translation. She said the claim was not accurate. That the Sanskrit word 'vrikshaniryasana' primarily means the tree's exudation, most likely sap or gum, from making a cut in a bark. It may or may not extend to mean a cutting or a graft.

This duality of nature and culture keeps springing up in varied ways. In July 2014, the Urdu poet Munawwar Rana, born in eastern Uttar Pradesh in 1952, circulated an anecdote on social media about a gathering where people were discussing the difference between table type mangoes and sucking varieties. Somebody said the cultured and civilized cut the mango fruit with a knife, while the uncultured and the uncouth suck it. Rana found it arrogant and offensive; he said those nursed on their mother's milk as babies prefer to suck mangoes, while those raised on packaged formula prefer to cut it. He posted this along with his couplet: 'Insaan ke haath ki banayi nahi khaate / Hum aam ke mausam mein mithaai nahi khaate' (That made by human hand, we do not eat / During the mango season, we do not eat sweets).

Some scholars have told me that there's much more to sucking type mangoes from beeju trees than we realize. I am confident of their reading. Besides, my taste buds get stimulated at the very mention of the juicy variability of the fibrous beeju mangoes of Malwa. 'Once famous mangoes of Malwa, Burhanpur, and Marathwada are not even mentioned today,' wrote Nene in his 2001 paper.[30] This one is on my to-do list. I owe it to my dead relatives as much as my living taste buds.

Chapter 3

Culture
Escaping Our Refugee Camps

'The mango is not just a fruit. It's a culture, a code of civility.'
—KHAN MOHAMMED ATIF

For consumers, the mango is a familiar fruit. They focus on its varieties and taste. This can be easily summed up. The mango can be sweet or sour. The pulp is either fleshy or juicy. The fruit can be plain or aromatic. The stone is either fibrous or fibreless, and the skin can be thin or thick. The size can be large or small. The same logic applies to mango pickles and chutneys. Yet, the edible mango in all its forms is only the tip of the iceberg: the visible side of what the mango means in India! Exploring the mango's cultural range took me down many rabbit holes. Or picnics!

I signed up for the one Sohail Hashmi organizes each summer. On a Sunday morning in July several years ago, a varied group of enlisters got into a bus in Delhi and headed 40 kilometres north to Rataul, a village in Uttar Pradesh, famed for its mangoes. Inside the bus, the chatter was upbeat. In Rataul, when we hit the orchards, the excitement gave way to serenity. We walked around, asking the hosts several kinds of questions. I felt the urge to climb a tree but did not; it seemed inappropriate. Besides, Hashmi was talking to some people and I did not want to miss out; he tells an interesting story.

Hashmi recalled, in another meeting, summer days in the 1960s when his family in old Delhi escaped the heat by driving down to the mango orchards in Mehrauli, near the Qutab Minar. Today, it is a busy patch called Andheria Modh on Delhi's southern fringe, on the way to Gurugram. They would get into horse-drawn tongas with packed food and a tub of ice. In the orchards, they'd take the mangoes from custodians and soak them in tubs. The food was always qeeme ka saalan with green chillies. When it got too hot, they had a cooled mango to douse the spice.

'Do you know why it got the name Andheria Modh?' Hashmi suddenly asked about the locality, not far from where he lives now. I did not know. 'That is because it is so dark under the mango trees. The canopy was so dense that it blocked all sunlight. It was then called Andheria Bagh.' It had been set up by Akbar Shah II (1760–1837), father of the last Mughal emperor Bahadur Shah Zafar II (1775–1862). Delhi historian Swapna Liddle has translated a Malhar composition Zafar wrote in the Braj language; it talks of the darkness under the trees of Andheria Bagh, the swings dangling from the trees, and the beauty of the monsoon there. Hashmi also remembers the 1970s, when Andheria Bagh and the surrounding orchards were cut down to make way for 'farm houses' for Delhi's rich and powerful. As remnants, some mango trees are still standing in Mehrauli's Desu Colony.

Andheria Modh is just one name that harkens back to a greener city. Several Delhi localities bear the names of orchards and water bodies that once stood there. Moti Bagh, now also a metro station, was once Mochi Bagh, an orchard planted by a cobbler (mochi) named Ramdas. The list is long: Jor Bagh, Karol Bagh, Maharani Bagh, Gulabi Bagh, Punjabi Bagh, Shalimar Bagh. There's also Raja Garden, Rajouri Garden, Tagore Garden....

The Hashmi family picnics in Delhi were not unique. Similar stories abound in many cities. My father often recounted the cookouts and picnics of his childhood and youth in Indore. Along with a friend who liked to bicycle, he used to go on rides extending for weeks. They cooked and slept in mango groves, with meagre supplies tied on to the cycle carrier. He never gave up that pleasure; he always looked for a reason to take family and friends out to a grove or a forest for a cookout. On my twentieth birthday, he took forty-odd friends to the forest of Sariska in Alwar, Rajasthan, where he cooked for everybody.

He spoke of the numerous mango groves that earlier surrounded Indore. To the city's southeast lies a locality called Navlakha; the name means nine lakh, which was the number of trees there. Was the count real or an approximation or imaginary? We don't know. What we do know is that a small river enters the city from that area. The orchards made a natural barrier against flash floods. This part of the city underwent the first land diversion and redevelopment by the East India Company after its victory against the Holkars, Indore's ruling family, at the Battle of Mahidpur in 1817. The adjacent locality is still called the Residency and it has some grand old mango trees.

Orchards and mango groves surrounded the ancient city of Varanasi. Its old names include Sundarban (beautiful forest), Anandvan or Anand Kanan (forest of joy). The city gets its name from the rivers Varuna and the Asi, tributaries of the Ganga. The panchkosi parikrama, a circuitous route around Varanasi walked by pilgrims, was formerly dotted not just with temples but stepwells and mango groves. Shyamsunder Jaiswal comes from an old business family of this city. He is the only remaining manufacturer of traditional zari (gold-plated silk thread) that goes into weaving the famed Banarasi brocades. His family owned an orchard north of the city. Twice a month on Ekadashi, the eleventh day of the lunar fortnight, the family business remained closed. The clan headed out to the orchard for picnics; special food items were cooked and consumed. During mango season, fruits were the main attraction for such picnics. 'Most well-to-do families had orchards that circled the city. The famous Banarasi Langda mango was grown in such orchards,' he said.

Unlike Varanasi, Jaisalmer, in the middle of the Thar Desert, has no rivers. But it does have a famed mango orchard, hostile conditions notwithstanding. If you head northwest from Jaisalmer along the Ramgarh road, in about 6 kilometres you can spot windmills and the cenotaphs erected in the memory of the royal family of Jaisalmer. These are approached via a road that runs along a dam. This embankment has created a tank to its north, irrigating the mango trees and a pleasure garden called Bada Bagh. The soil under the mango trees had been brought from the fertile Indus valley.

For a sense of scale, here's an example from M. S. Randhawa: 'The mango tree often attains the age of a hundred years or more. The biggest and oldest mango tree was discovered in 1949 by me in the village of Burail near the present city of Chandigarh.'[1] This tree gave shelter to the villagers for threshing wheat and was called chhapar, meaning roof.[2] Its trunk was nearly 10 metres in girth. Each of its nine trailing branches looked like giant trees themselves. The tree's crown covered 2,258 square metres. Each year, it yielded an average of about 17 tonnes of fruit, comparable to some orchards. In 1955, lightning struck and destroyed the tree.

It is difficult to find an old Indian city without any mango groves or orchards. To Patna's east is Digha, a locality with mango groves in ancient times and orchards till recently. Another feature of several old cities of North India was a Thandi Sadak, a cool road lined with trees, often the mango. Lucknow had one along the Gomti River, and some people say the road going towards Malihabad was also called Thandi Sadak. Such road names are found even today in Bhopal, Rampur, Hisar, Azamgarh, Haldwani, and Nainital, among other towns.

This cool road was common enough to get into film lyrics. The 1951 Hindi film *Jadoo* has a song titled 'Eji Thandi Sadak Hai', sung by Shamshad Begum. The lyrics came from the pen of Shakeel Badayuni, a poet from western Uttar Pradesh. In the song, the protagonist Sundari tells her love interest Pritam, a police constable: '*Akadh akadh ke chalo na babu, thandi sadak hai thandi sadak...*' (Relax, don't swagger, the street of love is a cool one...). In another 1951 film, *Sagai*, a song titled 'Bigad gayi bante bante baat' opens with '*Ek din Lahore ki thandi sadak par shaam ko...*' (One evening on Lahore's cool road...).

Andheria Bagh, Navlakha, and the many iterations of the Thandi Sadak all indicate a pattern of urban planning. Cities were surrounded by mango orchards, dotted with water bodies. Villages and cities without mango trees, if any, were looked down upon. The richest created pleasure gardens with spectacular animals and birds, sometimes in mini zoological parks; it was both a re-creation of the forest with plants and animals and an escape for recreation. In western Uttar Pradesh, when rich families discussed marriage proposals, one way to assess financial standing was to walk through their orchards and list old fruit-bearing trees, for they indicated old wealth. The finest mango varieties of India have come from such orchards because they were grown for the shauqeen who went to great lengths for quality.

The orchards of Rataul were set up along such a script. On Sohail Hashmi's mango picnic, I felt what it means to be rich and leisurely, even if momentarily. My breathing had slowed down along with the pace of walking; my mind too. A different kind of stimulation had set in. I saw a photographer from a Delhi publication climb a tree, not for fruit or a better angle; there was plenty of ripened fruit on the ground where he had left his camera. He just needed to climb. How did I know? Because I felt it, too.

Soon, several others began climbing, including those with unwieldy bodies. Their faces brimmed with excitement as they took their chances against gravity, their arms spread out for balance, not in the steady manner of tightrope walkers but like bird chicks awkwardly experimenting with nascent wings. The joy of climbing trees, like that of eating mangoes, reveals our primal self. It's hardly surprising the mango has us in its thrall. It is atavism at work. Evolution forged us in the trees. One part of us still lives up there. Trees call to ghosts of our pasts. In Hindi bhoot means both the past and a ghost. We can leave our past, but our past doesn't leave us.

Why were Indian cities nestled in mango groves? Not just for the fruit! An American psychologist's research offers a vital insight. Spoiler: Frances 'Ming' Kuo has not researched mango orchards. For over thirty years, she's studied how surroundings affect people. In particular, her studies explore how bad urban environments cause social problems like violent crimes. She had no interest in natural spaces or how they modify our behaviour. Except that today Kuo is an expert on the impact of urban greening on well-being. 'I have been dragged into this all the way kicking and screaming,' she told an interviewer.[3] Serendipity in science isn't valued enough!

Literature reviews had shown her that animals do not thrive in captivity, even when provided the best food and shelter: 'It turns out that zoo animals...are extremely expensive. They die at fairly alarming clips.' Yet, as a general rule, the animals thrive in their natural environment. This is called the 'habitat selection theory'; it implies that organisms break down socially, psychologically, and physically when they live in a habitat different from the one in which they have evolved. Some of these ideas begin with the celebrated biologist and naturalist Edward O. Wilson who died in 2021 at age ninety-two.

Kuo found that people living in buildings surrounded with green cover in Chicago were different in some ways from those who had no greenery around their living space. More residents of non-green buildings reported aggressive behaviour; they did not know or trust their neighbours.[4] Then, the psychologists measured their findings against crime data; it showed the same pattern. Localities where vacant lots were cleaned out and fostered as green spaces revealed a significant drop in gun assaults, with similar trends for crimes like burglaries. Kuo's literature review produced other studies, showing how greenery reduces levels of stress, making people friendly and open to others.

Since rich people tend to live in greener surroundings, she then checked if the better health outcome was determined by better access to healthcare. The results showed the improvement was not due to wealth but because of the greenery.[5] A 2015 study in London showed people living in greener areas were using fewer medicines prescribed for mood-related disorders like anxiety and depression.[6] Studies have shown that a three-day weekend in a forest preserve shores up the immune system by boosting natural killer cells, an effect not replicated in an urban surrounding.[7] Inside a laboratory, just seeing photographs of greenery or smelling fragrances associated with forests lowers the blood pressure and calms the nervous system. Therapists counselling people at the risk of suicide ask them to take walks in green surroundings. Gardening has long been linked to psychological health.

'When looking at urban environments the brain is doing a lot of processing because it doesn't know what this environment is,' psychologist Ian Frampton told a London newspaper about his study of MRI scans that showed how the human brain processes rural and urban environments.[8] 'The brain does not have an immediate natural response to it, so it has to get busy. Part of the brain that deals with visual complexity lights up: "What is this that I'm looking at?" Even if you have lived in a city all your life, it seems your brain doesn't quite know what to do with this information and has to do visual processing,' he said.

In 2007, for the first time in history, a majority of people began living in urban areas.[9] Today, that proportion is 55 per cent. Yet the best cities retain their greenery. New York City is unimaginable without its Central Park. Bengaluru, Pune, and Delhi were earlier reputed for their green cover. This idea went into the design of Chandigarh in the 1950s. But all planning has failed to cope with the tide of migration and urbanization.

Today, a sizeable proportion of the world's urban population, especially in developing countries, lives in deplorable conditions. The UN estimates the number of people living in slums at one billion.[10] Many of those who do not live in slums lack access to greenery and open spaces. They live like refugees, far removed from the forests and grasslands in which *Homo sapiens* evolved. Through gardening and landscaping and urban greening, we retain a sense of our natural habitat. In Indian cities, the mango groves and orchards were the refuge where people escaped to retain their natural humour.

The original idyll, a place of perfect comfort and peace, is the Garden of Eden in the Abrahamic tradition. 'From the soil, Yahweh God caused to grow every kind of tree, enticing to look at and good to eat.... A river flowed from Eden to water the garden, and from there it divided to make four streams,' says the 'Book of Genesis' in the Old Testament. The Bible is full of the imagery of gardening: '...And they shall say, This land that was desolate is become like the garden of Eden...' (Ezekiel, 36:35).

In Paradise, Adam and Eve felt no shame in their nakedness; the lapse occurred later due to the fruit of knowledge. People of the Book—Jews, Christians, and Muslims—have recreated the prelapsarian Garden in every way imaginable. It has inspired even the highly manicured gardens of the Palace of Versailles near Paris, the most ornate example of the French formal garden that sought to impose civilized, rational order upon wild nature.

The imagery of the natural world is found across traditions. The polytheistic Romans believed they were hunters descended from wolves. The Romans were preceded by the Etruscans, who were even more situated in the innate. 'There seems to have been in the Etruscan instinct a real desire to preserve the natural humour of life,' wrote novelist D. H. Lawrence in *Etruscan Places*, his 1932 collection of essays. 'And that is a task surely more worthy, and even much more difficult in the long run, than conquering the world or sacrificing the self or saving the immortal soul.' Lawrence had travelled in 1927 to Italy. The essays draw upon the contrast between the Etruscan/Greek embrace of nature and the Fascist order that Benito Mussolini wanted to impose.

The drive to impose order and control over nature has had drastic results. Europe may appear verdant but it has destroyed most of its wilderness through hunting and logging. 'Originally widespread in the Old World, sacred groves were reported in Greek and Sanskrit classics but were essentially wiped out in Europe by the arrival of Christianity and its attendant anthropocentrism,' wrote Madhav Gadgil, perhaps India's most feted ecologist, in a 2018 essay; he has researched and written on sacred groves since the 1970s.[11] 'The Christian church, with its towering pillars and soft light filtering in through colourful stained-glass windows, is said to evoke the sacred groves of yore.'

Desperate conservation efforts are trying to save the continent's last wilderness areas in Eastern Europe.[12] '...at least 58% (155 species) of all endemic European trees are threatened,' said a 2019 study by IUCN, the international conservation body.[13] Centuries of demonization led to the wolf being hunted out of most forests; in Christian lore, the wolf is a metaphor for Satan and enemy of the flocks. Now, several countries are reintroducing wolves in their forests, trying to bring back wilderness; it's called 'rewilding' and it is quite fashionable.

India has retained at least some healthy natural forests and wildlife despite the population pressure and the constant demands of industrial development and mining. Indians have also hunted wild animals and cleared forests for agriculture. But societies in

India have had a more complex attitude towards wilderness. One example is the ancient cult of the mango.

If India's rich had their orchards, the ordinary folk had groves. 'The mango was a wild fruit. Everybody ate it. Everybody had mango trees in the fields and in open lands. Now it is beyond the means of the poor,' said Prakash Bhatambrekar. Most villages in India had some form of wilderness close by—a pasture, a village forest, a sacred grove. While the vegetation varied from place to place, the mango was ubiquitous. Now it is difficult to find old people who remember witnessing this, while the newer generations cannot even imagine this culture of the mango grove.

They can catch a glimpse of it, however, in a book titled *Cities and Canopies: Trees in Indian Cities* (2019), by Bengaluru-based academics Harini Nagendra and Seema Mundoli. 'A number of British gazetteers from the early 1900s mention groves across north India. In Agra, 4 per cent of the land area was covered by groves of mango, jamun, bael, and other indigenous fruit,' they write, listing a rich selection of references from British travellers of the groves they saw, most commonly of the mango. And they noticed the same pattern: mango groves and water bodies went hand in hand. The book's fifteenth chapter is titled 'The Fellowship of the Grove'.

Pieces of the puzzle come from stray accounts. Despite having extended family across Madhya Pradesh, I'd never heard of land categorized as 'nistar van', meaning a forest for release. I came across this term in 2002 while researching evictions of forest dwellers. Such lands had many uses, from cremation grounds to sites for gathering food items, fuelwood and bamboo, among other things; in most villages, they had mango trees, if not mango groves. In the state, 10 per cent of each village's land was set aside for 'nistar'. Other states have examples, too. Rajasthan has community pastures called 'oran'. Gujarat has the salt marshes and grasslands of the Rann of Kachchh. Both oran and rann come from the Sanskrit aranya or the Pali araeen, meaning wilderness or forest.

These lands provided sustenance that couldn't come from agriculture. Several such fruits have just disappeared from northern India, like khirni, harfa revdhi, phalse, and numerous berries. A non-profit called Living Farms in Odisha studied wild foods harvested from forests by six tribal villages in two districts. It recorded 121 different kinds of uncultivated foods harvested between July and December of 2013.[14] Each household collected an average of 4.56 kilograms of such foods in each foray. Mushrooms and tubers figured prominently, along with the mango and jackfruit. One of the researchers in this study had documented in 2007 more than fifty fruits, including wild ones, gathered by Dalit communities in Sangareddy district of Telangana.

Such customs ran across social strata. In large parts of India, Rishi Panchami is an occasion for partaking in foods gathered from the wild; farmed foods are restricted on this festival that falls on the fifth day of the lunar month of Bhadrapada (usually in August). Even today, many people in cities follow the diet restrictions, especially in the Magadh region of southern Bihar. Villages here typically had two types of land outside the inhabited area: one is called chhavani, which literally means temporary shelter, and the other is baithaan, a resting space for grazing animals. Both of these were earlier dotted with mango groves. All the dried leaves of such trees were reserved for caste communities that needed fuelwood, like potters (kumhar) and bakers (bhadhbhunja). Only potters were allowed to cut diseased or dried-up branches.

The village and forestland connection often emerges on occasions featuring cookouts or the consumption of wild foods. One such observance is called laas, a contraction of the Sanskrit for felicity, ullaas (ut+laas, the ascent of joy). The root las yields more joyous meanings: beauty, grace, to appear in light and glow, play, gambol, lighten up, celebrate, engage in drama and sports. Lasya is the cadent dance of the goddess Parvati, different from god Shiva's frantic Tandav. Las lends itself to the name of the most famous mountain in Indian lore, the Kailas (spelt 'Kailash' now) Parvat, beyond the Himalayas in Tibet. It is the heavenly abode of Shiva, the deity of wild abandon; Pashupati, meaning the lord of the animals, is one of the forms in which he is worshipped. Kailas rises above the Mansarovar, the honour (maan) of a water body (sarovar). This mountain and the lake are not mere geographical locations; they betoken the state of being naturally joyous, of blissful innocence. Eden is no exception!

Everyone aspires to such bliss, both within and on the outside. Indian cities and villages did this by keeping the wilderness and water bodies close. Picnics and cookouts are escapes to a green haven. Across large parts of India, cookouts were a means for the upkeep of common resources. The festivity subsumed the labour and drudgery of repair and maintenance work. Anupam Mishra, a scholar of socio-environmental traditions of ordinary people, documented over several years the laas associated with water bodies; most of this work happened in the difficult summer months when the water bodies were dry.

Mishra was my mentor; I was still in school when I first travelled with him to see rural India. Among trees planted around water bodies was the mango; ancient texts such as Varahamihira's *Brihatsamhita* list the mango as one of the trees to provide shade around ponds. It's not just stray sightings or textual evidence. Older relatives have recounted to me the wedding parties of their childhood. The baraat, a contraction of the Sanskrit var yatra or the groom's travelling party, stayed in mango groves outside the bride's village. The lodging area was made comfortable and pleasant by the bride's family, with food and entertainment. 'They were charming, idyllic gatherings,' an uncle had told me.

Many tribal villages in coastal Andhra Pradesh still have localities called Vantalamammidi; it means 'mango kitchen'. 'These were mango groves under which travelling parties could take refuge and cook,' said Ravi Rebbapragada, tribal rights activist in

Visakhapatnam. Long before concrete banquet halls began to lend a grotesque touch to wedding arenas, the appropriate venue for a banquet was the mango grove.

In Amravati, Karuna Futane recalled her youth: 'When I got married, it was the norm for the bride to go back to her parents during the mango season; the festival was called maher vashin (literally, living at mother's house).' She usually spent the mango season at her mother's and recalled the excitement of plucking raw mangoes for pickles. 'Good pickling mangoes were shared and sent all over,' she said, her face flushed like a new bride's, even as her granddaughter tugged at her sari. Her husband Vasant is a campaigner for natural farming. 'In his youth, he used to plant mango trees by the roadside, saying it would give shade and fruits to passers-by,' she said. In Nagpur, reputed Marathi playwright Mahesh Elkunchwar recounted his childhood in Parwa village of Vidarbha, surrounded by mango groves. As an adult, whenever he went back, he noticed the groves were gradually getting cut down. Now, they are all gone.

Mishra had heard about a mango grove in the region south of the Narmada River that was a makeshift hockey stadium. The central part with the hockey pitch was open ground, surrounded by mango trees that provided shade to spectators. Enquiries after its location have been fruitless. He also spoke of a mango grove set up by a community of hijras (transgender persons). It was called Hijdon Ki Amrai near the town of Narshighpur in Madhya Pradesh. After I tracked down their residence near the old part of the town, the transgender persons took me north along the Jhirna Road. Here lies a graveyard surrounded by thirteen grand mango trees, where the community bury their dead.

One time, I happened to be travelling through Sohagpur, a town 120 kilometres west of here. I heard of a well-known face from the hijra community called Ghanshyam Guru. In 2010, Ghanshyam Guru embraced Islam and became Mohammed Haneef, undertaking

Haj a few years later. Hindu cremation grounds here do not allow last rites for the hijra community. Afraid that their souls will wander about after their bodies die, they embrace Islam for the dignity of a burial. 'The almighty is the one and the same, regardless of your mode of worship,' she told me. The community held large parcels of land in this region, granted to them for sustenance by big landholders or the village leadership. Such lands frequently had mango groves as common resources.

No community was untouched by the culture of the mango. Take Mumbai's Parsis. The Doongerwadi Tower of Silence in Malabar Hills has many mango trees. Such trees were grown from seeds. People took their chances with the mango's natural variability. If a tree gave sweet and palatable fruits, they were ripened and consumed. If the fruit remained sour, it was pickled. If the fruit was totally useless or the tree did not fruit at all, it still gave shade and timber. 'In the old days, it was a custom to bury every stone of a 'good' mango at any suitable place. "Good" mango trees were retained and the "inferior" ones were cut for timber,' wrote Yeshwant L. Nene of AAHF.[15] An investment in the mango could not go bad. 'Every native considers it incumbent on himself to perpetuate his name by planting a mango *bagh*, and if he can raise funds for the purpose he does so; and I am credibly informed that…even sweepers and Chumars have mango plantations,' noted the 1873 *Settlement Report of Budaon District* by C. P. Carmichael.

The acceptance of wilderness was not limited to wild foods. Agriculture itself was rooted in an appreciation of the nutrient cycles of the forest. It took a British soil scientist, Albert Howard, to document this approach. He wrote about this in his 1940 book *An Agricultural Testament*. Often called the father of the organic farming movement, Howard figured out Indian agricultural practices while serving as the Imperial Economic Botanist from 1905 to 1924. He then took over as the director of a government agricultural research farm in Indore. It was here that he figured out the compost-making method known as 'the Indore process' among organic farmers.

'The forest manures itself. It makes its own humus and supplies itself with minerals,' he wrote. 'Nature's farming, as seen in the forest, is characterized by two things: one, a constant circulation of the mineral matter absorbed by the trees; and two, a constant addition of new mineral matter from the vast reserves held in the subsoil.' The Indian farmer, he noticed, tried to mimic this: 'The agricultural practices of the Orient have passed the supreme test—they are almost as permanent as those of the primeval forest, of the prairie or the ocean.' The role of domestic animals in farming resembled that of wild animals in the forest.

A lot has changed, though. Urban sprawl has destroyed city greens; village groves have been felled mercilessly. Yet some things remain. Hashmi's annual mango trip is a new version of the laas, the cookout. The new sensations that urban people feel on these trips are actually craving for primeval joys. Then there are the Adivasi communities living inside forests or on their fringes without clearing them. The mango features prominently in the forest-oriented customs of several Scheduled Tribes from across India. Like the mango, this embrace of the wilderness is—or was—widespread. In ancient accounts, the rishis of the Vedic traditions set up their ashrams in the forest. The renouncers and sceptics of the non-Vedic schools also turned to the forest for the austerities of tapasya. 'It is not just a fruit. The mango is a culture, a code of civility,' Khan Mohammed Atif, a reputed scholar of Farsi in Lucknow, had told me several years ago.

'[The] ideal of perfection preached by the forest-dwellers of ancient India runs through the heart of our classical literature and still dominates our mind,' wrote Rabindranath Tagore, modern India's most famous poet. 'Our two greatest classical dramas find their background in scenes of the forest hermitage, which are permeated by the association of these sages,' he wrote in an essay titled 'The Religion of the Forest', compiled in the book *Creative*

Unity. Visva-Bharati, the university Tagore founded at Santiniketan, holds its annual convocation in a mango grove.

Tagore's life and work teem with a sensitivity rooted in the natural world. He wrote: '...in the level tracts of Northern India men found no barrier between their lives and the grand life that permeates the universe. The forest entered into a close living relationship with their work and leisure, with their daily necessities and contemplations. They could not think of other surroundings as separate or inimical. So the view of the truth, which these men found, did not make manifest the difference, but rather the unity of all things.'[16]

The mango tree has been a resource of wilderness at its munificent best. The fruit is one segment of a wider culture. The fruit of the summer is not why the mango has been driving India crazy for centuries. It's a different reason, a different season.

Chapter 4

Season
A Festival in Search of Revellers

'Oh, mango bud, I offer you
To Kama, grasping now his bow.
Be you his choicest dart, your mark
Some maid whose lover wanders far.'

—SHAKUNTALA

In a poem titled 'Nimantran' (invitation), Rabindranath Tagore said: 'No golden lamps or lutes are available now / But do bring some rosy mangoes in a cane-basket covered with a silken kerchief / And some prosaic food as well...' I visualized streaks of mango juice on the canvas of that iconic white beard! Before the fruit, though, comes the flower. In a poem titled 'O Manjari', Tagore turned mango flowers into a metaphor for a young woman. Manjari, from Sanskrit, means a cluster of blossoms. If the plant is not specified, it means, by default, the mango's springtime inflorescence. (Exception: in Marathi it stands for Tulsi flowering; mango flowers are called mohor.)

The mango fruit in the summer is like cooked food: it results from events of spring. The mango blossom is the fire that cooks; a fire that's been driving Indians crazy for centuries. Baurana is a verb in several languages of North India. It means to go mad with desire. Baur actually means the same thing as manjari: blossoms. Again, unless specified otherwise, baur or maur is the flowering of the mango. Spring is conjoined with the mango inflorescence in many parts of India. (Not all over India, though. For instance, on the south-western coast, the mango flowers in December and produces its fruit during the spring).

Since ancient times, people have been getting high just by hanging around flowering mango trees. Of the many accounts I've heard and read the most moving one came from Ravindra Sharma, an unusual scholar of folk art and crafts in Adilabad, Telangana. He was a repository of artisanal traditions in their lived context. The last time I met him, a year before he died in 2018, he spoke of going to meet master artisans in his youth. 'All around Adilabad were these amrais through which I had to ride my bicycle. During the spring flowering, the fragrance got so intoxicating that I'd often lose my way. I felt disorientated to the point that I'd forget what I set out to do. I'd just sit down, holding my head, like a junkie,' he told me, laughing loudly, sprawled on his easy chair in his house that he had turned into the Kala Ashram.

Spring is transformative, the world's most celebrated season. Both the English term 'spring' and the Hindi 'vasanta' mean to burst forth or materialize spontaneously; hence springs of water or a mechanical spring making boing-boing sounds. All through the year, plants are busy with workday tasks, like making food by photosynthesis, growing in size and girth. Come spring and plants (not all!) take a vacation to reproduce in an atmosphere of passion. This seduction has its own tools: colour, fragrance, and nectar. Plants are solar panels; they need sunlight to make a meal. This keeps them locked into the seasonal cycle. They are forced to modify themselves to the angle at which the sun shines on them. They reflect seasonal changes; animals follow the cue.

Human spring festivities are no exception. The season induces communal dance, music and feasts, merrymaking, and intoxication. It's not an accident that the springtime festival of Holi is celebrated with colour, cannabis, and communal revelry. 'Dance is the biotechnology of group formation,' Barbara Ehrenreich, American journalist and scholar, quoted a neurologist as saying in her book *Dancing in the Streets: A History of Collective Joy* (2006); she reviewed research on festivals from across the Western world for this book.

She showed similarities between the early worship of Jesus and that of Greco-Roman gods like Dionysus, Pan, and Bacchus. All of them are linked with wine, feasting, and ecstatic revelry. Ehrenreich goes on to catalogue the 'gods of ecstasy' from around the world. 'These itinerant musicians and masters of ecstatic ritual may well have been the prototype for the god Dionysus,' she writes. 'With his long hair, his hints of violence, and his promise of ecstasy, Dionysus was the first rock star.'

Ehrenreich's frame did not include Indian spring festivities and the mango. But the parallels are unmistakable. Consider the Sanskrit term madana or mad. It means intoxication; or going mad; or a male elephant in a rut; or the fluid that exudes from the temples of a rutting tusker. Madana is just another name of Kamadeva or just Kama, the ancient Indian deity of desire; his Greco-Roman counterparts are Eros and Cupid. Kama's business of love is transacted through his favoured tool, a bow made of sugarcane with a string made of humming bees. His arsenal contains five kinds of arrows, each with its own name and effect, representing a fragrant flower. There's ashok (*Saraca indica*), jasmine, lotus, lily and—his weapon of choice—the mango flower. It was with such a flower that he shot the most famous arrow of love, depicted in many tellings, at mighty Shiva.

Actually, Kama had been commissioned by other gods and Shiva's consort Parvati. They needed Shiva to wake up from his meditation and procreate with Parvati; only their offspring could destroy the powerful demon Tarakasur, who was giving them a hard time. Kama went in all guns blazing; accompanying him were his wives Rati (sex) and Priti (love) and his friend and associate Vasanta (spring). Kama's mere arrival transformed Shiva's habitat, turning the trans-Himalayan icy barren into a verdant garden with a pleasant southern breeze. He released an arrow at Shiva. The shootout did not go so well for Kama. Shiva, the fastest draw, opened his eye, the third one on his forehead, the destructive one you don't want open. It unleashed a death ray, instantly incinerating the god of love. The world turned loveless!

(It's not clear whether or not the arrow struck. But Shiva had noticed Parvati. She went on to give birth to their son Karthikeya/Murugan; he did kill Tarakasura. But Karthikeya was also the victim of the most infamous dodge by his brother Ganesha for the most famous golden mango. But that's another story! Back to the scene of the shootout!)

Parvati and Kama's wives pleaded with Shiva, asking him to revive the god of love. Shiva relented. But, mindful of Madana's mischievous ways, Shiva did this on the condition that he remains disembodied, existing only in spirit. He then diverted the fire of his death ray among other entities. They included: the mango, Vasanta, the Moon, flowers, bees, and cuckoos. Since then, Kama has moved out of sight into deep unconscious folds. Two of his names, Ananga and Atanu, mean body-less. Among his numerous names are Kandarpa, Darpak, Manmatha, Meenketana, and Mara. Each name comes with its own set of backstories. Each speaks of his former importance. The mango tree is a corporeal form of Kama.

Among his avatars is Pradyumna, the son of the ultimate singing–dancing rock star among the Hindu divinities: Krishna. In a memorable lecture on how religion shapes ideas, socialist leader Ram Manohar Lohia compared the influence of the three most popular Hindu gods: Rama, Krishna, and Shiva. Lohia was an atheist but possessed a nuanced understanding of society and religion. To him Rama is maryadit, or bound by ethics and codes, hence a social deity. Shiva is aseem or limitless, hence universal and all-encompassing. Krishna, however, is unmukt, liberated and unshackled, the ultimate rebel, hence a personal god.

It is no surprise that spring festivities, earlier linked to Kama, transferred gradually to Krishna. He describes himself thus in the Bhagavad Gita (10:28): '...I am Kandarpa, the motivation of procreation for begetting progeny'. Krishna is a revolutionary of the later Purana texts who helps people retain conscious agency in the face of life's big questions—he represents facing existential crises with joy and confidence. Kamadeva is an older Vedic deity; he prises open the unconscious, encouraging all to give up control, to surrender to our natural selves.

His buddy Vasanta is called Rituraj, meaning the king of ritus or the six seasons of India. The classical singer Kumar Gandharva paid a grand tribute to the king of seasons in the early 1970s. He created a three-hour live programme first called Geet Vasanta and then Rituraj Mehfil. When I first heard its recording a few years ago, I felt like mango sprouts were bursting out of my skin.

In traditional calendars, spring comprised the two lunar months of Chaitra and Vaishakha (falling in March–April). While other festivals have dates, spring was a festival in its entirety of two months. Called Madanotsava or Vasantotsava, it celebrated Madana's seasonal resurrection.[1] There are innumerable accounts of this festival in ancient and medieval texts. A few scholars have compiled them.[2] There's a bewildering array of deities, dates, names, locations, customs, and interpretations. One kind of fertility custom linked to Kama is the dohada, meaning 'pregnancy yearning'. It included a range of springtime practices like vriksh dohada, in which a young woman kicked an ashok tree to get it to flower. Another form had a young woman spitting a mouthful of wine on a maulshree or bakul tree (*Mimusops elengi*).

Like many festivals, it is enmeshed in seasonal and agrarian cycles. These 'gradually evolved into organized urban festivals. The culture of Kama, of amusement and merrymaking, appropriated these festivals as a mode of mass expression,' wrote Kanad Sinha, a Kolkata-based historian and Sanskrit scholar, in a 2013 research article on urban amusements, sports, and festivities of ancient India. '...the biggest urban celebration, with unlimited frenzy and ecstasy, was the Vasantotsava, the spring festival.... The culture of Kama was the heartbeat of urban India, and the festival of Kama must have been the high point of that.'[3] In medieval times, travellers and court chroniclers witnessed these celebrations and described them in their accounts. But Madana, already rendered invisible, was marginalized over the centuries. Spring festivities began to revolve around Krishna (and Saraswati, the Vedic goddess of speech, music, and learning). If the mango is Kama's reincarnation, the tree linked to Krishna is the kadamba (*Neolamarckia cadamba*).

There were signs of Madanotsava celebrations till late into the twentieth century. The dohada fertility ritual was in evidence till recently in the creation of orchards in and around Lucknow. Landholders, both Hindu and Muslim, requested new brides and daughters-in-law to plant mango trees on their land. The customary preparations for spring festivities began in Magh, the first month of Shishir or the winter season. The fifth day of Magh is called Vasanta Panchami or Madana Panchami; this is when people begin preparing for Madanotsava, forty days ahead of spring! The main offering on this occasion was mango flowers. (Southern India has its own seasonal cycle and festive traditions.)

In Varanasi, an old book provided a remarkable description of spring festivities. Its author was Rai Krishna Das, a prominent cultural figure who founded the Bharat Kala Bhavan. Titled *Prasad Ki Yaad*, it celebrates Jaishankar Prasad, Hindi's greatest modern romantic poet. It says the preparations got going in earnest on the fourteenth day of the second winter month of Phalguna, celebrated as Shivaratri. Instead of the traditional offering of bael (*Aegle marmelos*) leaves to Shiva, on Shivaratri he is offered mango flowers.

Processions were taken out with specific musical ragas sung to traditional lyrics. The book mentions 'Rangbhari Ekadashi', the colour-filled eleventh day of Phalguna. Deities were adorned in yellow and red, signifying flowers. Just as spring is the primary season, the unchallenged king among the Nava Rasas or nine essences of ancient aesthetics is Shringar or adornment. It is called Rasaraj, the king of the rasas. The king of the seasons thrives on the king of the essences, with a nod to the mango.

The Colourful Ekadashi was accompanied by a fair in Varanasi. The poet Prasad looked forward to the sugarcane juice prepared for this occasion. He used to ask for mango flowers to be crushed along with the cane, lending an intoxicating edge and fragrance. Das wrote that his muneem (accountant) got so high on this beverage that it became difficult to control him. And this, remember, is just the preparation; spring hasn't arrived yet! Intoxication is fundamental to the spirit of adventure and risk-taking, essential to spring. It drives people to set aside their inhibitions, their usual concerns. Madana's arrows of mango flowers are nature's venture capital, promoting projects beyond business as usual.

The year ends on the last day of Phalguna. The first day of spring ushers in the new year of several calendars including the Vikram Samvat, the most widely used traditional lunar calendar. (This, it is said, is why India's fiscal year begins in April.) The annual reboot in India has not been the winter solstice or the summer solstice, despite precise calculations of solar time since hoary antiquity. The spring has been India's annual restart. And the symbol of a new year of hope has been the mango flower. In numerous rituals, the mango flower is consumed on new year's day.

With the flowers fertilized in spring, the invisible god's annual assignment is complete. He returns to invisibility. The result of Kama's merrymaking appears in the summer: fruit! In Tamil, maamaikkavin is a term for a woman possessing mango-like beauty. Some of the most beautiful women in Indian lore are linked to mangoes. Amrapali was the most comely courtesan in the ancient republic of Vaishali in north Bihar in fifth century BCE. A gardener had found her as a baby, abandoned under a mango tree; her name means protected and raised by the mango. When she had grown up into a young woman, for a lark she entered and won a contest to be picked as the nagarvadhu, Vaishali's chief courtesan.

A highly accomplished dancer and musician, she had kings, merchants, and even monks infatuated by her beauty. In one account, Buddha warned his monks that Amrapali's beauty could corrupt the minds of the religious.[4] Likewise, Urvashi, the most beautiful apsara (celestial nymph), was created out of mango juice by the sage Narayana.[5]

It hasn't changed. In the movie *Bend It Like Beckham* (2002), a woman measuring the protagonist Jess for a dress describes her breasts as mosquito bites, adding that her design will make them appear like 'juicy juicy mangoes'. It gives a twist to the botanical name of the mango's plant family *Anacardiaceae*; it means outhanging heart. The mango's fleshy and drupe shape makes it a ready stand-in for human breasts; that the fruit is massaged and sucked loads up the symbolism.

In the Muria language of Bastar, marka dudo means breasts that are mango-like, hence beautiful. In the numerous articles about the mango that appear by the dozens each summer, the imagery is often sexual. In old Indian films, a man and a woman disappearing behind flowers or mangoes was the accepted innuendo for sex. It was common for pregnancy to be revealed through a woman craving for the sour taste of raw mangoes, a variation of dohada or pregnancy yearning. The fertility metaphors of the mango can sustain entire departments of cultural studies.

It might lean towards the feminine, but the mango does not overlook the masculine. The Gadaba and Kond tribes of Odisha find that the mango seed resembles human testicles.[6] The Gadaba say the first mango was created from the testicles of a hapless man called Gangu, who died a bachelor.[7] Vatsyayana's ancient Sanskrit treatise *Kamasutra* used the mango as a metaphor in only one instance. It describes the act of oral sex between a man and 'a person of the third nature', that is, a transgender or an intersex person. It lists the eight acts of fellatio; 'sucking the mango' is entry number seven, after the sixth that is 'polishing', and before the eighth and final, which is 'swallowing'. The *Kamasutra* repeatedly calls the mango a fruit fit for the amorous; several Ayurveda texts, including the *Charakasamhita*, mention it as an aphrodisiac.

One of the Sanskrit names of the mango is Chyuta/Chuta; it means to fall out or ooze out, drop by drop, or to get lost or to separate.... You do not need Sanskrit expertise to make the connection. In Carnatic music, some notes are described as chyut, leading to claims that they allude to the mango. But it is not so; they are 'dropped' notes, drawing from the original meaning, not the mango allusion. In Hindi/Urdu poetry, chuna or chu-aana is a verb used for the distillation of alcohol, as also the dropping of any valuable liquid or essence. The mango appears repeatedly with regard to alcohol in ancient texts; James McHugh, professor of religious studies, has numerous references to it in his 2021 book on alcohol in Indian history and religions.[8]

If you mention this Sanskrit name in North India, however, you will draw prurient laughter. Chut is by far the most common slang for female genitalia, drawing from the Sanskrit chyut. Hindi dictionaries have the word chutiya meaning a fool or an idiot, like the English cunt. But the tone in which it is employed on the street definitely has a ring of baurana or going mad. When he let Kamadeva exist in spirit, could Shiva himself have known how far and wide the god of love would permeate in flowery language!

꙳

Like Madana, the mango does not transact its business in language or literal meaning. It functions through more primal forms of storytelling, through the arts. The book to read for this is *Romance of the Mango* (2002) by art historian Kusum Budhwar, who grew up in Uttar Pradesh's mango belt. It documents how the fruit is all over the ancient and medieval visual arts of India, in sculpture and paintings. Several scholars have written a range of articles and book chapters documenting the mango's presence in the Indian arts—for example, in miniature paintings. The mango motif buta or keri is among the most common patterns in Indian textiles throughout history; there are claims that this was the origin of the paisley pattern but the evidence isn't clear enough.

Mystery also surrounds the bright pigment called Indian yellow, used on wall paintings in the country from at least the fifteenth century and, subsequently, in paintings in Europe. It smelled foul, it is said, because it came from the urine of animals fed on mango leaves. In 1883, at the request of the famed British botanist Joseph Hooker, the government assigned an official named T. N. Mukharji to investigate this. He went to Mirzapur near Munger to find that 'gwalas (milkmen)' indeed derived the pigment by boiling the urine of cows fed mango leaves.[9] The cows suffered much agony including stones in the kidney and the inability to pass urine. It is believed the pigment was banned sometime by 1908 to prevent such cruelty; synthetic dyes replaced it.[10] Art investigators have used the chemical analysis of this pigment to identify forgeries of older paintings in which the pigment had been used.[11] The matter remains mysterious, however; there is little evidence other than Mukharji's report.[12]

What we do know for sure is that the mango is always auspicious, signifying plenitude or fertility. And not just in Khajuraho! From yakshas and yakshinis under mango trees, to Shiva-Parvati, Ganesha, Rama, Krishna, Buddha...deities were often depicted with mango fruits and trees. Both revelry and penance are performed under mango trees in these sculptures. In later paintings, everyone from emperors and nawabs to saints was depicted under mango trees. There is a mid-eighteenth-century painting of Delhi's Nizamuddin Auliya under a mango tree; in front of him sits his most famous disciple, the musician–poet Amir Khusro, holding a stringed instrument that looks like a sitar.

Khusro is an important entry in the history of music and language of India. He is a link between older musical forms and newer ones like the khayal, ghazal, qawwali, and tarana. He is also said to have adapted older instruments into newer forms like sitar and tabla, though we do not know the exact details. There is evidence, however, that he loved the mango; he called it naghz tareen mewa-e-hindustan, the most supreme fruit of Hindustan. He especially liked the fact that, unlike other fruits, the mango was edible even when it was not ripe. He wrote sixteen riddles in a poetic form called Keh-Mukarni (meaning: say it, then deny it). In one verse, a woman asks another about something that appears year after year, is juicy to the lips, and for which she is willing to pay a price. When the other asks if she is talking about her lover, she denies it and then clarifies that it is the mango.

It is easy to find the mango mentioned in the lyrics or bandish used in ragas of Indian classical music. For a rendition of Raga Bahar, a springtime raga, the classical singer Kumar Gandharva composed a bandish titled 'Aiso Kaiso Aayo Rita Re'. It questions Vasanta about its tepid arrival, about the mango trees not flowering, about the bumble-bee not buzzing about. It then pleads the case of the pained koel or the cuckoo bird and the colourless flowers.

The koel is innately linked to the mango in songs, not the least because the bird's mating call is plaintive and piercing. It gets louder as mating season approaches in the summer, giving a sonic effect to the mango grove. The connection just keeps giving. Innumerable lyrics have riffed off koels and mango trees. Ustad Bade Ghulam Ali Khan sang Raga Malkauns to the bandish 'Koyaliya Bole Ambuva'. Begum Akhtar sang a Dadra composition to the bandish 'Koyaliya Mat Kar Pukar'. The bird is usually described in the feminine. As your neighbourhood bird watcher might tell you, it is actually the male who lets out that mating call to attract the female. It does so when it notices crows nesting, because koels are aggressive brood parasites. The koel is smart enough to deceive the crow, itself the smartest of birds.

Like the Antakshari, the Chahar-beit is a competitive lyrical form, performed as overnight Beit-Bazi. Its meter is from Arabic poetry but it came to India from Afghanistan via soldiers who used it to stay awake during campaigns. In peaceful times, Beit-Bazi shifted to mango orchards where Afghan musclemen were first employed as guards to protect the produce of prized trees. Later, they turned into gardeners and graftsmen. Unlike other lyrical forms, though, Chahar-Beit never got gentrified or reinvented into pop culture and is hardly known today.

It is in the well-recognized forms, however, that poets have left the most memorable tributes to the mango. The Sanskrit poet Kalidasa is the gold standard of mango writing. In his play *Shakuntala*, he compares the protagonist's fingers to the tender shoots of the mango. It is the flower, naturally, that really gets him going, preceding Tagore by centuries in celebrating Kamadeva's arsenal. The play depicts the custom of young women plucking mango flowers to offer them as rearmament to Kamadeva; just that here, his name is Manmatha for how he churns the mind. (Lock and load! Take cover, men, for you have no idea what's about to hit you!) Says Shakuntala in the play: 'Oh, mango bud, I offer you / To Kama, grasping now his bow. / Be you his choicest dart, your mark / Some maid whose lover wanders far.' A stately mango tree is a metaphor for a young and virile king, while the vines of jasmine entwining the tree become the female protagonist. Kalidasa made the mango blossom-spring-Kamadeva corollary his own. Innumerable mango articles get their jump-start from his verses.

The modern name bearing down upon mango writing is the nineteenth-century Delhi poet Mirza Asadullah Beg Khan 'Ghalib'. He was fixated on the fruit, however, not the flowers! He wrote an entire masnavi, an extensive poetic form derived from Persia, dedicated to the mango fruit; it is titled 'Dar-Sifat-e-Amba', or In the Praise of the Mango. In it, he calls the fruit sweeter than sugar from cane. The mango was the simplest of pleasures for this poet of riveting complexity. In perhaps the most famous modern couplet on the mango, he speculates on the fall of Adam from Eden after consuming the forbidden fruit of knowledge, which in Islamic canon is gandum or wheat. Ghalib questions Adam's choice of fruit:

> Firdaus mein gandum ke evaz aam jo khate
> Aadam kabhi jannat se nikaale nahi jate
>
> (If he had eaten the mango instead of wheat
> Adam wouldn't have been evicted from Paradise)

Three incidents of Ghalib's fascination with mangoes are repeated in many articles on the mango every season. All three come from the 1897 book *Yadgar-e-Ghalib* by Altaf Husain Hali, his student and disciple. One has a friend ribbing Ghalib when a donkey refuses to eat mango remains in the lane outside his house; only donkeys don't like mangoes, the poet replies. In the second, Ghalib is out inspecting a royal orchard in the company of the king Bahadur Shah Zafar. The king asks the poet why he was scrutinizing the fruits so closely. It is said, replied Ghalib, that each fruit has its consumer's name on it, so he was looking for his name. The king responds soon after by sending over a basket of choice fruit. The third anecdote is about a discussion among learned people on the most desirable qualities of mangoes. When it gets too intricate, Ghalib cuts it short; he says only two things are necessary in mangoes: they should be sweet and plentiful. Ghalib's letters are infused with his desire for the fruit and more than thirty-five varieties can be counted there.

One hears mango-related anecdotes of famous poets. Most are unverified and many are apocryphal, but even those ones tell a bigger story. After tasting a mango brought from his village in Unnao, Uttar Pradesh, the poet Suryakant Tripathi (1897–1961), known by his nom-de-plume 'Nirala', is said to have identified the tree on which it had grown. The Pakistani poet Faiz Ahmed Faiz (1911–1984) is said to have declared that only mangoes and grapes were entitled to be called fruit, the rest were all vegetables. Akbar Allahabadi (1846–1921) once wrote a letter in verse to Munshi Nisar Husain of Lucknow, asking him to send a consignment of mangoes.

Former journalist Himanshu Bajpai of Lucknow has gathered the poems and mango-related anecdotes of several Urdu poets, turning them into a live storytelling performance in the traditional Dastangoi form, revived in recent years after being forgotten. Bajpai gets several invitations to perform his *Dastan-e-Aam* during the mango season. He takes his time over Shabbir Hasan Khan or Josh Malihabadi (1898–1982), born in the land of orchards. In his later years, the poet moved to Pakistan and died there. His letters were full of nostalgia for the orchards of his village. He regretted that he was not going to be buried in Malihabad.

In the rural setting of Munshi Premchand's novel *Godaan* (1936, *The Gift of a Cow*), the mango appears repeatedly. It is there in the first chapter as the summer Sun rises from behind a mango grove. It is there in the last chapter; as the protagonist dies, his wife feeds him aam panna made of raw mangoes. Elsewhere is a description of raw mangoes in daal. Ripening mangoes become a metaphor for the youthfulness of his daughters. A discussion on philosophy and discrimination compares the mango's fruitfulness with thorny acacias and out-of-reach palms. The onset of the month of Phalgun at the beginning of a chapter is marked by mango flowering.

The most evocative prose on mango flowering I have found is an essay by Hazari Prasad Dwivedi (1907–1979), a giant of Hindi literature. Titled 'Aam Phir Baura Gaye' (Mango Trees Flower Again), it opens with a tale he had heard in his childhood: if a mango tree flowers before Vasanta Panchami, then one should take its flowers and rub them on the palm of the hand; this turns the hand into an antidote to scorpion stings...but only for a year.

He said that after growing up, he stopped rubbing the precocious flowers on his palms, but the mango flower and scorpion sting got clubbed in his imagination. From this point, in about 3,300 words, he takes the reader across the firmament of classical literature with the ease of a pollinator flitting from flower to flower, every now and then going back to the scorpion. Dwivedi concludes by saying that the mango flower is a gift of providence but the fruit is a result of human ingenuity.

Modern science agrees. The high-quality mango fruit is a result of generations of selection and breeding by humans. Yet, it is the flower that produces the fruit. Long before the setting up of market infrastructure in the late twentieth century, when people consumed only the mangoes grown in their region, a large part of India went mad with Madana's arrows to celebrate spring.

⁓

While Madanotsava has been losing currency since medieval times, spring celebrations continued till late into the twentieth century. But in a matter of two, maybe three generations, we have lost even the memory of the two-month festival. Today, it is difficult to find people who have even heard of it. In the towns of Mathura and Vrindavan, steeped in Krishna's lore, it is still a week-long celebration of colour.

For the rest of the Hindus, the two-month Madanotsava has shrunk to a one-and-a-half-day Holi. It is observed in large parts of India by the burning of the symbolic pyre of Holika. The next day, people engage in a token play of synthetic colours and consume celebratory sweets. That is that. At least the mango is still mentioned occasionally in association with the spring. Ashok, the other main tree of this season, has been entirely forgotten. In cities, the name now means the 'false ashok' (*Polyalthia longifolia* or *Monoon longifolium*), the ornamental value of which owes to gardening practices of the British Raj.

A study in contrast is a spring festival in Japan that has, in about 150 years, become the world's biggest spring bonanza. It's called hanami; hana means flowers and mi is to view. If the Indian spring was linked to mango inflorescence, the Japanese spring is associated with cherry blossoms. The arrival of spring covers the country in a sea of white and pink cherry flowers called sakura.

Other than all the usual implications of spring and flowering, in Japan it is also linked with the idea that we are made of several personalities, not just one. The blossoms signal transformation, the outward expression of the hidden self. It is associated with madness—baurana—and death, signifying the ephemeral nature of all life. The medieval Japanese term kuru'u means both 'to go insane' and 'to dance'. 'Cherry blossoms are intensely involved in conceptions and representations of the Japanese self,' wrote Japan-born American anthropologist Emiko Ohnuki-Tierney.[13]

Today, hanami is an international tourism event. A Japanese university estimated that sixty-three million tourists spend more than ¥300 billion (US$ 2.7 billion) each year on the cherry blossom viewing festival. There are 600 hanami locations drawn out on multimedia platforms; mapping companies make a killing. The cherry blossom festival was an important part of the plans for the 2020 Olympics in Japan, which were rescheduled due to the Covid-19 pandemic. No country in the world has deliberately created such an involved imagery around a plant or a flower.

When did this begin? In the 1870s, under the Meiji revival. Among other things, this revival set Japan on the path of industrialization. The cherry blossom came to symbolize imperial nationalism; universal conscription began in 1872. 'You shall die like beautiful falling cherry petals for the emperor,' it began to be said.[14] Soldiers were told, 'The obligation is heavier than the mountains but death is lighter than a feather.' It got amplified with the repackaged ideas of the samurai code bushido or the warrior's way. 'Chivalry [bushido] is a flower no less indigenous to the soil of Japan than its emblem, the cherry blossom,' opened a book titled *Bushido: The Soul of Japan* by Nitobe Inazo. He was a devout Christian in the US, who first published his book in English in Philadelphia in 1899. The book became very influential very quickly in a militarizing Japan.

At the end of World War II, with Japan staring at defeat, the air force formed squads called tokkotai to fly suicide missions. The symbolism surrounding kamikaze pilots was all sakura-driven. Many of the corps were named cherry blossoms; the glider plane was called the cherry blossom plane and had a cherry flower painted on it; the 'potential' or the weapon was called sakura-dan or the cherry blossom bomb. There are photos of pilots with branches of cherry blossoms on their uniforms and headgear. There is a photo of schoolgirls waving cherry blossom branches at tokkotai pilots queuing up for take-off and certain death. The attacks began in April 1945, when the cherry trees blossom.

State gifts of cherry saplings had begun much earlier. In 1912, the US president William Howard Taft's wife had requested such plants. The mayor of Tokyo sent 3,000 of those as a gift. The saplings soon became an instrument of diplomacy. Trees were planted in several foreign countries. Soon, minor cherry blossom festivals began providing visitors with trailers of the real thing in Japan. My first encounter with this was several years ago in Washington, DC. The east bank of the Potomac River was overwhelmed with white and pink blossoms the mayor of Tokyo had gifted a century ago. At that time, Cherry Blossom was merely the brand of shoe polish to me!

What the mango inflorescence has meant in India is much greater, much more complex and much older than what the cherry blossoms signify in Japan. In the duration that hanami went from a seasonal tradition to a symbol of chauvinist militarism to a $2.7 billion industry today, Madanotsava disappeared. Forget about being able to draw the rest of the world into a joyous celebration of nature and the seasons. We have exchanged two months of revelry and joy for a one-and-a-half-day observance of a funeral pyre, synthetic colours, and indifference.

The next time you notice the microblogging platform X, formerly Twitter, trending with #mangowars or #alphonso-vs-dashehri, pause a moment before joining the cause. Spare a thought for how little we know our heritage. For how Vasantotsava has disappeared in the absence of reinvention, even as festivals like Karva Chauth and Raksha Bandhan have gained traction via soap operas and pop culture. For the fact that Thandi Sadaks of our cities are now mentioned only in crime reports and horror shows on the television (look it up!). For Andheria Bagh that no longer shields the ground from the baking sun. For Navlakha that will not protect Indore from flash floods that are increasingly common due to the effects of climate change.

We have lost our spring!

Chapter 5

Religion
Bones of a Civilization

'We looked for saasandari (tombstones) and mango trees.'
—KUNWAR SINGH JONKO

That day in 1966, the barber was late. The ritual required that the first axe to fall on the mango tree had to be at the hands of the traditional barber. A. K. Raman Nair's cremation was held up. Some young men got impatient and asked P. Thankappan Nair, the nephew of the deceased, why the Nairs always felled a seedling mango tree for cremation. He did not know. Reminding the young men that such questions did not befit the sombre occasion, he promised them he would find out.

Such promises are made to be forgotten. P. T. Nair did not forget. He began his research on the mango, completing it in 1980. Its fruit appeared in 1995 in the form of a two-volume, 1,060-page monograph titled *The Mango in Indian Life and Culture*. The preface mentions how his curiosity began with his uncle's cremation in 1966. That 'the story of the mango dealing with its all-pervasive role in Indian life and culture has not yet been told by anybody.' Nair found that the use of mango wood for cremation is not confined to the Nair community but is widespread among others in Kerala. 'Mango seems to have been used for cremation all over India from the remotest past.' In Malayalam, mavin palakayil vayukuka (lay down on a plank of mango tree) indicates death. Sometimes, it is used as a curse.

The wood of a seed-grown tree is ideal for cremation in several regions of the country, including Mithila, in Bihar; across the state, in funeral customs, the four corners of a pyre are usually mango logs. Each time I visit a cremation ground, I enquire about the wood, including at the busy ghats of Varanasi. All I have found was that they get all kinds of wood, as long as it burns! I am still looking for that cremation ground savant.

'All accounts confirm that the mango is the tree of the departed.... No sacrifice is greater than felling a mango tree and using its wood for the cremation of the departed kin,' wrote Nair. There must have been a great demand for mango timber. Given how slowly trees grow, it would have taken long-term planning. Just as it was customary in many communities across India to plant a useful tree upon the birth of a child, some have planted mango trees in anticipation of deaths and cremations. Today, it is a big business. Beyond funerals, mango wood is the default samidh or fuel for the havan, the ritual fire. You can buy mango wood in small quantities from online retailers. Some offer the 'havan samagri kit' with mango twigs. Be it a wedding ceremony or an ordinary puja, fire is central to Vedic ceremonies. Fire of mango wood is considered the most auspicious to subdue unfavourable astrological alignments.

If the mango is so important for cremation and havan, it is even more central to Hindu weddings. A marriage can be seen as the formalization of desire—or the normalization of risk-taking. That is Kamadeva's turf! Besides, the mango stands for plenitude. It is only natural that mango leaves festoon the kalyana mandapa, a temporary space created for the wedding ceremony that centres on the fire from mango wood. In several parts including Tamil Nadu, Karnataka, and Maharashtra, banana stems form the four pillars of the mandapa, festooned with mango leaves. In parts of north Bihar, wedding rituals include a marriage between a mango tree, the stand-in for the groom, and a mahua tree, representing the bride. The actual bride and groom circle the trees in a ceremony that mimics the pheras around the fire. The toran or bandhanwar, mango leaves on a string, decorates many Hindu houses throughout the year. The commencement of any Hindu festivity, like the Diwali season, is often marked by stringing a fresh toran above the entrance.

Another common sight is a pot with mango leaves sticking out of its brim—or even a drawing of the same on a wedding invitation or a calendar or even the back of a truck, not to mention old sculptures. This arrangement has various names including purnakumbha or purnakalash; it means the 'full/complete pot'; its features vary from region to region. Three elements are universal, however: a pot of water, mango leaves lining the brim, and a coconut on top of the pot's mouth. In a puja, this kalash is placed on top of unhusked rice grains, embellished with auspicious signs drawn in vermilion, and sometimes a cloth is tied around the neck. Often, next to it are placed a lamp, incense, and a conch.

There are numerous explanations for these customs. Ravindra Sharma of Adilabad had an easy one. He said the kalash is set up before every puja as a sign of the five elements or the panch mahabhoot; it represents the entire universe. The rice beneath signifies the earth, the pot stands for water, the incense is air, the conch represents the sky, and the lamp indicates fire/light. The mango leaves stand for the splendour of life, for desire and fertility. The coconut represents the womb of creation.

The other signs of festivities give a clue as to the mango's role here. In most parts of India, the floor in front of the entrance is decorated with rangoli on festive occasions. In eastern India, this takes the form of alpana, made with soaked rice flour. In several parts, especially in North India, a related art form is chowk poorna; a square is made either on a wall or the floor near the entrance. This is filled with propitious patterns. A more classical form of such pattern-making is the mandala, which has travelled across Asia along with Buddhism. All these are invitations to the universe to arrive at the householder's door, bearing beneficence. Hence, the full/complete pot calls out to the five elements. These are means to connect, to drive away isolation and separation. In this scheme, mango leaves—and the wood, flowers, and fruit—attract all that is fulsome and fecund in creation, all that drives life.

Since its every part is so useful in Hindu customs, planting and raising mango trees was an act of punya, a good deed done at a cost without petty self-interest. The history of religious belief in India is often too complex even for the scholars who spend their lives studying it. Over millennia, varying schools and systems have argued, disputed, and influenced each other. Given the bewildering complexity of material, it is sensible to first acknowledge what we do not know. However, we do know for sure that the mango features in almost all schools and systems: be it the Upanishads and Purana texts, the Pali/Prakrit texts, or all the others too numerous to mention here.

During Samudra Manthan, a story that appears repeatedly across ancient epics and other texts, the Sumeru mountain was used to churn the ocean. Among the long and fascinating list of objects that emerged from this was the Kalpavriksha, the tree of wish-fulfilment. Because the type of tree is not specified, it is open to speculation. Sometimes, the banyan is depicted as the Kalpavriksha; its Sanskrit name vat is also a generic name for any tree.

The banyan and the sacred fig (peepal or *Ficus religiosa*) occur repeatedly in the earliest Vedic sources, but not the mango. The Rigveda does mention the saha; later texts use the term sahakara for the mango; but not all scholars are convinced the two are one and the same. Four kinds of compositions complement the Vedas: the Samhitas, the Brahmanas, the Aranyakas, and the Upanishads. Together they form the central canon of the Vedic faith: Shruti, meaning heard or revealed. Brihadaranyaka, among the oldest of the 100-plus Upanishads, cites the mango seed as a sign of rebirth.

The mango frequently appears in the texts of the later smriti or memory tradition; these are texts written by people as stories, not abstract philosophy. Several Puranas use the mango as allegory. The ancient grammarian Panini mentions it in his *Ashtadhyayi*; Kautilya refers to it in his *Arthashastra*. The epics Ramayana and Mahabharata have frequent references to the mango. It is highly unlikely that ordinary people read these texts and scholarly commentaries. Yet, many practices of ordinary people overlap with the instructions in these texts. It is impossible to say if a particular practice began after the texts were composed or if the authors recorded prevailing customs.

The mango idyll appears in three settings: van or forest; upavan or a planted forest; and vatika or a planted grove or garden. The first two are obvious locations for the mango. The third has a different role. The vatika is characterized by auspicious trees and medicinal plants. Most drugs used in ancient times were plant extracts; the pharmacological source needed to be close at hand. A vatika is incomplete without flowers and butterflies. The mango finds frequent mention in Ayurvedic texts for its medicinal values, so it naturally figured in the vatika of the classics.

Another term is the panchavati, meaning five trees. The count of five in Indian lore encompasses everything. The five elements (pancha mahabhoot) make the universe; there are five senses (pancha indriya); five fluids, like honey and milk, join to make the nectar offered to deities (pancha amrita); the primary unit for administration and dispute resolution had five members (panchayat); the subtle soul has five layers (pancha kosh); deities including Hanuman are depicted with five heads (pancha mukhi)...one can go on and on and on. (After all, five fingers of a hand make a fist.) Hence, the panchavati is a garden that encompasses five—hence all—trees. This is why the scriptures show a variety of trees in such gardens: banyan, figs, bael, ashok, gooseberry, blackberry, neem, and haritaki. But no vatika or panchavati is complete without the mango.

The mango has been highly useful since antiquity, with some sacredness attached to it. Yet, the mango is not as holy a tree as, say, the banyan or the peepal. It has been used in puja but it is also cut down, its wood burnt readily for a variety of uses. It is used in ritual but it is not ritualized beyond the reach of the ordinary. The mango does not have to bear sacred burdens. Rather, it is India's favourite festive tree!

In eastern Uttar Pradesh, there is the history-defying cult of Ghazi Miyan. While a ghazi in Arabic simply means a raider, the term has connotations of religious violence. Ghazi Miyan of Bahraich, however, is depicted as a protector of cows and cowherds. Till a century ago, many of his devotees were Hindus; their proportion has come down in recent decades due to socio-political changes. Each year, an urs or an annual festival related to him falls in the mango season, in which he is symbolically married under a mango tree; it is said he got up from his wedding ceremony to go protect cows, and was felled during these efforts, one of the many stories associated with the figure. Historians Shahid Amin[1] and Badri Narayan[2] have tried, separately, to figure out the history of this cult. They found no historical evidence to support any of the various claims associated with Ghazi Miyan and his purported rival Sohal Dev.

A more troubled religious enquiry into the mango came from Delhi's Syed Ahmad Khan (1817–1898), founder of the Aligarh Muslim University (AMU). In a meeting at the house of the mufti of Delhi, he deliberated on whether Islamic law permitted (halal) or forbade (haram) the eating of mangoes. Theologian and Jesuit priest Christian W. Troll published a biography of the founder in 1978. He wrote that Khan 'had defended his views, saying that to eat mangoes—though not a blameworthy action—was a matter of doubt, since the Prophet had not decided upon it explicitly, having himself never eaten them.' Troll quotes Khan's writing thus: 'I swear by God in whose hands rests my life...if a person does not eat a mango for the reason that the Prophet did not eat it, then the angels will kiss his feet at his [death] bed.'

In 1879, Syed Ahmed Khan recalled: 'This I had said with utter conviction and much fervour. The Maulana listened to it and remained silent. It was at this period of noisy and tumultuous Wahhabism that after the discussion just recorded I wrote this tract.'[3] Troll, a historian from the School of Oriental and African Studies in London, cites this as an example of the kind of questions weighing on Khan's conscience in the early 1850s: 'What he was later to call his "leaning towards Wahhabism", was basically his search for the authentic practice of the Prophet, and his determination to imitate it. Along with many contemporaries at that time, he became more and more aware of the gulf between Islam (as he conceived of it in its original purity) and the then state of Islamic practice in India.'[4]

Aligarh, where the AMU is situated, has been surrounded by rich groves and orchards for centuries. Many were encouraged and established by Muslim rulers from the fourteenth century. I've met several alumni of AMU who hold its founder in great admiration *and* own mango orchards. They are devoutly Muslim and passionate producers and consumers of the mango. The fruit has a way of joining the seemingly irreconcilable.

This ability to conjugate gives the mango a special role in all matters of matrimonial. In eastern and central India, there are many examples of the marriage of mango trees with each other, or with trees like the mahua, tamarind, and banyan. In some places, the wedding is between mango orchards, in others, it is between orchards and water bodies. These weddings are communal celebrations. The origin of this custom is perhaps from India's forest-dwelling tribal communities and their elaborate ceremonies of marrying trees.

The purnakumbha and the toran, however, are not merely about weddings or spring festivals. Again, Ravindra Sharma of Adilabad had an elegant explanation. Mango leaves, he said, remain green for a long time after they have been plucked from the tree. Any decoration with its leaves is likely to last longer because the leaves do not quickly dry out or get discoloured. A purnakumbha or a toran is often required to last a few days. Likewise, the resin in mango wood makes it burn quickly and vigorously.

Such everyday explanations reveal a duality. The mango bears fruits sacred and profane. It appeals to our senses and transcends them. Its wood creates the bed of fire for last rites while its flowers create the bed of seduction and procreation. The mango's sacredness might be marginal but it is central to India's earthly realities.

The heavily pregnant Maya was on the way to her parents' house in Lumbini, now in Nepal, when labour pains set in. She gave birth to a son in a garden, under a tree. Some sources say the tree was sal (*Shorea robusta*); others say it was the ashok or the mango or the plaksha (white fig). That's the nativity story of Siddhartha, better known as Gautam Buddha.

It is one of the most famous lives of ancient times. We have heard the story! Raised inside palaces by protective parents, married at sixteen, becomes a young father. Disillusioned with the pain and sorrows of existence, leaves the palace at age twenty-nine. Subjects himself to great penances to find answers to deep existential questions. Then, six years later, after a forty-nine-day spell of meditation, he attains enlightenment—bodh in Sanskrit—under a peepal tree at Gaya, now in Bihar.

Siddhartha became Gautam Buddha after realizing that (i) suffering is intrinsic to life, (ii) desire causes suffering, (iii) ending desire can eliminate suffering, and (iv) the eight-fold path can eliminate suffering. These are the four truths that launched a religion some 2,500 years ago, one that is followed by half a billion people today. Not to mention the influence on several other religions. Four truths packed within 140 characters! Eat your outhanging heart, X/Twitter!

Having attained inner peace, he could have retired to a life of blissful solitude. Instead, he gave in to the desire to spread his message by travelling across the Gangetic plains. Having renounced worldly things, Buddha was not going to lodge inside houses and palaces. So he stayed outdoors, under the shade of trees, often in mango groves that were ubiquitous in the Gangetic plains. He headed to Varanasi, a centre of learning and orchards and the mango. He went to a deer sanctuary ten kilometres north of Varanasi; it later acquired the name Sarnath. More than a millennium later, when the Chinese monk Hiuen Tsang or Xuanzang came to India, he went to Sarnath in 637 CE; he found a large monastery with 1,500 monks. 'In the great enclosure is a *vihara* about 200 feet high; above the roof is a golden-covered figure of the Amra (mango) fruit,' he recorded. Why, one might ask, did the summit of the vihara have a golden mango? In early Buddhist sculpture, Buddha was represented through symbols; the mango tree and fruit were prominent among these.

Buddha delivered his first major sermon in the sylvan surroundings of Sarnath. He spoke of Dharmachakra, the wheel of Dharma. That idea has travelled all the way across time to characterize India's national flag in the navy blue of the Ashoka Chakra, the Mauryan king's interpretation of Dharmachakra. Buddha's concerns were existential and universal. He needed an ordinary language to get across to ordinary people: the kind of metaphors that turn abstract ideas into imaginable forms. The mango must have been handy.

An apt example comes from Theravada, the oldest extant school of Buddhism. Its Pali texts contain recollections of Buddha's disciples of his teachings. In a text titled *Puggalapannatti*, four types of mangoes become an allegory for four types of humans, contrasting appearance with reality. One, a mango might appear ripe but actually be unripe; two, it might appear unripe but be ripe inside; three, a mango might appear unripe and actually be unripe; or, four, it might be ripe in both appearance and content.

Buddhist literature teems with mango references, most prominently the Jatakas or the 547 retellings of Buddha's recollections of his past lives. Some texts say Buddha told these stories in the mango groves of Jetavana, a monastery and vihara outside the ancient city of Shravasti, now in Gonda and Bahraich districts of eastern Uttar Pradesh. In the 539th Jataka, an ascetic asks Buddha the reason for his renunciation. 'The Great Being' replies in verse worth quoting:

> 'I wandered through my royal park one summer's day in all my pride,
> With songs and tuneful instruments filling the air on every side,
> And there I saw a Mango-tree, which near the wall had taken root,
> It stood all broken and despoiled by the rude crowds that sought its fruit.
> Startled I left my royal pomp and stopped to gaze with curious eye,
> Contrasting with this fruitful tree a barren one which grew close by.
> The fruitful tree stood there forlorn, its leaves all stripped, its branches bare,
> The barren tree stood green and strong, its foliage waving in the air.
> We kings are like that fruitful tree, with many a foe to lay us low,
> And rob us of the pleasant fruit which for a little while we show.
> The elephant for ivory, the panther for his skin is slain,
> Houseless and friendless at the last the wealthy find their wealth their bane;
> That pair of trees my teachers were—from them my lesson did I gain.'[5]

Take a moment to digest that: the mango tree as Buddha's teacher! Then there's another story of a magnificent mango tree, perhaps in Vaishali. After eating a particularly delicious mango, Buddha asked his disciple Ananda to plant it at an appropriate spot. Then, he washed his hands above that spot. A grand white mango tree is said to have grown out of it.

The republic of Vaishali produced the most famous story linking the mango and Buddha. It has to do with Amrapali, the stunning courtesan we met earlier. She had heard of Buddha and his teachings and was present at his sermon in Vaishali; it made a deep impression. One version of this story says she sent him a parrot that talked to people in a human voice. It extended Amrapali's invitation to Buddha. Impressed by the bird, he accepted the invite and actually stayed at her mango grove. She invited Buddha over to her house for a meal; he accepted.

On the way back, her carriage collided with another. It had rich and influential men from Vaishali going to invite Buddha. They rebuked her as the 'mango woman', a euphemism for slut. She got her revenge by telling them Buddha was coming to her house for a meal, not to theirs. At her palatial house, she offered Buddha her mango orchards for the cause of propagating Dharma. Some versions say she offered everything she owned. Buddha accepted. Later, after women came to be allowed inside monasteries, she is said to have renounced the world for the life of a bhikkuni, attaining the peace she could not find in all the power and attention that her charms commanded.

The parrot's mango connection crops up in the lore of another non-Vedic religion: Jainism. Mahavira, Buddha's contemporary in the same region, is believed to have been the last of the twenty-four Jain tirthankars (founders and teachers). Jainism has its own cosmology and epistemology. In a Jain story, a parrot brought the 'fruit of immortality' from Harimela, an island in the middle of the ocean. In its 'north-east corner stood a large mango-tree, bedewed with ambrosia; and that the fruit of this tree restored youth by curing deformities, diseases, and old age.'[6] In another story titled 'The Adventures of Princes Amarasena and Varasena', a parrot couple leave behind the two fruits from two magical mango trees on a mountain called Sukuta.[7]

Ambika is a Jain yakshi/goddess who protects and attends to the twenty-second tirthankara Neminatha. Also called Amba or Amra, she sits on a lion, under a mango tree, with a mango fruit in hand; even her names come from the mango. She is the yakshi of—you guessed it—prosperity and fertility. The Kalpavriksha, too, is not restricted to the Vedic schools; it is also a symbol of munificence in Jainism and Buddhism. Their literature described mango varieties as well as practices like ripening mangoes.

It was in a mango grove that Buddha spent his last night in Kushinagar, now in eastern Uttar Pradesh. He stopped to rest in the metalsmith Chunda's grove. It is said he was given a meal of pork that was on the turn. A bhikhu or renouncer was required to eat whatever the householder gave him in alms. After Buddha had moved on, he was laid low by severe gastric distress. He had a message sent to assure Chunda that the meal he served was not the cause of his illness. Buddha died soon after and was cremated under a sal tree. Tathagata is one of Buddha's many names; it means 'he who thus came/went'.

'Gautama was born under an ashoka-tree, received enlightenment under a peepal-tree, preached his new gospel in mango groves and under shady banyans, and died in a sal grove,' wrote M. S. Randhawa. 'Never before or after has a religion been so much associated with vegetation.'[8]

Kunwar Singh Jonko, a farmer and tribal rights activist in Jharkhand's Chaibasa town, was telling me the story of getting back the land of his ancestors. I asked him how his father knew which part of the forest was theirs. He said they had two ways to tell: saasandari (tombstones) and planted trees not part of the natural sal forest. What kind of trees, I asked. Trees like the mango, he replied. An old mango tree inside the dense forest here can only have come from a seed planted a long time ago by those who lived nearby, who expected to eat its fruit. If there is a tombstone, that seals the matter.

I had met Jonko in 2003. He belonged to a Scheduled Tribe called Ho. (In some tribal languages of Jharkhand, 'Ho' means human; the elephant is called 'Marang Ho' meaning big man.) Jonko had taken me 60 kilometres inside the forests west of Chaibasa in Jharkhand's East Singhbhum district. His village Katamba—the name hints at amba or the mango—had been set up in 1978 as part of the Jungle Andolan, a movement to reclaim land from which the colonial government had evicted their ancestors a century ago. The British wanted to cut trees and obtain timber to make railway sleepers. So between 1878 and 1891, they notified the region as a forest reserve. This was a clear violation of the rules the British had themselves set up in a treaty that followed a violent uprising in 1831, called the Kol revolt.

In 1978, some cousins invited Jonko's father to reclaim their lands by force; that was the Jungle Andolan, another battle in the old war for a tribal heartland. But nobody was old enough to remember and identify the location of their former village. They went searching. 'We looked for saasandari and mango trees,' he said. Where they found both of these in proximity they re-established their village without asking the forest department. It was a violation of forest laws set up in 1871 and 1927; tribal people say the colonial laws violated their customary laws. They say the British forest laws were created to justify the injustice meted out to the tribal people by cutting down their forests.

This story is common to indigenous communities across the world. There's a saying attributed to Native American tribes: Before the White Man commits a crime he makes a law to justify it. Forest-dwelling communities were not the lawless heathens they were made out to be. Their customs worked through convention, as does the unwritten constitution of the UK. They are embedded in religious practices that interpret the natural world. Several beliefs and practices of India's Scheduled Tribes are ingrained in the mango.

The earliest textual accounts of tribal customs come from documents of colonial administrators and anthropologists. The mango features frequently in their creation stories, in the accounts of the first beings. The names of several villages of Scheduled Tribes such as Kondadora, Bhagatha and Nukadora in Vishakhapatnam district, Andhra Pradesh, have the suffix '-mamidi', which means the mango in Telugu; for example, Pulusumamidi. Some Gond communities in Odisha consider the mango a divine gift that sustains them in drought years; the mango fruits in the hottest time of the year. The Gonds eat not only the fruit but take out the seed kernel and grind it into flour in the summer. From a fleshy fruit of indulgence, the mango turns into a means of surviving scarcity and drought.

In the world of journalism, though, the mango kernel has acquired a negative connotation, especially with reference to the drought-prone tribal areas of Odisha. Every now and then, some news report or the other equates it with famine and starvation deaths. A few reports have traced fatalities to food poisoning after consuming mango kernel cakes or suchlike. It's not the mango kernels that cause the deaths, however; that has to do with eating stale or rotten food. When food items prepared with mango kernels are stored in the heat for a day or two, they attract harmful bacteria and fungi that can cause food poisoning. People eat stale foods made of mango kernel flour either because there is nothing else to eat, or because they are unaware of the danger. In one news report, the deceased tribals had malaria, but the newspaper headline mentioned only the mango kernel. It has become a tick, invoking the mango kernel to lend the report an alarming overtone of starvation.

Yes, starvation deaths point to the ineffectiveness of food security schemes. No, it is not down to some inherent evil in the mango kernel, which has long been a basic ingredient of the diets of the tribal people. The mango kernel's link with starvation is merely another example of how stories from tribal regions are reported: with more alarmism than understanding. There are far greater dangers to the Scheduled Tribes that are underreported, such as land alienation due to rampant mining. The mango kernel provides allegorical compensation for lacunae in research. The forest dwellers are not readily accessible to urban reporters; besides, they are not represented by spokespersons and lobby groups. As Jonko explained to me, the evidence is not to be found in books and experts. Often, it is found in tombstones and trees.

Chapter 6

History
The Great Leveller

'On account of the hardy nature of the tree, low cost of maintenance, and profuse yield, it has come to be known as the poor man's fruit, and thus possesses mass appeal.'

—LAL BEHARI SINGH

In 1949, socialist leader Ram Manohar Lohia was arrested for violating curfew. He was leading a protest against the ruling dynasty of Nepal in front of the Nepalese embassy in New Delhi. He spent a month and a half in prison; along with him was a young activist named Rajinder Sachar. While in prison, Lohia received a basket of mangoes, Sachar was to later write: 'It was a gift of goodwill from Prime Minister Jawaharlal Nehru.'[1]

Vallabhbhai Patel, the deputy prime minister, was upset about the message this sent. Nehru replied that Lohia was not a criminal but a political activist and that personal and political relations should not be mixed, noted Sachar, who many years later went on to serve as the chief justice of the Delhi High Court. Nehru had always looked upon Lohia with soft eyes; after all, Nehru was one of the earliest promoters of the young firebrand in the 1930s, when, along with some others he had formed the Congress Socialist Party. As the years rolled by, Lohia turned into Nehru's staunch critic. In sending the mangoes to Lohia, Nehru was remembering the old warmth, not the new hostility.

The mango is a gesture of detente in statecraft. For example, whenever India and Pakistan resume dialogue after a period of hostility, mangoes are summoned. A basket of mangoes is a tried and tested way to disarm an antagonist and appear in a benevolent light, and it has been so for centuries. Gifts of mangoes have influenced geopolitics, history, and the maps of empires. Some of the biggest wars in India's history were fought in mango groves; the results of wars have been attributed to mangoes.

'History, as we know it, is a record of the wars of the world, and so there is a proverb among Englishmen that a nation which has no history, that is, no wars, is a happy nation,' wrote M. K. Gandhi in his 1909 manifesto *Hind Swaraj*. 'How kings played, how they became enemies of one another, and how they murdered one another is found accurately recorded in history, and, if this were all that had happened in the world, it would have been ended long ago.' The mango fruit fuelled Gandhi during his campaigns for non-violent mobilization. 'At one time in Bengal when I was working very hard I lived entirely on mangoes,' he told Louis Fischer, an American journalist who went on to write Gandhi's biography.[2]

Fischer first met him in 1942 when he spent a week with Gandhi at the Sevagram Ashram in Wardha. He described seeing Gandhi at the crack of dawn each day, eating mango pulp. 'I again found him scooping mango sauce out of a deep glass,' he wrote on 7 June 1942. Two mornings later, he found 'Gandhi was having his mango meal'. In his 1891 speech to the London Vegetarian Society, Gandhi had called it 'the most delicious fruit I have yet tasted'. Gandhi's fondness for mangoes was such that the fruit escaped his notoriously quirky experiments with diet.

The complicated relationship comes out in 'Mangoes and Mahatma', the concluding chapter of American historian Nico Slate's 2019 book on his diet. 'Mango is a cursed fruit. It attracts attention as no other fruit does. We must get used to not treating it with so much affection,' Gandhi wrote in a 1941 letter. Beginning in 1906 in South Africa, he had set about turning his life into that of an ascetic. He wrote about his human struggles to master his senses, at least by the superhuman standards he had set for himself.

'The ambivalence in Gandhi's relationship to mangoes reveals a larger tension at the centre of his diet, a tension between seeing food as an instrument of service and celebrating food for its own sake,' wrote Slate. Gandhi's mango travails reveal his struggles. 'At stake was his ability to live a life that was passionate but not selfish, joyful but not lustful. The difficulty of discerning healthy desire from unhealthy lust is revealed by the most important detail in Gandhi's encounter with the mangoes he yearned to share: their intended recipient.'[3] It was Sarladevi Chowdhrani, a forty-seven-year-old activist he had met in 1919 in Lahore, when he was fifty; they kept their intense affection platonic. In a 1920 letter to her, he wrote: 'Revashankerbhai came in this morning. He brought some luscious mangoes. I fretted to find that you were not here to share them.'

His ambiguities and uncertainties, wrote Slate, turned his diet experiments into a source of connection: 'That is the lesson of the luscious mangoes. Our struggles with food are never ours alone.'[4] Growing mangoes is its own way of establishing human connections. In 1932, during one of his nineteen imprisonments, Gandhi planted the seed of a mango he had eaten in Pune's Yerwada prison. Upon getting released, he took the sapling with him in a pot; he was practically homeless now, having vowed not to return to Ahmedabad's Sabarmati Ashram till the country was freed; he never did go back. The sapling is now a tree in Hyderabad in what was once the house of Congress leader Sarojini Naidu.

India remembers its historical figures by their association with the mango. Yet that's no easy task; leaders do not document their lives as rigorously and honestly as Gandhi did. Historian Damodar D. Kosambi complained in 1956 about the lack of material despite India's great literary heritage: 'Many Indian kings of the middle ages (e.g. Harsa circa 600–640) were incomparably superior in their education and literary ability to contemporary rulers in Europe; they had personally led great armies to victory in heavy warfare. Nevertheless, not one seems ever to have thought of composing a narrative like Caesar's Commentaries.'[5]

Many historians have worked to fill the gaps apparent in the 1950s. Their research shows how many of India's historical events and figures were linked to the mango. The fruit becomes a means to tease out the political attitudes and economic policies of rulers. How administrators approach the mango reflects their views on public welfare or natural resource management or their cultural connections with the people they govern. This can throw up one or two surprises.

The fruit's social role kept surfacing in my conversations with old people in Lucknow. Amid over-the-top stories of the mango passion of Awadh's nawabs and Pathan warriors, somebody or the other pointed out that the mango is a qaumi or awaami fruit. Qaum means common property in Arabic, but can also mean nation, community, caste, or clan. Awaam, also from Arabic, means common. The mango was sustenance for the poor, I heard from at least three people in Lucknow. When the range of food items dried up during the summer, poor people ate their roti with mangoes.

Experienced people in Bihar echoed this observation. 'It was the poor man's fruit. But now there are no mangoes for the needy,' said Binodanand Jha, retired director of Bihar's education department and a source of customary knowledge. 'On account of the hardy nature of the tree, low cost of maintenance, and profuse yield, it has come to be known as the poor man's fruit, and thus possesses mass appeal,' wrote agriculture scientist Lal Behari Singh in a widely cited 1960 book.[6]

As mentioned earlier, ancient scholars like Varahamihira and Parashara encouraged the planting of useful trees like the mango.

This was a matter of policy early on, at least in the 3rd century BCE. The edicts left by the Mauryan emperor Ashoka illustrate this. One such testimonial was inscribed in a pillar edict erected at Topra, near present-day Ambala; in the fourteenth century, it was brought to Delhi's Feroze Shah Kotla, where it stands today. It mentions that the king had mango groves planted along the highways. The pillar erected at Allahabad, now Prayagraj, includes his queen Karuvaki's edict. It says: '...the officers everywhere are to be instructed that whatever may be the gift of the second queen, whether a mango-grove, a monastery, an institution for dispensing charity, or any other donation, it is to be counted to the credit of that queen...'

The mango has long been an element of statecraft. Mauryan ambassadors helped spread both the mango and Buddhism. Monks took the Indian mango to parts of Southeast Asia. Subsequent rulers kept following the policy of mango propagation. Various Gond rulers of central India planted mango trees along their roads. This policy is revealed in the chronicles of the Moroccan Ibn Battuta, who travelled to various parts of the Indian peninsula in the 1330s and 1340s. The mango is the first entry in his account of Indian trees and fruits. In his *Rihla* he described Koil (today's Aligarh) as 'a handsome city possessing gardens. Most of the trees are mango trees.'

Sher Shah Suri, a sixteenth-century ruler from Sasaram in Bihar, had mango trees planted along the Grand Trunk Road, the ancient Mauryan-era highway that he had rebuilt. Historians have noted that during the sixteenth and seventeenth centuries 'fruits of the common kind, mangoes, melons, berries, coconut, etc., were available to the poor in season'.[7] Historian Lallanji Gopal wrote that in Mughal times, 'anyone who converted his cultivated land into an orchard was entitled to get all his revenue remitted.'[8] In Bihar's Mithila region, mango orchards were not taxed even in the twentieth century.

Horticulture also became a means to increase revenue in Maharashtra. 'During the Peshwa period when financial difficulty became acute and the revenue rate was raised, the cultivation of wasteland and of cash-crops such as mangoes, coconuts, and betelnuts was specially encouraged by the state...such a promotion became a duty of local officials,' said an economic history of India.[9] The mango has been a component of welfare and economic policy for all kinds of rulers: from large empires to small estates, ambitious military adventurers to debauched nawabs, and enlightened governors to powerful bigots. These included a number of rulers except the British. To be sure, they did like the mango fruit. British travellers tasted and admired the mango, penning accounts of its popularity. British botanists wrote books about it and played a hand in creating grafted varieties. British land surveyors noted its economic worth. 'Mango on the roadside is a profitable source of income,' said the 1892 *Report on Arboriculture* from Gurdaspur, Punjab. The 1899 *Report on the Land Revenue Settlement, Nagpur District*, signed by one R. H. Craddock, said: '...it is a meritorious act to plant a mango tree on account of the food and shelter it yields'. The mango kept giving.

But the British land policy had no room for the well-being that the mango generated. With the Permanent Settlement formalized in 1793, the new zamindari system had begun to tell on the well-being of ordinary people by the early nineteenth century. 'The common lands of the village (shamilat) also suffered invasion by "the newly-created zamindars, who have all a propensity to cutting down mango topes, and appropriating to themselves tanks, wells, and grazing lands",' wrote historian Eric Stokes, formerly a subaltern or a junior army officer, in a 1980 book. It quotes a British official in Kanpur as saying the 'new zamindar was "very seldom...so imprudent as to live within the village he has acquired", but employed an agent "who, backed by the authority of Government, is able to realise revenue, and seize upon everything visible and tangible".'[10]

There was a shift in attitude. The qaumi or awaami fruit disappeared from state policy. When the British began setting up railways in India in the 1850s, there was no effort to plant mango trees along the railroad. Quite the opposite took place: widescale logging began for railway sleepers. 'A great deal of India's deforestation can be traced to the colonial period. It is also important to study the ecological impacts of colonialism such as deforestation because these colonial empires were the forerunners of contemporary globalisation. One of the most important causes of deforestation in India was the building and expansion of the railways,' wrote historian Pallavi Das in her 2015 book on the railways.[11]

This unprecedented deforestation disrupted the hydrological cycle of several regions, leaving them vulnerable to drought and famine. It also led to conflicts with forest-dwelling tribal communities, causing a series of hools (revolutions) in the tribal heartland. Even Indian princely rulers switched priorities to align with the colonial development plan. In Alwar, the maharaja seized control of community forests and sold them to railway contractors; villages increasingly began suffering drought.[12]

In 1911, the British decided to build a new city near Delhi to move their capital. There was much discussion on the trees that were to adorn New Delhi. 'The archival record is rich with reports and memos by foresters, horticulturists—even civil servants with strong opinions—who argued for or against candidate species,' wrote author and naturalist Pradip Krishen in his 2006 book.[13] 'One factor, above all, stands out—all their lists contain a clearly stated prejudice against deciduous trees that go bare and remain unsightly for some period in the dry season. That is clearly why the amaltas, mango, siris, and shisham, for example, did not make the cut.'

A policy of indifference felled the tree of desire. What if the British had taken to the awaami fruit like previous rulers! Would it have made their rule less destructive? It is a question worth asking, because the mango has helped localize foreign rulers for long.

Awaam is the plural of the Arabic word aam; it means that which is widespread and general, marginal not central. In most of North India, the common word for the mango is aam from the Sanskrit aamra. This leads to a surfeit of puns on aam aadmi or the common man. The aam was central—not marginal—to India's aam aadmi.

⁂

'Sikandar ne Porus se ki thi ladhai, jo ki thi ladhai, to main kya karun...' ran the refrain of a popular song from the 1962 Hindi film *Anpadh* (Illiterate). The lyrics ask, rhetorically: 'If Alexander fought a battle with Porus, how is it any concern of mine!' The song played often inside my head during history classes in school. India has a lot of history, way too much for your garden-variety non-historian. Yet there are rewards, like reading about Alexander's soldiers eating green mangoes and getting dysentery. Alexander the great, undefeated general, laid low by the mango loosies! Delhi Belly, 326 BCE!

The reference occurs in *Historia Plantarum* (Enquiry into Plants), an important classical source on natural history that influenced modern botany and taxonomy. Its author was Theophrastus, probably a disciple of Plato and a contemporary of Aristotle. His observation has been translated thus: 'There is also another (tree) whose fruit is long and not straight, but crooked, and it is sweet to the taste. This causes griping in the stomach and dysentery; wherefore Alexander ordered that it should not be eaten.' Translators believe this fruit was the mango. *Hobson-Jobson* mentions one or two other sources as saying that Alexander prohibited his soldiers from touching the fruit.

There's another reason this caught my attention. I've heard similar anecdotes, all fantastical, about other battles; the soldiers on the losing side are always said to have had either too many mangoes or been denied mangoes. (Yes, Alexander did not actually lose, but his victory was Pyrrhic; he ended his campaign in Punjab and headed back.) I heard a similar story about the Battle of Chausa, fought on 26 June 1539, between the armies of Sher Shah Suri and Babur's son Humayun. Sher Shah's forces surprised their opponents at night, driving Humayun out of India.

I've heard jokes in Lucknow and Varanasi mocking Humayun's soldiers for consuming too many Samar Behisht mangoes, a name sometimes used for Chausa, a prized northern variety. There are several references to Sher Shah naming the Chausa variety to commemorate his victory—that the mango variety was irrigated with the blood of the soldiers who fell at Chausa. There's a temptation in folklore to ascribe loss in battle to the bewitching appeal of mangoes.

A full 218 years later, on 23 June 1757, a battle fought in a mango orchard changed Indian history. This one occurred in a Bengal village called Palashi, after the palash tree, the flame of the forest, Anglicized to Plassey. The British East India Company had brought its forces from Kolkata, about 160 kilometres south. They were up against Siraj-ud-Daulah, the nawab of Bengal, based in Murshidabad, 50 kilometres north of Palashi. The well-appointed Bengal army had more than 50,000 infantry and cavalry, supported by the French East India Company and nearly sixty field guns.

The British contingent numbered less than 3,200, all included, with nine field guns. Their forces had set up near the Hooghly River in Laksha Bagh, a mango grove spread over almost 50 acres. But the British had a covert alliance with several of Siraj's associates, including his main general Mir Jafar. Still, what reversed the odds was a monsoon downpour at noon. Accustomed to English weather, the British had covered their gunpowder under the additional cover of the mango trees. The Bengal army did not. Their gunpowder turned to soggy gloop and the British guns fired grapeshots at charging soldiers.

This British victory was the beginning of India's imperial colonization. Historians tell us the British won because of a complex range of reasons, including Mir Jafar's treachery. Historians with a sense of humour say it was the lack of cover. But it's not difficult to find non-historians who say it was all down to the mango. That the British soldiers were satisfied after eating mangoes of Laksha Bagh, while Siraj-ud-Daulah's soldiers watched them from a distance, desperate for mangoes, unable to focus.

That Siraj lost near a mango orchard had a ring of retribution. One legend, almost certainly apocryphal, says Siraj once had a servant beheaded because he cut a special mango badly. It had to be cut with a bamboo knife so that the flesh did not get tinged with a metallic taste. (Modern experiments show the mango tastes best when handled with gold cutlery.) This refined mango sensibility came to an abrupt end when Mir Jafar's son had a metal weapon put into Siraj's flesh.

Siraj was not the only ruler with mango quirks. Ibn Battuta picked up a story in the Malabar region in the fourteenth century: 'I was told that the raja of Quilon (Kollam) rode out one day in the environs of the city. His way led between gardens and with him rode his son-in-law, who came of the royal stock. The latter picked up one mango fruit which had fallen out of the orchards. The raja gave him a look and straightway pronounced the sentence of death. He was cut and divided into two halves; one half was nailed to the right of the road on the cross, the other to the left. The mango fruit was likewise cut into two halves, and one half of the fruit was placed on each half of the corpse. Thus he was left as a warning to the people.' Ibn Battuta did not specify the nature of the warning. What did it say? 'Do not touch my mangoes, even when they fall out of the orchard'? We do not know. That is the trouble with history. The important questions never get answered.

Such stories show that in India, mangoes comprise the pinnacle of temptation that lead men astray, like alcohol in other places. I have noticed this repeatedly in social settings. When rebuking somebody for making a mistake or losing concentration or falling asleep, a person might ask: 'What happened? You had too many mangoes?' If nothing else explains distraction and failure, military or civil, bring on the mango. No excuse is greater!

The First Battle of Panipat was fought near a mango orchard in 1526. Delhi's sultan Ibrahim Lodhi lost to Babur. The Third Battle of Panipat in 1761 is believed to have taken place in Kala-e-Aam, supposedly a grove with black mangoes. There are stories claiming that the mangoes turned black due to the blood spilled in the battle between the Maratha confederacy and Ahmed Shah Abdali. On the outskirts of Panipat in Haryana, stands a memorial to the battle, called the Kala Amb Park. A red obelisk has been placed at the spot where a black mango tree is said to have once stood. Under that tree, the Maratha commander Sadashivrao Bhau is said to have died. The website of Haryana Tourism says 'people come here for a stroll in the peaceful surroundings'.

It is not that the mango incites war. It is unlikely that mango groves called out to belligerent rulers looking for a suitable site to battle. More likely, mango groves were so commonplace in India that you were never far from one, whether you were fighting a battle or not. It is just that wars get recorded in history. A bunch of people going to a mango orchard for a cookout is not history. No, that is life.

Babur was nuts about fruits. His memoir *Baburnama* begins with a description of his hometown, now in Uzbekistan. 'Fergana is a small country, abounding in grain and fruits,' says the second paragraph.

'It produces...fruits in abundance, excellent grapes and melons. In the melon season, it is not customary to sell them out at the beds. Better than the Andijan nashpati, there is none,' reads the fifth paragraph. 'It has running waters, beautiful little gardens and many fruit-trees but almonds for the most part in its orchards,' it says a few paragraphs later. '...there is Marghinan...a fine township full of good things. Its apricots and pomegranates are most excellent. One sort of pomegranate, they call the Great Seed; its sweetness has a little of the pleasant flavour of the small apricot and it may be thought better than the Semnan pomegranate. Another kind of apricot they dry after stoning it and putting back the kernel; they then call it subhyani; it is very palatable,' it says after a few pages.

It seems ridiculous, how he keeps going on like a gardener. Babur's plantmania might be exceptional but it has a context that is worth a closer look. It did not just lead to the form of the Mughal garden; it also initiated a new culture of fruits in India, adding a layer to the culture of the mango. The gardens surrounding Humayun's Tomb in Delhi or Taj Mahal in Agra follow a geometric plan. 'Gardens of the Mughal period in India belong to a historical tradition of formal gardens extending over three continents, and at least five centuries. From West Asia and Persia to North Africa and Southern Europe, i.e., Moorish Spain, and in the East to Central Asia and the Indian subcontinent,' wrote Mohammad Shaheer, a reputed landscape architect and scholar.[14]

The big influence behind the four-fold layout of the Mughal garden is the Persian chahar bagh.[15] '...while Persian tradition has been one of the main forces in the evolution of the paradise garden, its origins are far older than the Persian empire. It is indeed described in the Book of Genesis. 'And a river went out to Eden to water the garden: and from thence it was parted and became into four heads,' wrote Sylvia Crow in her 1972 *Gardens of Mughal India*.

Gardens have been central to life in Central Asia with its dry terrain and extreme weather. In parts of Afghanistan, temperature can vary up to 40°C within a day. The lowest temperature recorded is −54°C; the hottest summer temperature recorded is 54°C.[16] It is not amenable to agriculture. Yet some regions here are ideal for horticulture, especially in the northern parts. Afghans specialized in the crops they could grow: pomegranate, grapes, apricots, almonds, and apples. Besides, on both sides of this region lay ancient civilizations that were gardening powerhouses: China to the east and Persia to the west. Just throw in the Silk Road, revived in the thirteenth century under the Mongol empire, and you've got yourself a horticultural superhighway.[17]

This is where Babur came of age. How did he find time from his fruity obsession to recruit an army and indulge his military ambitions? Poorly, it must be said. Born on 14 February 1483 in the Fergana Valley of what is now Uzbekistan, Babur became a king at the age of eleven. His father had died in a freak accident; he was attending to pigeons in the dovecote at the edge of the palace; it collapsed and flung him to his death into a ravine underneath. (Since Babur's son Humayun was to die sixty-two years later after slipping on the stairs of his library in Delhi's Old Fort, perhaps Babur was better off admiring melons in open gardens.) Three years later, Babur seized Samarkand, only to lose Fergana to ambitious relatives. When he went back, he lost Samarkand; this yo-yo continued for a while.

He moved to Kabul, turning his ambitions southwards to Hindustan. After his son's wedding festivities, he slept in the garden. In the morning, as light broke, he offered his prayer and asked for a sign: 'God! if the government of Hindustan is destined to be given to me and mine, let these productions of Hind be brought presently before me, betel-leaves and mangoes, and I shall accept them as an omen.' The sign arrived soon after in the form of half-ripened mangoes, soaked in honey for preservation, all the way from Lahore, in the hands of Dilawar Khan, son of the governor of Lahore, Daulat Khan Lodhi. He was conspiring against his cousin Ibrahim Lodhi, who was not very popular because of his inclination to torture and behead people randomly.

'When Babur's eyes fell on the fruit, he arose from his throne, and prostrated himself before the Almighty, who, he was persuaded, of His boundless generosity, had granted him the sovereignty of Hind.' This oh-so-perfect story is not from Babur's memoir but from *Tarikh-e-Salatin-e-Afghaniyah*, attributed to one Ahmad Yadgar, a courtier of Jahangir's. Its content is copied from earlier works. Even if entirely fictitious, the account does reveal a desire among Babur's descendants to colour his turn towards India with something essentially Indian: mangoes.

Once in control of Delhi, Babur was struck with meloncholia, a yearning for melons. He wrote in a letter: 'How should a person forget the pleasant things of those countries, especially one who has repented and vowed to sin no more? How should he banish from his mind the permitted flavours of melons and grapes? Taking this opportunity, a melon was brought to me; to cut and eat it affected me strangely; I was all tears!' He noted on 24 June 1529, in Agra: 'A Balkhi melon-grower had been set to raise melons; he now brought a few first-rate small ones; on one or two bush-vine I had had planted in the Garden-of-eight-paradises very good grapes had grown; Shaikh Guran sent me a basket of grapes which too were not bad. To have grapes and melons grown in this way in Hindustan filled my measure of content.' He went to great lengths to score Central Asian fruit. On 7 July 1529, he wrote: 'Today wages were given to 150 porters and they were started off under a servant of Faghfur Diwan to fetch melons, grapes, and other fruits from Kabul.'

People knew the new sultan was cuckoo about fruits. If you wanted time with Babur, you did not have to be imaginative to think of a gift. 'The Mughal upper stratums spent extravagantly to grow or import exquisite fruits,' wrote researcher Anku Bharadwaj in a 2015 paper on food symbolism among the early Mughals.[18] 'Gifts of fruit were a matter of protocol in the upper stratums of Mughal society, and were an implicit language of diplomacy. For example, Babur gave away oranges as gifts; he bestowed the fruit of two trees on Shah Hasan, gave the fruit of one tree each to several baigs and to some, he gave one tree for two persons.'

This became the standard for Babur's descendants. 'The entire Mughal court was conversant with the political language of fruit,' wrote Lizzie Collingham, a University of Cambridge historian 'interested in linking the minutiae of daily life to the broad sweep of historical processes'.[19] Babur begat Humayun, who was driven away by Sher Shah. His son Akbar set up an 'imperial fruitery, staffed with horticulturists from Persia and Central Asia', wrote Collingham. Akbar begat Jahangir who 'wrote at tedious length on the merits of apples from Samarkand and Kabul', on 'exactly how many cherries it was possible to eat at one sitting', on 'his uncle's apricot trees', and 'the astonishment of the sheikhs of Ahmadabad at the superiority of Persian melons over those grown in their native Gujarat'. Jahangir was followed by Shah Jahan who, like his father, 'took particular delight in having fruit weighed' in front of him. Jahangir begat Aurangzeb, who was presented with '100 camels loaded with fresh and dried fruit and nuts' when the ruler of Balkh in Afghanistan was trying to impress him.

Foreigners were astonished to discover the proportion of their income that the Mughal noblemen and administrators spent on fruit. Thomas Roe, English ambassador who visited Jahangir's court, failed to recognize the compliment he was paid when Prime Minister Asaf Khan sent him a basket of twenty musk melons. Collingham quotes him complaining, curmudgeonly, that 'all I have ever received was eatable and drinkable,' and that the Indians must 'suppose our felicity lies in the palate'. Roe didn't realize he'd been sent the highest gift of the Mughals.

Several historians have pointed out that fruits were a way for the Mughals to feel connected with their homeland in Central Asia. Persian melons and apples from Samarkand weren't just fruits; they marked a culture. Apart from the gold and silver, the monsoon rains and 'unnumbered and endless workmen of every kind' that Hindustan provided, Babur did not really take to his new territories in the four years he spent in India. His remains were later taken from Agra to be laid to rest in Bagh-e-Babur in Kabul. 'The towns and country of Hindustan are greatly wanting in charm. Its towns and lands are all of one sort; there are no walls to the orchards,' Babur observed. 'Hindustan is a country of few charms.'

There was only one fruit of Hindustan he admired, though: 'Mangoes when good, are very good, but, many as are eaten, few are first-rate...Taking it altogether, the mango is the best fruit of Hindustan. Some so praise it as to give it preference over all fruits except the musk melon, but such praise outmatches it.' And this is where a shift happens, because even as his descendants paid homage to Central Asian fruit, the mango was their favourite. Akbar remarked: 'This fruit is unrivalled in colour, smell, and taste; and some of the gourmets of Turan and Iran place it above musk melons and grapes.'

This is when a mango orchard of one lakh trees was set up in Tirhut called...Lakhi Bagh. There are references to orchards with one lakh mango trees in Samastipur (now in Bihar) and Palashi. Skilled horticulturists were invited from Persia and Central Asia and the post of Darogah-e-Bagh (inspector of gardens) was created for the administration of royal gardens.[20] Orchards began to get tax breaks.

Jahangir made it plain that 'notwithstanding the sweetness of the Kabul fruits, not one of them has, to my taste, the flavour of the mango.' His patronage of quality mangoes resulted in mango cultivars being collected from far and wide and grown in orchards of what is now western Uttar Pradesh.[21] There is an account of a grower surprising him with mangoes late in September and then in October.

By the time it came to Shah Jahan's reign, Mughal nobles running his military campaign in Uzbekistan felt like foreigners. This was no homecoming to the land of Babur; in fact, they disliked Central Asia and spoke longingly of home back in Hindustan. While Babur paid couriers to bring fruits from Kabul, Shah Jahan had a courier service set up to bring quality mangoes from coastal Maharashtra to Delhi.[22] In 1663, the best mangoes in the Delhi market came from Bengal, Golconda, and Goa, wrote French traveller and physician Francois Bernier, who had attended on Aurangzeb and his brother Dara Shukoh.[23] Shah Jahan accused a young Aurangzeb of eating the best mangoes from the emperor's favourite tree in the Deccan, when duty required him to send the fruit to his father.

Historian Audrey Truschke wrote a biography of Aurangzeb in 2017. It has an account of the twenty-five-year-old prince visiting his aunt in Burhanpur, now in Madhya Pradesh. There, in 1653, he saw Hirabai Zainabadi, a singer and dancer, playfully pluck a mango in an orchard.[24] (The situation makes one recall the 1975 blockbuster *Sholay*, in which the protagonist Veeru helps Basanti pluck raw mangoes.) The two fell in love. It is said the young prince was even willing to drink wine for her. She died within a year. Aurangzeb is described as a vegetarian who ate millet bread, drank only Ganga water, and kept long fasts. After seizing power, he imprisoned his father, telling Shah Jahan that he would have to eat just one dish for the rest of his life. On his cook's advice, Shah Jahan asked for dal, which could be prepared in a different way throughout the year.[25]

Aurangzeb's austerity did not apply to mangoes. He often demanded he be sent baskets of the fruit. He is said to have named two mango varieties with Sanskrit names: Sudha Ras (nectar) and Rasna Vilas (luxury of the tongue). He wrote frequently about mangoes, complaining if the consignment got spoiled in transit.[26] Wrote Truschke: 'Aurangzeb...was a man of studied contrasts and perplexing features...He was angered by bad administrators, rotten mangoes, and unworthy sons.'[27]

Following Aurangzeb's death in 1707, his successors wilted in the face of growing Maratha military power to the south and the British East India Company to the east. But a new culture of fruits had taken hold in India with formal eating and ornate gifting of fruits. 'The ruling class also invested directly in...horticulture. Everyone from "the emperors to rich peasants" had orchards producing for the market. The fact that India began to grow a variety of fruits adopted from Central Asia, Iran and the New World owed much to this aristocratic enterprise,' said an economic history.[28]

The Mughal obsession with the mango radiated to two great centres of its refinement: Tirhut (later called Raj Darbhanga) and Murshidabad. The Lakhi Bagh of Darbhanga initiated a new emphasis on horticulture in a region steeped in stories of Buddha's mango groves. In 1882, the Darbhanga ruler wrote to the botanist Joseph Hooker for help with setting up their gardens. Hooker recommended Charles Maries, who then became the superintendent of gardens, setting up the famed Anandbagh Botanical Gardens surrounding the Darbhanga Maharajah's palace. The horticultural practices he compiled and created are still in use.

Murshidabad's mango history is not yet documented. Faruque Abdullah, a lecturer in history at a college in the city, has painstakingly gathered written and oral accounts from the descendants of the nawabs, and also people from other prominent families. It appears that the second nawab of Bengal, Shuja-ud-Daula (r. 1727–1739), was especially fond of mangoes. It was under his rule that gardeners and skilled horticulturists were invited for refining mango cultivation. He commissioned a research station called Ambekhana or House of the Mango under the supervision of a mango manager called Aam Tarash.

The nawabs used to invite special guests for their elaborate mango parties. If a certain mango became perfectly ripe in the middle of the night, the nawab and the guests woke up at that hour to enjoy it at its best. While trying out several varieties in one go, a palate cleanser was required; this was a variety called Anaras, slightly sour and pineapple-flavoured. We have already considered the delicacy of Kohitoor. This obsession with fine mangoes influenced the courtly patronage of horticulture in several parts of India known for fine mangoes: Lucknow, Hyderabad, Junagadh. Among the wealthy merchant families and princely states, the mango manager became a common position.

The influence of Central Asian fruit culture goes unnoticed. Arun Kumar, a journalist and socio-political activist, opened my eyes to it. He was born in 1941 to a prosperous family in Kandla, a town in western Uttar Pradesh, in the heart of rich orchard territory. He once told me about his uncle Gyan Prakash, a firebrand local leader of Hindu nationalist outfits. 'When it came to selling the crop of his mango orchard, however, he refused to give it to Hindu gardeners,' said Kumar, recalling an incident in 1948. 'He didn't say no to the man, just something innocuous. After he left, he said Pathan gardeners of Afghan extraction had raised the orchards with love and care, so he was only going to sell the crop to them.' It went further. Arun Kumar recollected, with great excitement, this one time when a cow had given birth to a calf; his uncle refused to give it to Hindus who came asking because they needed a bullock. 'I'll give it to my Pathans. They look after animals very well,' he remembered his uncle saying.

He was a treasure trove of anecdotes, revealing forgotten aspects of social life. 'The Pathans in our region were very good with making grafts of trees and they were committed horticulturists, with a passion for fruit-bearing plants,' he once told me. It is a stereotype but there are cultural reasons behind stereotyping. The most popular image of the Afghan fruit merchant features the actor Balraj Sahni in the 1961 film *Kabuliwala*, based on Rabindranath Tagore's 1892 short story of the same title. The protagonist, a dry fruits seller, misses his daughter in Kabul, so he brings fruit for the young girl Mini and doesn't accept payment.

Across the country, the Muslim presence in the fruit trade, especially the mango, is much bigger in proportion to their population. As it happens in India, Muslims have castes too. One such caste is Rizwan, a name used for those who tend to the gardens of the biblical Paradise. I've met Rizwan families that were invited to form nurseries in western Uttar Pradesh and Bihar. I found corroboration from somebody who knows the sector inside-out. At Mother Dairy, Omveer Singh had handled the operations of fresh fruits and vegetables; when I met him in 2019 he was the managing director of the National Dairy Development Board. He said that when he entered this sector in 1986, more than 90 per cent of fruit wholesalers were Muslims. He said that currently, the figure must be 70–80 per cent: 'Not just fruits, about 50 per cent of the wholesalers of fresh vegetables we deal with are Muslim.'

The spread of mango husbandry is a metaphor for a sort of cultural joining, like grafting together two varieties. It shows how communities brought together by awkward historical events learn to tolerate each other, finding some use in each other. This does not happen through cliched mystical-religious experiences but by sharing mutually admired objects like the mango. Consider the current customs of celebration. It is now common across northern India to gift boxes of dried fruits on Diwali. This is a Punjabi custom owing to the region's proximity to Afghanistan.

Little facts remind us of such links. Actor Dilip Kumar was the son of a fruit merchant from Peshawar. Music industry mogul Gulshan Kumar Dua assisted his father in running the family fruit juice shop in Delhi's Daryaganj, before hitting paydirt in the devotional music wave of the 1980s. The fruit juice corner became a feature of Indian cities only after the Partition in 1947, with the coming of refugees. I've noticed the name Chaman on some such shops. In Farsi, it means a garden. Traditional horticulturists use it often. 'They used to say bagh chaman ho gaya (the garden is thriving),' Arun Kumar had said. It was a common name among Hindu and Jain business communities of northwestern India.

Chaman is also the name of a town along the Afghanistan-Pakistan border; it is known for fruit trade. But Afghanistan's horticulture is in shambles now. 'In 1972, horticultural commodities supplied 40-60 per cent of all export earnings. Dried fruits from Afghanistan once accounted for 60 per cent of the world's market,' said a 2016 research paper.[29] 'Horticultural production is now estimated at less than 30 per cent of 1978 levels. Many fields are abandoned, many orchards destroyed; tree nurseries, seed sources, water, input, and knowledge are limited or non-functional.'

The Afghan Civil War damaged the Gardens of Babur in Kabul. Since 2008, it has been restored with the help of the Aga Khan Trust for Culture (AKTC). A few years ago, I met the director of its Delhi office, conservation architect Ratish Nanda, a former associate of architect and scholar Shaheer. The trust has worked to restore the Humayun's Tomb complex in Delhi. Completed in the 1570s, the tomb is the prototype of what went on to be called Mughal architecture; its most famous example is the Taj Mahal. Nanda said Mughal architecture was not based on Iranian influence, as is commonly believed. He had travelled through Iran and parts of Central Asia, looking to trace its roots. He found some influence, but also found signs of Mughal influence on Iran. He explained how the tomb's grand scale, the site plan, the material used, the four-dome layout of the edifice...it was all unprecedented.

As we walked about the tomb, flocks of rosy starlings appeared like dark clouds in the sky, conducting their aerial dance that ornithologists call murmuration. It was March; the colours of spring were taking over. The rosy starlings, long-distance travellers, were headed to their summer refuges in the Himalayas and Central Asia. It occurred to me that they might stop by the Gardens of Babur in Kabul. Newspaper features say the greenery of the gardens is a great respite for residents of the city ravaged by decades of war and violence.

The mango's political utility is old but not outdated. Each prime minister of India must have his or her mango anecdotes. Ajit Prasad Jain, India's agriculture minister in 1957, wrote: 'They sometimes called mango the bathroom fruit, but one has to see Jawaharlal Nehru deftly slicing the mango on the banquet table—two pieces apart with stone removed. Ladies with delicate fingers are often seen envying him.'[30]

On foreign travels, Nehru took cases of mangoes along with him. He gifted such cases to writer George Bernard Shaw and to the US President John F. Kennedy in 1961. He handed out lessons on eating the mango to Soviet leader Nikita Khrushchev, who 'enthusiastically adopted the squeezing and sucking style of eating mangoes', wrote Vikram Doctor.[31] '...but Chinese premier Zhou Enlai was shown the more formal slicing and spooning method. A correspondent of *The Times of India* reported in 1955 that as Zhou ate his mango 'his beetling brow relaxed, his lips rippled into a smile.... He had entered a new world of sweetness and goodwill. Thereafter, he ate out of Mr. Nehru's hand and signed the famous joint declaration.' Seven years later, China was to invade India and start a war, becoming a strategic ally of Pakistan's.

Pakistan was not to be left behind. In 1968, its foreign minister gifted a crate of mangoes to Zhou Enlai's boss, Mao Zedong. Chairman Mao was not interested, so he passed it on to a propaganda team. The fruits became Mao's avatars, symbols of the Great Leader in a communist version of idol worship. When they began to decay, the mangoes were boiled in water and all the workers of certain factories had one teaspoon of the divine elixir. The mango was part of the political spectacle in China's National Day Parade of 1968. Wax copies of the mango began to sell; the fruit featured in the propaganda posters of the Cultural Revolution. Alfreda Murck, a historian of Chinese art at the Columbia University in New York, wrote a book about this weird cult of the Mao-mango in 2013, titled *Mao's Golden Mangoes and the Cultural Revolution*.

Really, name a prime minister! Lal Bahadur Shastri gifted mango baskets to the Soviet premier. P. V. Narasimha Rao and Atal Bihari Vajpayee continued to use the mango in foreign relations. In 2006–07, when Manmohan Singh was prime minister, the mango played a dramatic hand in diplomacy. Indo-US relations were thawing after a period of awkwardness and trade sanctions following India's 1998 nuclear tests. A civilian nuclear deal was under negotiation, among other matters. The US was looking for a way to get India to allow the import of Harley-Davidson motorcycles, which did not meet India's vehicular emissions norms. Some Indian-origin entrepreneurs in the US found out; they were desperate for Indian mangoes, which were banned from import into the US to keep out insect pests. Their lobbying led to a trade deal under which India lowered its emission standards for the motorcycles and the US lowered its phytosanitary regulations to allow the import of Indian mangoes.

Mango diplomacy with the US and Soviets/Russia is nothing compared to the fruit's role in Indo-Pak relations. Each time a side makes an effort to resume dialogue, baskets of mangoes cross the international border. In 1981, Pakistani President General Zia-ul-Haq sent mango baskets to his Indian counterpart Neelam Sanjiva Reddy and Prime Minister Indira Gandhi. A newspaper cartoon dated 14 July 1981 has a mango crate being loaded into an F-16 fighter craft; in the foreground, Zia stands next to a bomb, instructing the pilot: 'Deliver the mangoes first—and the bomb later!'[32] The mangoes were of an exquisite Pakistani variety called Anwar Rataul. It did not take long for some mango growers from the village of Rataul to land up at the prime minister's residence bringing with them the variety named after their village; some relatives had taken it with them to Pakistan at the time of the partition and renamed it Anwar Rataul.

When Narendra Modi became the prime minister in 2014, he invited his Pakistani counterpart Nawaz Sharif to attend the ceremony of oath-taking. One year later, in July 2015, Sharif sent Modi a box of mangoes as an Eid gift. Five months later, Modi paid Sharif a surprise visit on Christmas Day for his granddaughter's wedding. Before the 2019 general elections in India, Bollywood actor Akshay Kumar met Modi for a 'non-political interview'. The high point of that interaction was when he asked the prime minister how he likes to have his mangoes.

The mango has to carry burdens, political and non-political. It is both our great distraction and a reliable means of conversation. It is to Indians what the weather is to the English. The more we talk about it, the more it becomes our mirror. When people talk about the mango, they end up revealing something about themselves. Our mangoes define us.

Chapter 7

Origin
A Unit of Diversity

*'...the Indian plate was...a biotic ferry during its
northward voyage from Gondwana to Asia...'*

—RAKESH C. MEHROTRA

India belongs to the mango like it does not belong to much else. Does the mango belong to India, though? This question has two parts. Where did the mango originate? What is India?

The first question is for botany to answer—actually, palaeobotany, its antiquarian branch. Paleobotanists dig for fossils through layer after dirt layer, hoping to find fossilized remains of plants and animals. It is a piece of life, petrified by time, with a story to tell. Fossils are not juicy and colourful like the mango blossom or fruit. They do not inspire poets. There are no spring festivities, no collective intoxication and dancing, no playing with colours. This is the boffin's borough. Yet, this is what we must delve into if we wish to know the mango's origin.

The second question is even more complex. There is an India defined by current political boundaries. Outside this lies Bangladesh, which has the mango in its national anthem; Pakistan, which is as mango-mad as India; and Sri Lanka, which has its own proud mango traditions. They also have a claim on the mango.

Do we say South Asia? Physical geography does allow us to look beyond political boundaries. That's a different place, contained within the mountains to the north and the east, the desert and the Iranian plateau to the west, and the sea to the south. There's an ancient name for it: Jambudwipa, the island of the jamun or the Indian blackberry. (Why, for the sake of all that is sweet, not 'Amradwipa' after the mango!) Even this subcontinental idea has a term limit. Since we are looking at the past, we will have to keep rolling back. If you travel back in time to more than 50 million years, there is no South Asia, no Himalayas. There's only an island, a renegade tectonic plate, drifting around what we call the Indian Ocean today. The Tethys Sea lay between the Indian plate and the northern supercontinent Laurasia, containing Asia, Europe, and North America.

Go back 140 million years and it's no island. It is joined with Antarctica, a part of a continental conglomeration. Geologists call that supercontinent Gondwanaland or just Gondwana, after the forest-dwelling Gond tribe of central India.[1] Based on that, our greater undivided India can stake a claim upon Antarctica, Australia, Africa, South America, and even the Arabian Peninsula. Any takers for a 'Pan-Gondwana' movement? Akhand Gondwana?

The origin of the mango lies in such antiquity. There was no South Asia, no India, no Indians nor people of any persuasion. The idea of humans was a distant possibility among our ancestors who, for all we know, might have been enjoying the fruits of the mango's ancestors (more on that in the coming chapter). It requires that we set aside all our calendars. And that we set aside all cultures and identities, all ideas of India and non-India, all notions of nation. Lift ourselves out of human solipsism and join biology's dance to the music of deep time.

This is inconvenient. Today, we are the world's most dominant large-sized animals. Our ability to remodel our habitat has taken away the capability to imagine the world without ourselves in it. And yet for most of our species' existence, we have been unremarkable; we did not build villages or cities till about 11,000 years ago. 'If some creature had come from outer space 150,000 years ago,' said author and biogeographer Jared Diamond in an interview, 'humans probably wouldn't figure on their list of the five most interesting species on Earth.'[2]

The mango has been interesting for a lot longer. Nobody disputes where the mango's cultivation began. There's evidence the people of the Indus Valley Civilization were consuming the mango. We know this from objects dug up from a site dated to 2600–2200 BCE in Farmana, Haryana, about 70 kilometres northwest of Delhi.[3] In 2010, two researchers carried out a laboratory analysis of food remnants stuck to pounders and grinding stones. It showed traces of a brinjal curry made with ginger, turmeric, and mango—from about 4,500 years ago! But the mango is millions of years old! Where is its natural home?

The question is frivolous at one level; its answer will settle merely the bragging rights over the mango. But there is a lot more to it. The mango's origin speaks to deeper questions about the natural world that created us. Tugging at the string of the mango's story has helped me crawl out of the confines of my imagination. It has afforded me a peek into life's complex tapestry, a fabric so rich it would have been difficult, nay, impossible to conceive. Science also refreshes that most valuable of gifts: stories. Just that this is non-fiction. These stories are based on research, as far as it is humanly possible to find the truth.

'I represent dumb life forms, the ones who cannot talk, cannot tell their stories,' said Gaurav Srivastava, a young scientist who studies flowering plants and the dynamics of the Indian monsoon at the Birbal Sahni Institute of Palaeosciences (BSIP) in Lucknow. We had a long chat and I asked him how he got interested in botany. 'I had no option. My father was a botanist. My lessons began during my childhood when he took me to the subzi mandi, shopping for fruits and vegetables. He asked me botanical questions about each item.' I could relate to that. My interest in fruits and vegetables also sprang from visiting the mandi with my mother or my uncle. Just that I did not study botany.

⌁

'Do you want to see it?' Rakesh C. Mehrotra, emeritus scientist at BSIP, asked me. It took me by surprise. I had enquired about the mango fossil he had found, angling for an explanation only. 'Yes, please,' I said, 'if it is not inconvenient'. He got up from his chair, walked around his table strewn with papers, along some cabinets with scientific equipment and sample boxes, to a metal almirah. He reached inside and pulled out something the size of a large pineapple wrapped in protective plastic. He put it on the table in front of me.

It was a piece of shale, black rock, with an imprint of a mango leaf. 'How old is it?' I asked. 'About 25 million years old,' he said. 'May I touch it,' I checked. 'Yes, of course,' he replied. I put both my hands on the cold stone and felt my pulse quicken. Here was a great monument in a small stone, a high-definition imprint of a mango leaf, readily recognizable even to an eye untrained in botany. Mehrotra had found it in a coal mine in Tinsukhia, Assam. The same mine had yielded a fossil of a coconut.

Mehrotra has often used his year-end vacations to go digging in the dirt. When most people are drawing up their party plans or recovering from hangovers, he and his associates bundle up their equipment into a train and head to the Northeast. There, in coal mines, they look for patches to excavate. There are no guarantees. It takes years of experience to figure out where to dig and how. Finding a fossil is only half the job. Mehrotra has to look at its characteristics with forensic care, looking for clues that can link it with existing life forms. He is a qualified taxonomist; his eye is trained to catch details of a pattern or a minor variation. All science is drudgery and requires interminable patience. Palaeoscientists need it in spades.

Then, he has to compare those features against what other scientists have already found. He has to look for reliable collaborators willing to go through the material and the conclusions. Then a paper is written and sent to scientific journals, where it undergoes peer review. Other scientists scrutinize each detail, probing and questioning every premise. Then, and only then, after they are satisfied, is the paper published. A small contribution logged into the colossal body of science! To be pulled up in search engines by other researchers!

It was in one such paper that I first read Mehrotra's name. Published in 1998, it reported a leaf fossil found in Damalgiri, about 16 kilometres southwest of Tura in West Garo Hills, Meghalaya.[4] In this instance, Mehrotra collaborated with David Dilcher, a renowned American botanist-geologist. In the paper, Mehrotra gave a new name to the species found: *Eomangiferophyllum damalgiriensis*. How old is this? Geologists have dated the sediment containing this fossil to the 'upper Palaeocene'. That is, the upper layer formed during the Palaeocene epoch, 66–56 million years ago or MYA.

This makes the fossil 59–56 million years old. At this time, the Indian plate was an island south of the equator. What is Meghalaya today was then a part of the Indian plate. It collided with Asia about 50 MYA. After reading the paper, I was eager to meet its author and ask him: Does this prove that the mango originated on the Indian subcontinent?

'Yes!' Mehrotra replied, emphatically, when I met him. Has anybody found an older fossil of the mango anywhere else? 'No,' he said, explaining the complexities of natural history. In an earlier dig, he had found fossils of eucalyptus wood in Rajasthan and Gujarat from a time before India's collision with Asia. The fossils Mehrotra spoke of were from 66–59 MYA. A dig in Argentina yielded an even clearer eucalyptus fossil from 52 MYA.

The eucalyptus, however, is not found naturally in either South America or India. We know the eucalyptus as an Australian tree today. But the oldest eucalyptus fossil found in Australia is no older than 23 MYA. We can infer that the eucalyptus originated in some part of Gondwana but we cannot exactly point to the location. This was news to me. The introduction of eucalyptus by the Indian forest department has adversely affected groundwater in several regions since the 1980s. It has become an example of unthinking foreign introductions. While Mehrotra's findings do not change the effects of the eucalyptus on groundwater, it helped me overlook the fundamentalist rhetoric of native *vs* foreign.

He also gave me a crash course in plate tectonics. While science agrees that the Indian plate collided with Asia about 50 MYA, the two plates did not unite immediately. India remained an island and the Tethys Sea stood between the two land masses for a lot longer. Why? Because the collision happened deep beneath the ocean floor, along the edges of the two tectonic plates. What we see of the continents is barely the tip. The continental plates run much deeper into the Earth's crust. Much deeper than the deepest oceans! The plates get driven around because of upheavals in the Earth's lithosphere, that is, its uppermost crust and solid mantle. The oceans keep spreading around these changing tectonic positions.

After India had collided with Asia about 50 MYA, it took another 27 million years for the Tethys Sea to drain, allowing the joining of land above water. Land-based plants and animals could have migrated only after this continental conjunction.[5] This means even the mango fossil from 25 MYA, which Mehrotra found in Assam, existed two million years before the land connection. What is the oldest mango fossil found outside India? It was found in Thailand from rocks dated to the Miocene-Oligocene; that's 23 MYA or about the time of the India-Asia land connection.[6] Thailand, along with the rest of Southeast Asia, was a part of the Eurasian plate before the collision.

I asked Mehrotra if I could see the older fossil from Damalgiri, Meghalaya. He said that it was in the institute's collection and that

I could see it another time. I went back a year later just to see and touch the leaf fossil from 59-56 MYA; it has a leaf gall, a globular protuberance that happens due to certain insects. I took a photo of this leaf-in-stone and hung it as a toran on the gate of my mind. It will never wilt.

⁂

Taxonomists have identified about seventy species of mango and are undecided about another ten. Southeast Asia has the largest number of wild mango species. Indonesia has thirty-three, Malaysia twenty-nine, Thailand thirteen, and Vietnam has nine. India has nine species, including *Mangifera indica* and all its 1,000-odd varieties; five of those species are from the Andaman and Nicobar Islands that are close to Myanmar, Thailand, and Indonesia. While the subtropical mangoes of India tend to have one seed inside the fruit (monoembryonic), Southeast Asia's tropical mangoes tend to contain multiple seeds (polyembryonic).

The mango has a clear Southeast Asian slant. That's where lies its 'centre of diversity', where most of its wild species or 'landraces' are found. For a comparison, the apple has a similar connection with Central Asia. The fruit's diverse wild forms are found in Kazakhstan, where the apple originated; from here it travelled along the Silk Road to Europe in the Middle Ages. Likewise, the Central and South American region has about 200 wild species of the potato and thousands of varieties.

Origin is equated with diversity. It is a common belief that where a plant appears in its most diverse forms is where it must have originated. For example, if you search for 'centre of diversity' on the internet, say on Wikipedia, you are redirected to 'centre of origin'. The United Nations' Food and Agriculture Organization (FAO) has different definitions of the two. But it sticks them together, as seen in its literature on the 2001 International Treaty on Plant Genetic Resources for Food and Agriculture. One is often taken for the other.

Given the number of mango species in Southeast Asia, it appears likely that it originated there. It is one of Earth's grand theatres of tropical life, comparable with Central Africa and Central America. Six of the world's twenty-five biodiversity hotspots are in Southeast Asia, more than any other region. It harbours about one-fifth of the planet's plant and vertebrate species. And new species are discovered routinely. The mango could have evolved here in the plant family *Anacardiaceae*. In 1876, British botanist Joseph Hooker supposed that India was not the cradle of mango but was a later recipient of the tree; he was the director of the Royal Botanical Gardens, Kew, and had travelled in the Himalayan region in search of plants.[7]

The story is not that simple, though; it seldom is, as Mehrotra's eucalyptus fossil had shown me already. Let's leave aside the mango and its family *Anacardiaceae* and consider another plant family: *Dipterocarpaceae* (meaning two-winged fruit). It dominates the forests of Southeast Asia. A bulk of the timber trees of the region are dipterocarps. The island of Borneo, which has the greatest diversity of the mango, has more diversity of dipterocarp trees than any other place on the planet. More than 280 species of dipterocarps comprise 80 per cent of the trees in the canopy of the tropical forest here. We can conclude that the origin of the *Dipterocarpaceae* family is in Southeast Asia. And so it is believed.

Yet recent evidence has raised doubts. It just might be that dipterocarps originated in the Gondwana supercontinent, travelled north on the Indian plate, and then jumped over to Eurasia after the collision, diversifying later in Southeast Asia. The sal tree (*Shorea robusta*) is a dipterocarp common to Indian forests. Besides, some recent studies have linked this family to another called *Sarcolaenaceae* that is unique to Madagascar, the island east of Africa that was joined with India till 88 MYA.

A 2004 study showed the family from Madagascar shares a common ancestor with the dipterocarps.[8] Since then, international groups of plant systematics have been debating whether they should combine the two families into one.[9] Dipterocarps are spread across the tropical rainforests of South America and Africa, both of which were parts of Gondwana, like India and Madagascar. If the two plant families have a common ancestor, and one is found only in Madagascar, it seems likely that they both evolved in Gondwana.[10] Because India took the life forms of Gondwana to Asia, it seems the logical conclusion. Then, more evidence appeared to back this theory. In 2010, a team of mostly Indian scientists reported a major find of fossils in amber in Gujarat from 50–52 MYA; it contained plant fragments resembling dipterocarps.[11] Yet the matter is far from settled.

Mehrotra does believe that the mango spread from India to Asia, as indicated by the identifiable 'megafossils' he found. But he maintains that dipterocarps came the other way after the collision. His trained taxonomist's eye is unwilling to accept that the sample in the amber from 50-52 MYA is actually that of a dipterocarp. No other fossil of the family has been found in India in sediment older than the collision, while there are older fossils of the family from Southeast Asia.[12]

Some scientists have a different explanation for the relationship between Madagascar's species and Asian biota (biota or biome is the sum of all organisms from a place or time). They put it down to a chain of 'stepping-stone' islands that exists along the path the Indian plate took northwards, from the Reunion Islands through Seychelles, Mauritius, and the Maldives to Lakshadweep. A group of scientists wrote: 'Clearly, a former stepping-stone island chain could have served in transoceanic dispersal of lineages not only travelling from Asia to the Madagascar region, but also those travelling in the opposite direction.'[13]

Mehrotra lamented the fact that geologists are taking over palaeobotany. 'Geology treats it as just one of the branches of palaeontology,' he said. This means animals and plants are handled by the same discipline. Animal fossils are always more exciting;

which plant can stand up to a T. Rex in popularity! Palaeontologists are not trained in plant taxonomy and plant systematics, Mehrotra pointed out.

At the same time, taxonomists cannot find jobs in research institutions; genetics is now a big influence in plant systematics. This results in few students opting for taxonomy now, said Srivastava, who did his PhD under Mehrotra. Plant taxonomy is a difficult discipline in any case because of the immense fieldwork. 'Forest-dwelling tribals identify plants better than several taxonomists,' Mehrotra said.

Palaeobotany is complicated. Just as one of its branches looks at megafossils to identify them through taxonomy, another branch studies fossils of tiny grains of pollen. Some who study fossilized pollen maintain that dipterocarps crossed over to Asia from India. Such stand-offs are common in the sciences. In fact, this is the way the scientific method works: conflicting theories co-exist till somebody finds the evidence to settle it one way or another.

'As it is assumed that *Dipterocarpaceae* are of Gondwana origin (no scientific proof), we may as well assume that the *Anacardiaceae* are of the same origin, but this remains unproven,' wrote Andre Kostermans, a reputed Indonesian botanist, in his 1993 book *The Mango*. 'I fully agree with (Lal Behari) Singh's (1960) belief that the theory of the maximum number of species of a certain genus coinciding with the site of the origin of the genus is a fallacy and not based on fact,' he added.[14] That a well-regarded Indonesian botanist is inclined to accept India and Gondwana as the origin of the mango counts for something.

A life in science is a series of experiments in uncertainty. The words of two masters are worth repeating here. 'Science is a very human form of knowledge,' said Jacob Bronowski, a towering historian of science, in 1973. 'We are always at the brink of the known, we always feel forward for what is to be hoped. Every judgement in science stands on the edge of error, and is personal. Science is a tribute to what we can know *although* we are fallible.'[15]

Physicist Richard Feynman was the epitome of charm and glamour. He described science as 'a satisfactory philosophy of ignorance', saying: 'The scientist has a lot of experience with ignorance and doubt and uncertainty.... We have found it of paramount importance that in order to progress we must recognize the ignorance and leave room for doubt.... Scientific knowledge is a body of statements of varying degrees of certainty—some most unsure, some nearly sure, none *absolutely* certain. Now, we scientists are used to this, and we take it for granted that it is perfectly consistent to be unsure—that it is possible to live and *not* know.'[16]

The palaeosciences are no exception. Megafossils are not easy to find, although they tell a story more precisely and can settle a matter locally and in high resolution. Pollen fossils are plentiful but pollen can travel long distances; they tell a regional story in lower resolution and a broader sweep. Both have their uses and their limitations.

Traditionally, most scientists have acknowledged the lack of clarity about the mango's origin. Wild mango trees are found in the forests of Northeastern India, Myanmar, and Bangladesh, even if not in Borneo-like diversity. Apart from the megafossils that Mehrotra found in Northeastern India, other evidence also explains how the mango might have evolved on the Indian plate before the collision.

Vandana Prasad, also of BSIP, has excavated and studied pollen fossils extensively. 'Vastan in Gujarat has yielded pollen of several *Anacardiaceae* species, including *Mangifera indica*, in a good state of preservation,' she told me. In her published papers, she has argued that the vegetation India brought along from Gondwana underwent rapid transitions during its northward journey.[17] It was a long trip of about 9,000 kilometres, beginning around Antarctica. 'India moved so fast up the latitudes that the clades (branches on the tree of life) of Gondwana adapted and became distinct,' she explained.

How fast? Since we are dealing with supercontinental real estate, we can examine this from the banal maxim of property dealers: location, location, location. Around the time that the mango evolved, India was a turbo-charged island in equatorial waters. It was moving northwards at such speeds that a scientist called it a 'cheetah plate' or a tectonic Ferrari in 2007.[18]

How fast is that? Around 100–200 mmpy (millimetres per year) or 4–8 inches per year; this might well be the land speed record of a tectonic plate.[19] Africa and Australia, India's siblings from the joint family of Gondwana, were moving at far slower speeds of 20–40 mmpy. The Indian plate's continental roots were melted by hot rock upwelling from inside the Earth's mantle. (This reduced the Indian continental plate's thickness to about 100 kilometres. Africa, Australia, and Antarctica are all about 180 kilometres thick.)

Lightened by the melting of its deep roots, the Indian plate began rapidly moving north. The plants and animals living on it underwent dramatic transitions across a range of climatic conditions. Location is only half the story, though; the other half is timing.

About 55 MYA, the northern edge of the Indian plate had crossed the equator. Wet tropical rainforests and swamps overran the Indian landmass. This is when the Earth's climate changed dramatically. Scientists call this rapid global warming the Palaeocene-Eocene Thermal Maxima (PETM). For about 200,000 years, the Earth turned into a greenhouse. Of this, about 20,000 years were sheer pandemonium. The planet's average temperatures increased from 5°C to 8°C. The ice caps melted; forests emerged around the poles with crocodiles roaming about. (PETM has become the focus of new research because of some parallels with the present. Burning coal and petroleum over the past 200 years has increased the average temperature by 1.1°C. PETM might provide the hindsight scientists need to anticipate the effects of climate change.)

Here, we need context. PETM occurred about 11 million years after the great cataclysm that wiped out the dinosaurs not just those living on land—all dinosaurs. This suddenly left a lot of room for mammals. Life began a slow recovery but was in the doldrums. The fossil record does not yield anything spectacular. Until PETM, that is; it brought in dramatic changes. Suddenly, mammals began to adapt and diversify.[20]

Three mammal orders really took off around this time. One, odd-toed ungulates (*Perissodactyla*) that today include horses, zebras, asses, and rhinoceroses.[21] Two, even-toed ungulates (*Artiodactyla*) that today feature pigs, deer, cattle, giraffes, hippos, camels, and their cousins who took to the sea, whales and dolphins. Three, the primate order that produced lemurs, monkeys, and apes including us. Several plant species also underwent rapid changes in this period.[22]

Most of India was already under evergreen forests. Vandana Prasad has written about the southwestern tip of Kerala that retains a few isolated species of this period. Conditions in India then were comparable to present-day tropical regions of Central Africa, South America, and Southeast Asia. 'Canopy plants need high temperatures, high humidity, and high rainfall,' Srivastava had told me. Mango is a tropical canopy tree.

It seems viable that the rich tropical forest that covers Southeast Asia was populated by plants and animals that arrived on the tectonic raft that was India. Mehrotra transcends of his usual understated tone and bursts into metaphor in one of his papers describing this: '...the Indian plate was not only a biotic ferry during its northward voyage from Gondwana to Asia but also a place for the origin of several plant taxa.'[23] Long before the Columbian Exchange, the continents were making global journeys and exchanging species.

Was the mango one of these plants? Existing evidence does allow us to claim that the mango evolved and came into its own on the

Indian plate. A piece of Gondwana that risked tectonic copulation with Asia! It brought the gifts of rapidly transforming plants and animals from the southern supercontinent to the northern one. It set off a planetary biotic exchange, with a prominent side-effect called the Himalayan range. Such pressures—tectonic and climatic—drive life's evolution. The mango is one unit of life's bewildering diversity.

Caveat: The natural world does not fit neatly into human comprehension. As and when new evidence emerges, we must be ready to rewrite what we know today, reminding ourselves that the new understanding might be rewritten in the future. There is no saying what fossils lie buried in which part of the world, waiting for the right taxonomist to come along. Even our ideas of landmasses and continents are deeply influenced by culture and politics. In 1997, geographer Martin Lewis and historian Karen Wigen published a book titled *The Myth of Continents: A Critique of Metageography*. They warn of unconscious frameworks that often shape the social sciences. That all geographical divisions like East and West, even the seven-fold continental framework, are oversimplifications; they misrepresent the world.

Whichever way the fossil evidence points, wherever the mango's natural origin might lie, nothing can change our unique cultural relationship with the fruit of India. Besides, to understand our natural connection with the mango, we need to look beyond every cultural association.

BOOK TWO

The Fruit of Wilderness

Chapter 8

Angiosperms
An Abominable Mystery

'The ability of the angiosperms to accommodate and maximise benefits from animal behaviour has been responsible for the evolutionary success of the group.'

—DAVID DILCHER

How far are you willing to go to find yourself? On 4 July 1997, NASA landed its Mars Pathfinder on the red planet; it unleashed the rover Sojourner. After years of preparation, six months in flight, and a budget of US $150 million, we finally set our electronic eyes on Mars. Live pictures were beamed on TV screens across the world. NASA turned it into a popular event. When the rover ran into rocks, they were named after cartoon characters—Barnacle Bill, Scooby-Doo, the dog, and Yogi, the bear. 'Someone high up in NASA must have issued a firm directive: "Keep it cuddly, guys. We don't want Mars to seem like, you know, outer space",' commented author Barbara Ehrenreich.[1] 'But cuteness short-circuits the whole process of learning and discovery. When we turn the Martian terrain into a comic strip...we are making things seem tame and familiar before we even know what they are.'

Our stories thrive on the conquest of the familiar over the unfamiliar. In science fiction, characters reach outer space to find emotional succour. Joseph Cooper in the film *Interstellar* (2014) goes through a wormhole near Saturn to find his daughter beyond a blackhole; in *Contact* (1997) Ellie Arroway moves through space in an elaborate and expensive spacecraft to find her father; at the end of *2001: A Space Odyssey* (1968), David Bowman goes through a vortex of light near Jupiter and sees...himself as a baby. We go to exotic locations and return with selfies for display pictures of social media accounts.

And so it is with quest for the mango. From the annual mango writing season to our books, our stories about the fruit are actually tales about ourselves. These are passionate excuses to talk about kings, nawabs, and prime ministers; prophets, philosophers, and travellers; poets, connoisseurs, and celebrities. Archaeology can take us back 4,500 years to the Indus Valley Civilization for the mango. Palaeobotany takes us back millions of years; yet all it says is that the mango tree's ancestor was present during the Upper Palaeocene in what is now Meghalaya.

All our mango practices—cultural and horticultural, traditional and modern—result from tinkering with a fruit that our ancestors found in the wilderness. How did the mango occur in the wild? That question grew on me. Botanical literature offers a clue. 'The mango tree is believed to have evolved as a canopy layer or emergent species of the tropical rainforest of South and Southeast Asia,' said a scientific review.[2] Other than 'canopy' and 'tropical', there is not much to go by. The choice is between stopping here and stepping into the unfamiliar. When I mentioned such choices to Subhash Chacha, he said: 'Play your hand.'

I began sampling research by scholars from diverse disciplines. They do not speak to the mango directly. But they do offer pieces that, when put together, reveal a picture, albeit a hazy one, not a precise portrait. They offer tantalizing peeks into why we go crazy over fruits and flowers. The mango in this picture frame is not the fruit we know and admire. It is but one of the many quirks of evolutionary biology. The path to this fruit of wilderness goes through a cat's cradle of timelines; its basic unit is MYA (million years ago). Fasten your seatbelts!

Usually, such accounts begin with what we know and then go back towards the unknown. That did not help. So I hacked it the other way—beginning with the origins of life, travelling along the path of emergence, evolution, genetics, ecology, horticulture, and then to markets. From the unfamiliar to the familiar! And I still found my ancestors on the other side.

⁂

In biology, you are not merely what you eat; you are also what eats you. Pop culture has a handy maxim for the three primal urges: food, sex, danger. Each organism looks for sustenance, tries to reproduce, and avoids becoming somebody else's food. All three needs must be met right here in the biosphere. Nothing comes from outside except sunlight. The story of life also revolves around the Sun, our real star. My favourite teller of this story is Nick Lane, an evolutionary biochemist at University College London, who is a leading authority on the basics of life. A lucid speaker and author of five books, he has a way of illustrating the world outside the human imagination in familiar terms.

Here's a summary: the Earth was formed out of stardust about 4,500 MYA, after the Sun became a 'main sequence star', turning hydrogen to helium. Water arrived early, perhaps on comets and asteroids that crashed into this young planet. The best current estimate is that life had begun in a deep-sea vent by 3,800 MYA. The earliest forms of life were microbes, now classified as archaea; they breathed and cooked by fermentation. Then came bacteria. Some figured out autotrophy, which is Greek for 'self food'. The autotrophs made their food themselves using sunlight. Photosynthesis initially used hydrogen and sulphur. Gradually, some green-coloured bacteria figured out the chemistry of splitting water and combining hydrogen with carbon dioxide to make sugar. Their waste was oxygen, a gas so reactive it was not found in its free state. Bacteria perfected this skill and became the emperors of the blue planet. Their success and mounting waste began oxidizing the air, sea, and earth, which peaked 2,400 MYA. Lane explained this in a 2002 book titled *Oxygen: The Molecule that Made the World*.

Archaea and bacteria continue to thrive. (Evolutionary biologist Stephen Jay Gould has said that our planet has always been in the 'Age of Bacteria', not the age of reptiles or mammals.) Then, mergers and acquisitions of cells took place. Two such events went on to shape all complex life; they happened around 1,500 MYA, give or take a few hundred million. In both, a small cell began living inside a bigger cell; both worked for each other without destroying each other. This is endosymbiosis. One cellular entrant became mitochondria and the other chloroplast. The first took over energy functions, allowing the bigger cell to develop a nucleus and special organelles for a range of new functions. The second contributed to the evolution of larger, more complex autotrophs: plants. Both mitochondria and chloroplast maintain their own DNA separately.

That's for food. For reproduction, early microbes just divided themselves into two. But this simple asexual duplication does not drive complexity. Diversity results from sexual reproduction. Sex mixes and remixes genes, providing variation in subsequent generations. Successful variations take, unsuccessful ones get selected out. Complex life underwent a dramatic expansion and variation around 541 MYA. It is called the Cambrian Explosion, as seen in a range of fossils from around the world. Fossils of tube-like marine animals suddenly appeared in the sediment of this era.

Plants had already made landfall; they were rootless and moss-like. In about 100 million years, some had begun to show a vascular system to transport juices of life across much larger and vertical bodies. These had roots, stems, and leaves for more specialized work. Trees had arrived. One big reason for their success was a partnership their root system had struck with fungi and microbes. The plant world still thrives off this partnership, keeping our world green.

The land offered advantages like more sunlight. Soil offers rich nutrients if you can send down roots. There are disadvantages, too. Rooted autotrophy fixes a plant to one spot, severely limiting the scope of spreading their progeny. A successful solution came to seed-bearing plants (spermatophytes). They produced small, motile seeds that unzipped into new plants. Such plants went on to acquire many superlative forms.[3] They include trees that live for up to five millennia, like the bristlecone pine of America; a seagrass spread across 200 square kilometres of sea floor off Western Australia over the past 4,500 years; trees taller than 120 metres, like the eucalyptus of Tasmania or the giant sequoia of California; and trees with seeds that weigh 20 kilograms, like the double coconut of Seychelles. Some of these giants begin life in seeds weighing a few thousandths of a gram.

They have tenacity, too. The ginkgo is a living fossil; its close relatives appeared 270 MYA in what is now China. A ginkgo tree in Hiroshima survived the atomic bomb in 1945 even though it was less than a kilometre from the point of explosion.[4] Five other ginkgoes in a 1.5-kilometre radius of the explosion survived.

The seeds were bare, hence the name 'gymnosperms', which is Greek for naked seed. (Gymnasts were so named for performing naked in ancient Greece.) This category includes ginkgos, conifers, and pines. Gymnosperms were well established by 380 MYA and were very successful 360–300 MYA in a period called the Carboniferous; their burial created the largest coal deposits that we mine today. Their success had an undesirable side-effect. Their large vascular bodies were full of the juices of life. They attracted animals or heterotrophs, Greek for 'other food'. For mobile animals, a self-reliant vascular autotroph was a sitting duck.

After they were established on land, animals made dramatic reproductive adaptations. Fish and amphibians, both early forms, reproduce via soft eggs laid and fertilized in water. Their eggs cannot survive on dry land. Water is essential to life. Scaly-skinned reptiles adapted to enclose their eggs in hard and dry shells, keeping them moist inside. The earliest creatures related to mammals emerged 325 MYA and small, arboreal mammals were established by 225 MYA. They had adapted to retain the egg inside their body till the foetuses were more developed.

With diverse animal predators crawling all over them, trees had to contend with a lot. They did not suffer passively. An arms race unravelled. For example, castor beans contain ricin, which is one of the deadliest poisons known in nature, more toxic than cobra venom (which still lacks an effective antidote). Many plants make cyanide; the seeds of apples, peaches, and almonds contain this poison.[5] The mango tree produces the irritant urushiol, although its cousin poison ivy is infamous for it. Several plants, especially grasses, have serrated knife-like edges on their leaves to ward off browsers; some plants have prickles and thorns. A few turned carnivorous, like the Venus flytrap.

In response, herbivorous animals developed multiple stomachs to break down cellulose in leaves that is tough to digest. Leathery lips and tongues tolerated serrated edges and thorns. The largest animals on land tend to be herbivores; giraffes developed long necks, and elephants have powerful trunks. Animals also have biochemical responses to neutralize plant poisons.

Insects were among the early plant predators and faced defensive adaptations. As happens in such conflicts, hostile parties forged alliances. Fungi and pathogens became a part of this conflict, being used as deterrents. Some plants attract insect-eating ants. Herbivores in grasslands had to deal with some of the largest carnivores like sabre-toothed smilodons, which are now extinct, and tigers that are verging on extinction. Such complex relationships make up the web of ecology.

The plant-animal relationship was not entirely antagonistic.[6] There were opportunities in animal mobility. If they could be manipulated into providing transport services, their predation could be tolerated. Making a virtue out of a vice, plants and animals entered partnerships. Insects began carrying pollen for plants.[7] Over time, a different kind of seed-bearing plant had emerged: angiosperms or 'vessel seed', with enclosed seeds. Their lines diverged from gymnosperms a long time ago, although their fossils begin to appear only from about 140 MYA.

Angiosperms really took to animals. Pollinators increase the chances of out-crossing or cross-pollination. This drives sexual experimentation and variety. Flowers were the great angiosperm adaptation to lure insects. Over time, the size, shape, and colour of flowers became customized to the vision and taste of the insect of interest. Insects developed eyes that could discern the brilliant colours of flowers; birds did the same. Flower petals became both billboards and landing strips. When plants needed to repel animals, they produced bitter poison; when they needed to attract them, their flowers offered sugary nectar. Flowers became portals for evolution's WiFi syncing. (Kamadeva'a arrows, remember, are made of flowers!)

The seed resulting from this pollination was encased inside colourful and sweet fruit. If nectar is the incentive for pollination, fruit is the incentive for animals to carry away the seed so that it takes to new lands. Plants offer animals multiple payment plans to suit every situation; evolutionary ecologist Jonathan Silvertown listed these in a remarkable book.[8] Nuts and grains are a kind of 'prepaid scheme' in which the plant provides animals a portion of its seed as food; this ensures that some of the seeds are taken away to germinate at a distance. A 'post-paid' plan is to get burrowing animals to store seeds for a rainy day; for example, squirrels hoard acorns. Fleshy fruits are a 'pay-as-you-go system' with an in-built deterrent. Under this, the flesh is the reward for carrying the seed, but the seed itself is inedible, for example, due to irritants and poisons.

The flower-fruit schema was a runaway success. 'The ability of the angiosperms to accommodate and maximize benefits from animal behaviour has been responsible for the evolutionary success of the group,' wrote David Dilcher, a leading biogeographer and palaeobotanist.[9] Flowering plants diverged quite late; yet they form the largest group of plants today, with about 300,000 living species. In contrast, gymnosperms are considered 'highly conservative'; their species today number between 900 and 1,200.[10] Of the plants we see around us today, 80–90 per cent are angiosperms, including almost all important agricultural plants.

This rapid diversification of flowering plants has long intrigued naturalists. Their sudden success boggled the mind of Charles Darwin, challenging the kind of 'descent with slow and slight successive modifications' (later termed evolution) that he had proposed in his 1859 book *On the Origin of Species*. It was a source of great and repeated frustration, right down to his final years. In 1881, the year before he died, Darwin wrote a letter to his botanist friend Joseph Hooker, director of the Royal Gardens, Kew, who had first said that the mango originated in Southeast Asia.

Darwin wrote: 'Nothing is more extraordinary in the history of the Vegetable Kingdom, as it seems to me, than the apparently very sudden or abrupt development of the higher plants.' In another letter to Hooker, this one in 1879, he wrote: 'The rapid development as far as we can judge of all the higher plants within recent geological times is an abominable mystery.'[11] There was 'nothing...more extraordinary', he wrote, calling it the 'most perplexing phenomenon.'

Evolutionary biology has come a long way since Darwin. Nevertheless, 'Darwin's Abominable Mystery' has become a trope. It was the title of a special issue of a botany journal on Darwin's 200th anniversary and the 150th anniversary of his path-breaking book.[12] 'Our results uphold Darwin's suspicions that simple explanations for the mystery of angiosperm diversification are inadequate,' said a 2004 review paper.[13]

Now, science has more tools to explain the diversity of flowering plants. What were called 'higher plants' earlier are now categorized as dicotyledons or dicots.[14] The pathbreaking research has not come from older fields like phylogeny, taxonomy, or the fossil record, although these have seen great advances, too. It has come from genetics.

Several early naturalists had speculated about biological inheritance. Unbeknown to them, Austrian friar Gregor Mendel, in the 1850s, had begun documenting hereditary traits in successive generations of pea plants. Also a mathematician, Mendel kept elaborate records of the thousands of plants he bred, creating a database. Mendel's work gained currency only after his death in 1884. Gradually, improvements in microscopy gave scientists a closer look inside cells. The ideas of genetics were established by the 1930s in mathematical models. By 1943, DNA was identified as the material that transfers biological inheritance. Ten years later, the structure of DNA became known.

It emerged that organisms that appear closely related can be genetically distant, even as closely related organisms differ in appearance. Earlier, taxonomists had to spend years to observe physical features of organisms and record them painstakingly; Darwin spent years studying barnacles, describing their traits in excruciating detail. Advances in genetics made it possible to look inside the genes responsible for those traits. After recording the genes in DNA, scientists began doing exactly what Mendel had done: turning their descriptions into computations and data. Computers and the internet have allowed people across the world to share genetic descriptions in the form of statistical data. A whole new world opened.

Geneticists can test the speculations of the early naturalists. They have proved or disproved a lot of what early naturalists like Darwin had theorized through their taxonomic labours. Consider that the seed of a plant contains the ingredients for creating another plant. Genes work in a similar way: hereditary traits are folded into functional genes, which are part of an entire library of DNA. This library is the genome. The chromosome is a condensed form of DNA. The full set of chromosomes is called ploidy, which is Greek for folded form. Another metaphor for it is installing a software program on a computer or a phone: a small installation file unzips into a full-blown app.

Not all organisms need this. Many microbes, for example, reproduce simply by duplicating themselves. Their DNA gets passed on directly in uncompressed threads; they are monoploid. It's a simple copy-paste job. Other organisms reproduce sexually, including humans. This is quite complicated. The DNA gets compressed before getting passed into sex cells called gametes. These contain only one copy of chromosomes that fuses with another gamete of the opposite sex to restore the chromosomal set. Thus, fertilization restores two ploids, making them diploid. For example, humans have forty-six chromosomes; the egg and sperm of each parent contain one set of twenty-three chromosomes; after fertilization, the foetus has a full complement of forty-six.

If only it were this simple! There are organisms with three or more sets of chromosomes, hence polyploid. This multiplicity of ploidy occurs in certain conditions. During reproduction, for instance, an egg might fertilize with two sperms (polyspermy). Or errors might occur during the segregation of chromosomes into sex cells (meiosis). In such a situation, a gamete goes through with both its sets of chromosomes. If such an egg or sperm gets fertilized, it contains an extra set of chromosomes from one parent. (A human with 46+23=69 chromosomes will be a triploid. One with 46+46=92 chromosomes will be tetraploid.) Such a progeny has more than two sets of chromosomes. Boom, polyploidy!

Among humans, such fertilization is not viable. Even if it occurs, the mother's body aborts it right away. Polyploids are rare and unstable among animals; exceptions include some frogs and salamanders.[15] But plants play a completely different game; they have a much greater tolerance for polyploidy. An ancient tropical fern called adder's tongue has two sets of 720 chromosomes or a total of 1,440.[16] If genetically viable, such plants can survive. They cannot breed even with their own type. But that is not always a problem. The flowers of most plants are hermaphrodite; many can fertilize themselves. This creates a new species. Polyploid implies that whole genome doubling (WGD) has occurred. The duplicate genes, freed from their old functions, can risk undertaking mutations, permutations, and combinations. Polyploidy boosts speciation, that is, the creation of new species. DNA analysis has explained this.

The genome of each organism is a genetic record of its ancestry. The genomes of all seed-bearing plants show polyploidy in their distant ancestry.[17] 'Most flowering plants have been shown to be ancient polyploids that have undergone one or more whole genome duplications early in their evolution,' wrote a plant geneticist.[18] His team went a step further and traced the age of polyploidy of flowering plants; it found that 'many different plant lineages seem to have experienced an additional, more recent genome duplication'.

The team cites a number of studies showing how polyploid species adapt to tolerate changing environmental conditions: 'Due to advantages such as altered gene expression...polyploid plants might have been better able to adapt to the drastically changed environment 65 MYA.' A majority of these 'independent genome duplications' happened about 66 MYA. This was a cataclysmic period, a mass extinction. It wiped out at least half of all species on Earth; some estimates say the loss was up to 90 per cent. All land-dwelling dinosaurs disappeared, and so did marine and flying reptiles.

It is called the Cretaceous-Paleogene extinction event, or the K-Pg boundary. Its exact cause is not known but there are two likely suspects. One, large-scale volcanic eruptions began around 67 MYA along the Western Ghats of the Indian plate, which then was situated about where the Seychelles islands are today. The lava created the Deccan Plateau. The eruptions spewed volcanic gases containing sulphur, carbon, and ash, blocking sunlight and unleashing a volcanic winter. Two, a large asteroid hit southeastern Mexico about 66 MYA; the impact added a massive cloud of dust to the volcanic ash already floating in the atmosphere. Global wildfires and dust-cloud winters inhibited photosynthesis, driving numerous plant species to extinction. Mayhem ensued, altering the biosphere dramatically.

Yet many lines of flowering plants not only escaped this upheaval, they adapted, diversified and began a conquest of the world that has continued to the present day. Angiosperms had been around a lot longer.[19] Their oldest fossils are from around 140 MYA, the beginning of the Cretaceous period.[20] 'The fossil record strongly suggests that the first major diversification of angiosperms took place during the Cretaceous period,' said a 2011 review, which is largely where the matter stands.[21]

Flowering plants began to diversify during 130–90 MYA, leaving their mark on the ecosystem.[22] The climate of this period was quite warm, greenhouse conditions prevailed, and the polar ice caps did not exist.[23] This diversity came in handy at the K-Pg boundary. The cataclysm speeded up this diversification, helping angiosperms become the dominant group of plants today. Scientists are increasingly finding much higher polyploidy in domesticated plants as compared to their wild relatives. The polyploidy did not occur after the plants were tamed; instead, it was the polyploidy that made them more pliant for domestication.[24]

Genomics is a new field. While it has a method to test and prove several theories, it has also thrown up a lot of new information that is poorly understood. For example, most of an organism's DNA was earlier called 'Junk DNA'. Now, scientists merely say they have not yet figured out their function. It is known, however, that polyploidy is more common when organisms are under environmental stress.

Climate change, for example, is a great source of stress; organisms that cannot cope with it go extinct.

Environmental disturbances force organisms to change how they reproduce. They bring together species that otherwise live apart, leading to hybridization. A telling example comes from 'facultatively' sexual species. These can reproduce both sexually and asexually. Science has known that such species tend to switch to sexual reproduction when under stress and to asexual reproduction when the stresses disappear.[25] They try out new genetic combinations to deal with new stresses. (Do remember this when we tackle the mango's crazy genetics!) When the stress disappears, the survivors can slide back into genetic stability. This explains how polyploids become diploids or even monoploids and vice versa. Humans are diploids but our genome reveals polyploid ancestors. When the going gets tough, the tough...just...diversify genetically.

Polyploidy very likely played a hand in the evolution of moss-like vegetation into vascular plants about 400 MYA. But the gymnosperms have not diversified like the angiosperms. Flowering plants are the progeny of genetic speculators, handy at dealing with the unknown consequences of a mass extinction. When business-as-usual ceased, the gamblers found divergent success. I wonder if Darwin would have found closure.

The theatre in which the mango evolved did not exist before the K-Pg cataclysm. There were no tropical rainforests with flowering plants. Insects, old pals of the flowering plants, had done well to survive the extinction event. The terms of this partnership were established long ago. Now, in a nascent world, flowering plants had needs that the insects could not service. They needed new recruits from among the animals.

The animal world had seen a major turnover. The dinosaur success had been honed over tens of millions of years; the most ferocious, the largest, the airborne...they were all reptiles. Mammals had emerged at about the same time as them. But early mammals were small, evasive, and nocturnal creatures. They hid in the shadows during the day and hunted at night, eating even baby dinosaurs.

The K-Pg boundary was as devastating to animals as to plants. The survivors, however, had a post-apocalyptic world full of opportunities. Now, mammals could roam around in broad daylight, holding their heads high. The oldest fossils of *plesiadapiforms*, a primate-like group, begin to appear about 65 MYA, right after the K-Pg boundary. This is also the period from which we see both megafossils and pollen fossils of the mango. It was a new dawn of fecundity.

'The scale of biological turnover between the Cretaceous and Paleogene is nearly unprecedented in Earth history,' wrote a team of forty-one scientists about the time between 66 MYA and 23 MYA.[26] Within this period of 43 million years, one phase is peculiar.

As we have seen previously, about 55 MYA came the Palaeocene-Eocene Thermal Maxima (PETM). This period of climate change required plants to adapt their shape as well as their modes of seed dispersal. In open plains and deserts, plants rely on the wind to carry their seeds.[27] But the warm and wet tropics of PETM were ideal for rainforests with dense canopies that block air flow.[28] Besides, even if a seed gets to the forest floor, very little light goes past the canopy. One adaptation was to make larger seeds, packed with more nutrition. That has its own problem: large seeds cannot be transported by the wind or water or insects.

Birds were excellent seed transporters because their flight took them to distant places, increasing the range of a tree that employed their services.[29] Even today they remain the preferred seed dispersers of many plants that produce small fruit and berries, especially in temperate regions.[30] But the payload birds can carry is limited for the same reason airlines limit the weight and size of baggage on airplanes: aerial transport is expensive. Flowering plants in the tropical rainforest with larger seeds needed a different kind of seed carrier in their canopies.

As we have seen already, two orders of hoofed mammals really took off around the PETM, along with the primate order. This is about the time when solitary primates began merging into more complex social groups. Having descended from small arboreal mammals, they lived in the trees. They provided an option for trees looking for fresh talent for seed dispersal. (Another order begins to get noticed about this time in the fossil record: Chiroptera. It includes fruit-eating bats.)

New employees require training. Timing is of the essence. A large seed needs time to grow and mature on the tree. If the fruit gets plucked before that, the tree loses its investment in reproduction. Any plant nutritionist will tell you fruits and flowers are expensive projects. In the interest of efficiency, trees create a signal system, a way to deter the seed carrier from grabbing a fruit when the seed inside is immature. By the same logic, the tree needs to broadcast and advertise the fruit when the seed is ready.

One solution lay in scent. Plants release a range of volatile chemicals that disperse as aerosols. Some smells are irritating, meant to deter unwanted animal attention. Some are meant to attract, as in the case of flowers and ripe fruits. A team of scientists at the German Primate Centre in Gottingen recruited botanists and studied how the smell of ripe fruit attracts birds and primates. First, they tested whether fruit-eating spider monkeys could smell the difference between ripe and unripe fruit. Yes, they could, even in difficult conditions.[31]

They then checked for differences in the smell of fruits. They found those meant for birds did not release such volatile chemical signals, even when the plants were more closely related to those eaten by primates. The researchers concluded that the smell of ripe fruit is an adaptation of the plants to attract primates, not a random event that the primates exploited.[32] It was the plants that began hitting on animals! (Think of this when you see someone pick up a mango and smell it to detect ripeness.) Scientists have found more than 300 volatile compounds in mangoes; not all of them contribute to the aroma.[33] The particular combination of these volatiles gives each mango variety its distinctive aroma and taste.

Smell, though, does not go too far with primates. Their sense of smell is nothing compared to, say, dogs and cats, which have a patch of furless wet skin at the tip of their nose. Called a rhinarium, it is a refined smelling sensor. Primates lost their rhinarium long ago. Only one branch has retained it: the wet-nosed (*strepsirhini*) prosimians or pre-monkeys, who diverged from the dry-nosed (*haplorhini*) simians. That wet-nosed branch today has lemurs of Madagascar, bushbabies of sub-Saharan Africa, and lorises of South Asia and Southeast Asia. (Lemurs remained obscure till the *Madagascar* movie franchise made them famous in 2005, particularly in the character of King Julien, whose Sri Lankan/Indian accent comes from the voice of British comedian Sacha Baron Cohen. While we are on lemur-comedian trivia: the Bemaraha woolly lemur gets its scientific name *Avahi cleesei* from the lemur-loving British comedian John Cleese of Monty Python fame.)

We can smell a mango when we pick it up and bring it close to our nose. But without a rhinarium, primates cannot smell ripe fruit from a distance. And canopies of tropical rainforests are not exactly cosy spaces. No, scent was not going to cut it! A primate-customized communication code was required. Plants did possess an older visual signalling system for this, a colour code they had evolved for birds. While the seed matured, the colour of the fruit skin remained green like the leaves. After the seed had matured the skin of berries and fruits changed colour, telling the birds their snack was ready.

This colour code continues today. Many bird-dispersed seeds come in small sizes inside berries and fruits coloured red or black.[34] This has to do with the capabilities of the bird eye. Birds can tell four different elements of vision: brightness, blue, green, and red. Many can see in the ultraviolet range also. They have descended from aerial reptiles. All reptiles have eyes with a full colour range, as do many fish, pointing to the common ancestry of all vertebrates. (Some snakes also possess heat-seeking infrared sensors made famous by the aliens of the *Predator* media franchise.)

Mammals and reptiles, however, had diverged much earlier, perhaps more than 300 MYA. Since mammals were largely nocturnal for millions of years, they developed good night vision. This rich nightlife had a cost: they lost the gift of full colour vision. A team of scientists studying lemurs in Madagascar realized that nocturnal species enjoyed an advantage if they did not possess the ability to see red; it improved their ability to discern fruit just by brightness.[35] The study provides an idea as to how nocturnal mammals might've lost the ability to see red; it is useful only for daytime feeders.

Today, most mammalian lines are 'bichromatic', they can tell two colours, blue and green, but not red. Marine mammals like whales have, in fact, gone full 'monochromatic': they can only tell light from darkness, with zero colour perception. The colour code that plants had developed for birds was not useful with mammals. There was no scope for a coevolutionary partnership. And then a miracle happened. Even as their sense of smell became compromised, primates gained the ability to see red.[36] The trees with colour-coded fleshy fruits had punted on the appropriate candidate!

Chapter 9

Primates
And Then Happened a Miracle

'...primates acquired their suite of diagnostic features through "diffuse coevolution" with angiosperms...increasing fruit and seed size at this time might have been central to shaping the origin of the group.'

—JONATHAN I. BLOCH AND COLLEAGUES

As it happens, the person who figured out a key part of the fruit-primate relationship was an anthropologist opposed to racism and bigotry. Robert Sussman's (1941–2016) life was an example of cross-disciplinary partnerships. Born in New York, Sussman studied sociology and anthropology at the University of California in Los Angeles, obtaining his PhD in 1972.[1] He grew up in the 1960s, a period punctuated by the civil rights movement in the US. He became interested in racism and eugenics.

He looked for answers in physical anthropology, the evolutionary study of human behaviour. This required studying both our extinct hominin ancestors and extant primates. He went on to become an international authority on the behaviour and ecology of primates. His life's work resulted in *The Myth of Race: The Troubling Persistence of an Unscientific Idea* (2014), his book examining the social construct of race. In the forty-three years between his PhD and his last book, Sussman made memorable contributions. He challenged existing concepts of how our primate and hominin ancestors evolved and proposed new ones. Two of his ideas bear upon the mango story.

One, our ancestors did not evolve as fierce hunting animals. Depicted in popular culture as 'Man the Hunter', the stereotype has been used by assorted ideas of supremacy. (Never 'Woman the Hunter'!) During the Cold War, it became a cornerstone of American jingoism.[2] '...there has always been a lot at stake in concepts of human ancestors—more than objective science, more than impartial pursuit of truth,' he wrote in a book with conservationist Donna Hart, titled *Man the Hunted*.[3] 'Especially in the late 1800s and early years of the twentieth century, we—humans and our ancestors—had to be on the top of the species heap. We had to be the smartest species. We had to be special and powerful and above other animals. And very importantly, humans had to be ranked in a hierarchy of races with European humans at the apex.' Along with several scientists, Sussman argued that our ancestors spent a lot of their time avoiding becoming a snack to predators like hyenas, leopards, and large birds of prey. The primate need to stick together in social groups stems from finding strength in numbers in the face of danger. This idea went into *Origins of Altruism and Cooperation*, a scientific series on primatology he co-edited in 2011.

Two, Sussman proposed a cohesive theory about how primates and fruit-bearing plants struck a partnership of coevolution. He created the framework for studying how fruits in trees defined primates, not hunting for insects or small prey. This mutualism resulted not only in trees producing fleshy fruits but also in the creation of one of the greatest theatres of life: the tropical rainforest.

'Bob Sussman achieved legendary status early in his career, with diverse and brilliant contributions that defied arbitrary boundaries between academic disciplines,' said fellow primatologist Crickette Sanz in a 2016 obituary. 'He proposed the reigning theory of primate evolution, championed the role of cooperation in sociality, and educated a generation about the dangers of race—just to name a few of his many contributions.'[4] Sussman collaborated widely. In the 1970s, a primatologist friend took him to Madagascar to study lemurs, our most distant cousins in the primate order. He then diversified into macaques and became a well-regarded conservationist, working in Mauritius, Costa Rica, and Panama, among other places.

Primate evolution was largely a mystery in the 1970s. Beginning in 1913, a string of scientists had proposed that the hallmark primate characteristics were adaptations to the need to move in a complex, three-dimensional space in trees.[5] These include: forward-facing eyes with stereoscopic vision; a reduced sense of smell; grasping hands and feet; large brains; and a long juvenile stage. Several scientists had speculated about what had shaped these signature traits.[6]

In 1972–74, Matt Cartmill, a biological anthropologist with important contributions to the field, arranged all available evidence. He proposed that the earliest primates were shaped by hunting for insects at night.[7] Only nocturnal predators such as owls and cats had forward-facing eyes like those of the early primates, he argued. Fruit-eating arboreal mammals and day-time predators had 'orbital' eyes—located in the 'orbs' or the side of their heads for a more panoramic field of vision, like in goats and horses and cattle. The large primate eyes, said Cartmill, helped spot their quarry, while grasping hands made it possible to catch insects in the lower canopy and the undergrowth of the forests.[8] It fit the 'Man the Hunter' stereotype.

In the 1980s, while researching lemurs, Sussman had begun looking into flowering plants; he had already begun a partnership with Peter Raven, who went on to become a celebrated botanist. In a 1991 paper Sussman cited plant fossils to say that there was an upswing 56 MYA in flowering plants; around the same time primates and fruit-eating bats emerged.[9] He emphasized the connection. The following year, Cartmill responded with a proposal to settle this conundrum.[10] He said scientists needed to find fossils of early primates that showed they possessed fruit-eating teeth and grasping hands but *not* forward-facing eyes. This would prove that the change in eye position followed a diet of fruits instead of hunting insects. Sussman persisted with his theory. Three years later, he wrote: 'These organisms were linked in a tight coevolutionary relationship.... The first known primates...were a product of a long coevolutionary interaction with the angiosperms.'[11] There was no evidence to settle it.

Then, in 2002, two palaeontologists discovered in Wyoming, US, a well-preserved fossil from 56 MYA of a mammal not seen before.[12] It was not exactly like a primate because it lacked forward-facing eyes but it did possess grasping hands. And its blunt teeth were typical of a fruit diet, not insect predation.[13] Sussman argued that if primate evolution was driven by hunting, then grasping hands and orbital convergence would go together. One of Sussman's collaborators had shown previously that opossums, insect-hunting arboreal marsupials in the Americas, had orbital eyes, not the forward-facing eyes of primates and bats.[14]

Cartmill's theory is still alive, given the complexity of the subject; it could be a combination of hunting and fruit-eating that drove primate evolution. But evidence has been mounting in favour of Sussman's hypothesis.[15] Scholars have been gravitating towards the fruit theory, and not just primatologists but scientists from other disciplines, too. A widely cited 2007 paper said this about Sussman's work: 'Thus, the sequence of characters added in early primate evolution suggests that primates acquired their suite of diagnostic features through "diffuse coevolution" with angiosperms through the Paleocene and that increasing fruit and seed size at this time might have been central to shaping the origin of the group'.[16]

Sussman's theory currently provides the more elegant framework to explain what shaped primates: eating fleshy fruits on trees. This partnership also drove the success of the angiosperms and their fruit-bearing ways.

Primate colour perception is key to understanding fleshy fruits. Its peculiarities are worth a closer look. Science divides mammals into twenty-seven orders.[17] They are further sorted into three cohorts: placentals (most of the ones we know), marsupials (those with pouches found in Australia and South America), and monotremes (weird ones like the platypus). Among placentals, only the primate order has trichromatic vision.[18] All Afro-Eurasian or Old World simians can tell red from green. But not the prosimian tarsiers, lemurs, galagos, and lorises. Some can, most cannot. The Americans are the exception. Some American simians can see red, some cannot. Among several species, only the females are trichromatic. A nocturnal species called the owl monkey—yes, it does look like that!—has no colour vision at all, seeing the world in black and white.

The New World monkeys are exceptional in other ways, too. They originated in Africa, splitting from Old World primates 40 MYA.[19] Because their flat nose opens sideways, they are called *platyrrhini*, Greek for broad-nosed; Old World simians are *catarrhini*, meaning down-nosed. The best estimate in science is that the broad-nosed ones used natural rafts to cross the Atlantic Ocean 36–40 MYA.[20] (Much before Eurasians walked across the Beringia land bridge about 15,000 years ago; much much before Viking Norsemen crossed the northern Atlantic in the eleventh century; much much much before the Columbian Exchange began at the end of the fifteenth century. A hoot in respect!)

Scholars have speculated for a long time on the colour vision of primates. In his 1879 book titled *The Colour-Sense: Its Origin and Development*, Canadian author Grant Allen wrote: '...the taste for bright colours has been derived by man from his frugivorous ancestors, who acquired it by exercise of their sense of vision upon bright-coloured food-stuffs; that the same taste was shared by all flower-feeding or fruit-eating animals; and that it was manifested in the sexual selection of brilliant mates, as well as in other secondary modes, such as the various human arts.'[21]

The number of scientists investigating the origins of primate colour vision has boomed.[22] It's not just a question of scientific curiosity! One motivation is to find a cure for colour-blindness in humans. About 8 per cent of all men (one in twelve) have some deficiency in perceiving colour; among women, this figure goes down to 0.5 percent (one in 200).[23] One reason behind this difference is that the genes responsible for differentiating red from green are located on the X chromosome. Females get two copies of the X chromosome; if one is deficient, the other makes up for it. Males get only one copy of the X chromosome; if that is deficient, then that's it.

In 2009 a team of American scientists took two colour-blind squirrel monkeys and injected their retinas with a virus that introduced the human gene for the red-detecting pigment in cone cells. Twenty weeks later the monkeys began distinguishing red from green.[24] This could help cure colour-blindness.[25] Yet we do not know exactly 'how' simian trichromatic vision evolved.[26] The best we know is that the gene that coded for green duplicated and mutated to perceive red. But such mutations occurred more than once, which is why some American monkeys, and some prosimians, see colour but others cannot. It will take a while to understand this clearly.

As for 'why' primates regained the capacity to see red, there are three hypotheses. The first is the 'bare skin theory'; it argues that colour vision results from the complex social relations among primates.[27] All old-world primates tend to have bare faces and many have bare rumps. Their colour vision is near-optimal for telling changes in skin colour, which in turn reflects levels of oxygen in the blood. Their faces and rumps change colour in response to each other, for example, while blushing or signalling sexual availability. (Dating apps are mere adaptations!) Simians have highly expressive faces and complex social systems, beginning with a social turn during the PETM. Reading each other well can be a matter of not only life and death but also food and reproduction—to tell anger from excitement, repulsion from sexual attraction, and cooperation from aggression. Hence, colour vision. Of course, humans take face-reading to another level. It is not just photographers and magazine designers looking for a cover-worthy face; there are innumerable self-help books, software programs and websites dedicated to face reading. Erotic art and pornography are examples of the vividly visual nature of our perception and imagination.

The second theory is the 'folivory hypothesis'; it puts primate colour vision down to the need to discern nutritious leaves (foliage, hence folivory).[28] Some primates get a substantial part of their diet from leaves; a few subsist entirely on leaves. They prefer younger leaves of reddish colour, containing more protein. Leaves are available all the time in the tropics, whereas fruits are seasonal and scarce.[29] Central American forests contain the loud mantled howler monkeys (*Alouatta palliata*). Most of their diet comes from the tender leaves of particular trees. And they have full trichromatic vision.

It's the third theory, however, that has the most takers: that colour vision resulted from 'frugivory' or eating fleshy and colourful fruits, plentiful in the tropical rainforest.[30] Studies show that fruits eaten by primates are often coloured orange, yellow or brown, all three requiring the ability to see red. This is different from the red–black scheme that plants had devised for birds.[31] This worked well for the trees because not only did primates carry away the fruit, they also tended to discard the seed in gaps in the canopy, where there was light for the seedling to prosper.[32] Why? Because the primate home range is relatively large; they move across long distances and in varied habitats. Primate teeth and grasping hands are exactly suited to fruit eating.[33] A 2013 review of all scientific material on seeds says fruit is the most widely used type of food among primates, found in the diets of 91 per cent of the species examined. It goes on to describe how frugivory is widespread in nature; the seeds of 70–94 per cent of woody plants in tropical forests are dispersed by animals; in temperate forests, this figure is 30–40 per cent.[34]

Living on trees and searching for colour-coded ripe fruit has shaped the primate body in several ways. Hanging from branches, the simians have rotating shoulder joints; apes have remarkable shoulder joints that enable fast brachiation or swinging from branch to branch. The most elegant brachiators are gibbons, the lesser apes. Moving through three-dimensional space in the canopy has given primates a good perception of depth, powered by the convergence of orbital eye sockets into forward-facing, stereoscopic eyes. Fruit-eating bats are the only mammals which can be compared to primates for their complex visual system. Insect-hunting bats, equipped with superior echolocation faculties, do not have the same visual apparatus, however.[35] This visual faculty of the primates expresses culturally in a myriad ways, even in us humans—from cave paintings to neon advertising billboards to the video games we play like addicts. The influence of fruit is very deep. Plants invented the advertising industry from flowers and fruits.

Some scientists argue that leaves and fruits have a combined effect in helping primates see red. I'm inclined to stick with the fruit club for more than one reason. To fruits, we owe a lot more than our RGB (Red, Green, and Blue) colour vision. It has a sneaky role in yet another standout feature of the higher primates: their large brain. The body-to-brain size ratio of higher primates is greater than all other mammals. The primate brain is large at birth and develops slowly thereafter; the playful juvenile stage of primates is much longer than other animals. All higher primates have a long childhood; they play and experiment and learn. None more so than humans; our babies are hopelessly helpless at birth, lacking even the strength in their necks to hold up their outsized heads. This large brain has given us culture, language, and civilization.

How did primates get this large brain? Diet, said the first line of argument. The dietary habits of most primates are diverse; the rapid growth in brain size of our hominin ancestors is put down to the dense nutrition of meat; back to 'Man the Hunter'.

Then a theory emerged that put it down to the 'social brain', overlapping with the 'bare-skin theory' of colour vision. Complex primate societies required the kind of cognitive skills that drove a larger brain size.

Three scientists questioned both these theories in 2017, pointing out that the brain is an energy-intensive organ that grabs ready-to-digest, high-density nutrition that sugary fruits provide. 'Specifically, frugivores exhibit larger brains than folivores,' they wrote; 'frugivores exhibit 25 per cent more brain tissue than folivores of the same body weight'.[36] Omnivores who add meat to their diets do as well or a little better than those who eat both fruit and leaves; both do much better than those who are restricted to eating leaves. Finding quality fruit requires greater brain processing, so it's really a sweet circle; the social networking comes later, as a by-product. This study was widely reported in the scientific world. It has tilted the brain size debate in favour of the primate desire for fruit.

What moved my mind has to do with intoxication, specifically, ethanol. It's called the 'drunken monkey hypothesis' and it's attributed to Robert Dudley of the University of California, Berkeley. An evolutionary physiologist, he saw a monkey eat an overripe fruit in 1999 while on a research visit in a rainforest in Panama. It occurred to him that the primate brain might have evolved the capacity to respond to alcohol, that the taste and odour might 'stimulate modern humans because of our ancient tendencies as primates to seek out and consume ripe, sugar-rich, and alcohol-containing fruits'. He floated the idea in an academic paper the next year.[37] A few years later, it became a book titled *The Drunken Monkey: Why We Drink and Abuse Alcohol*.[38]

'The argument here is that our attraction to alcohol goes back about 18 million years, to the origin of the great apes, if not 45 million years with the origin of diurnal fruit-eating primates,' Dudley said in a 2014 interview.[39] When microscopic yeast feed on overripe fruit, it converts the sugar into ethanol to keep away competing bacteria. Animals that routinely eat fruits end up consuming small amounts of alcohol, too. 'Chimps, our closest relatives, are getting about 90 per cent of their caloric expenditure from ripe fruits; and where there is sugar in the tropics, there is alcohol,' he added, putting a new gloss on Jamaican rum.

There are clear advantages: the caloric value of ethanol is nearly twice that of carbohydrates, that is, sugars. Apart from its gustatory charms, alcohol whets the appetite, the well-known 'aperitif effect'; one of the by-products of fermentation is glutamate that frees up umami, the attractive savoury sensation known as the fifth fundamental taste.[40] But ripe fruit is not easy to find in the tropical rainforest. When it is available and fermented, the smell of alcohol wafts about, tempting frugivores with calories, taste, and intoxication. Dudley proposed that celebratory feelings of socializing and sharing food and alcoholic drinks are a continuation of the quest for fruit in the forest. But his arguments were theoretical.

Just a year after Dudley's book was published, a team of seven American scientists released the results of a genetic analysis that gave his theory an empirical leg to stand on. The team looked into the DNA of eighteen primate species for the sequences that code for an enzyme; called ADH4 (alcohol dehydrogenase class IV), it breaks down alcohol in the body. The team sought to understand its evolution. It found, one, that in most primates the enzyme does not act on ethanol. Two, it found a single genetic mutation (called A294V) that increases the ability to metabolize ethanol by forty times. And, three, that this mutation occurred independently in two distant primates; the first is a gremlin-like lemur called aye-aye that feeds on beetle larvae, but might be getting its ethanol from fermented nectar; the second is the common ancestor of all African great apes.[41]

That is the progenitor of gorillas, chimpanzees, and the hominin line that produced us. This ancestor lived in Africa about 10 MYA. This was a cooling phase in the Earth's climate, which must have changed the vegetation and available food sources. African apes spend plenty of time on the ground, even if they are not bipedal like humans. The ability to metabolize fermented fruit would have helped hominins walk on two legs by providing a valuable source of calories on the ground. Without the A294V mutation in ADH4, they would have been inebriated by consuming fermented food, stumbling about on two legs, out of control. This mutation gives us the ability to hold our drink!

A few months after this mutation was reported, an international team published a study. It reported long-term and recurrent ethanol drinking by wild chimpanzees in the West African Republic of Guinea. In the seventeen years between 1995 and 2012, a total of thirteen chimps had ethanol from the sap of the raffia palm tree that the Guinea residents tapped with plastic containers tied to trees, much as toddy is tapped in India.[42] In each of the fifty-one drinking events spread over twenty sessions, the chimps crafted leaves into drinking implements; they turned them into scoops, folds, and sponges to draw out the ethanol from the plastic containers. And there it was: tools devised with dextrous hands, the taste for ethanol, the ability to metabolize alcohol...it spoke of hallmark human characteristics. No creature is as closely related to humans as chimpanzees.

'Our desire for alcohol is not an evolutionary mistake. There are good reasons for why we get drunk,' wrote Canadian sociologist Edward Slingerland.[43] 'No informed decision, at either the individual or social level, can be made without a better appreciation of the role that intoxication has played in creating, enhancing, and sustaining human sociality, and indeed civilization itself,' he said in his 2021 book *Drunk: How We Sipped, Danced, and Stumbled Our Way to Civilization*. Barbara Ehrenreich explains how the subversive deities—from Pan, Bacchus, Dionysus to Jesus—offer wine to the weary; how alcohol is deeply tied to with communal revelry, dance, and carnival. Alcohol consumption is nearly universal in human societies across the world to the extent that its addiction is a disease with serious social and public health consequences.

No, our genes or our desire for fruit cannot explain alcoholism. It is a complex problem with varied psychological and sociological factors. Yet the drunken monkey hypothesis does shed new light on why so many people find themselves so helpless in their alcohol addiction. In fact, Dudley had a personal motivation to explore this theory over so many years. '...my father was an alcoholic who drank heavily, and whose premature death was in part caused by his unsuccessfully treated addiction,' he wrote in his prologue. 'Our family...experienced first-hand the sometimes violent and dangerous consequences of life with an alcoholic. I well remember as a child being simply puzzled as to why anybody, let alone a parent, might engage in such self-destructive and socially damaging behaviour.'[44]

I never saw my father drink but I have felt the damage of alcoholism. Along with my cousins, I watched Subhash Chacha, my uncle to whom I owe my interest in the mango and plants in general, spiral into alcoholism. It took away his finer side, his gentle and artistic bent, his sunny outlook, and his sporting nature. His fierce temper got worse; he turned impossible. His considerable physical strength kept him going till long after the doctors had given up. He still demanded foods and fruits, but alcohol had colonized his taste buds. We began avoiding him, unable to reconcile the drunken outbursts with the loving patriarch who kept our clan together. The name that first comes to me in reference to mangoes is the same one pops up first with regard to alcoholism.

I have tasted wine made from Alphonso mangoes in Mumbai. It tasted experimental. I was happier with the mangoes. Without the ferment, the fruit has everything to hold us in thrall. The fragrance and aroma of ripeness, which even our suppressed sense of smell can detect at close range. The hallmark orange-yellow colour that our trichromatic eyes recognize. But primates, spread across the tropics, have varied diets. Some have toughened their digestive system to tackle cellulose. Some eat insects, meat, and organs. Yet, fruit is an important ingredient in the diet of almost all primates.

That is because fruits deal in the primary currency of desire: sugar. Its addiction is far greater than alcoholism. 'From birth, humans are attracted to sweet-tasting foods, and for good reason: all 10,000 taste buds in the mouth have special receptors for sweetness. Sweet foods cause the taste buds to release neurotransmitters that light up the brain's pleasure centres,' wrote food historian Andrew F. Smith in *Sugar: A Global History*.[45] In reaction, they release endocannabinoids that whet the appetite.

'This may have an evolutionary explanation: about 40 per cent of the calories in breast milk come from lactose, a disaccharide sugar that is readily metabolized into glucose, the body's basic fuel. The sweetness leads infants to eat more, making them more likely to survive.' Sweet foods promise the safe nutrition of mother's milk. Bitter foods signal toxins and danger; which is why we are warned to stay away from bitter people because they are toxic. 'Once we become conditioned to consume sweet foods, even the sight of them will cause us to salivate; the saliva will help begin the process of breaking down the carbohydrates, signalling to the digestive system that nutrients are on the way,' wrote Smith.

Honey, the purest sugar in nature, is not readily available; bees mount a stinging defence against marauders. Fruits, meanwhile, are custom-made for animals, a worthy substitute for mother's milk when the baby is ready to wean. Naturally, a baby's first real food is sugary: honey and sugared rice or kheer are commonly used in India. If the season is right, it is the mango. The sensation of sweet runs deep in most languages. From 'sweets for my sweet', the metaphor is all over our languages—honey, sweetie, sweetheart, and names like Madhu, Madhuri, and Mishti. Musicians talk of a sweet sound. In cricket, batsmen practise to hit the ball with the sweet spot of the bat. A sweet dish is what ends a good meal; in Rajasthan, a good meal begins with five sweets. Sweetness is joy.

Anil Agarwal, a widely respected journalist and environmental activist, was my editor for four years. He fought a long and heroic battle with cancer. He once described how each round of chemotherapy took away the sensations in his mouth, making everything taste metallic. As the measured poisons in medicine left his body, the four main tastes returned slowly; first bitter, then salt, then sour and, in the end, sweet. His face became animated just at the memory of recovering the taste of sweet; it was the taste of hope, the taste of life. Other cancer patients have told me they recover their taste in a different order. Oncologists say the tastes return in the inverse order in which a patient loses them during chemo.

Not just diabetes and heart disease, our sugar addiction has driven us to historic horrors. James Walvin had a taste of it in the 1960s while researching slavery in a Jamaican sugar plantation. The British historian wrote about it in his 2017 book *Sugar: The World Corrupted From Slavery to Obesity*.[46] It is a sweeping view of how our desire for sugar has shaped history. 'For the best part of four centuries, sugar was cultivated in Brazil and the Caribbean by enslaved Africans and their descendants born into slavery. Sugar and slavery went hand in hand, and it seemed to most of the people involved in the system—except, of course, the slaves themselves—that there could be no sugar without slavery. The crudest measurement of sugar's corrupting influence was that the Western world devised, perfected, and justified that most brutal of systems for its own pleasure and profit. What greater corruption could there be?'[47]

Our industrialized world has a new form of slavery, deployed by the fast food industry. 'Very few people suggested, say in 1970, that sugar posed a global health problem. Yet, today, sugar is regularly denounced as a dangerous addiction—on a par with tobacco— and is the cause of a global epidemic of obesity.' Some countries now have a sugar tax, especially on sugary carbonated drinks. Our sugar addiction is changing the nature of fruit itself. Since the dawn of agriculture about 10,000 years ago, all manner of breeding techniques have been used to produce ever sweeter fruits. What we buy in the market today is nothing like the wild fruits from which they come. Mango varieties of North India, for example, are too sweet even for South Indians.

'People do not realize that everything that we're seeing in the supermarket, they're like superstars. They're like supermodels and elite athletes like LeBron James,' said food writer Frederick Kaufman.[48] They contain a lot more sugar; are either seedless or have few seeds; and have pulp that is creamy and free of fibre. The situation has gotten such that the Melbourne zoo stopped giving fruits to its animals after they began gaining weight and showed signs of tooth decay.[49] The monkeys stopped getting the extra sweet bananas because they stopped eating other foods that were important for nutrition. We get our sugar addiction from our primate ancestors. We are spreading obesity among our primate cousins.

A few years after my father's death I met a friend of his, a journalist from Kashmir. He recalled travelling with my father in rural Kashmir one time, through apple orchard country. It was harvest season and my father felt like eating fruit. The friend had the taxi stopped and was about to look for the caretaker. My father stopped him; he said fruit obtained by permission does not taste as good as that plucked from the tree, without a say-so. They walked in and helped themselves to a couple of apples. Right on cue, the gardener arrived. He saw two primates, approximately seventy years of age, munching on his produce. The friend told him, in Kashmiri, what my father had said. He laughed. Then he gifted them some fruit.

Fruits drive us crazy; none more so than mangoes in India. A fruit vendor in northern Delhi left fifteen crates of mangoes unattended during an altercation with somebody three months into the Covid-19 pandemic, on 23 May 2020. Passersby pounced on the crates and looted mangoes worth ₹30,000. Ordinary people, not criminals! Bystanders shot videos; it became a story read and watched widely. Then, those who had seen the report on TV poured in with offers of financial help. The vendor later said he was overwhelmed by the support.[50]

I witnessed a similar scene in Varanasi. In front of the Central Office of the sprawling campus of the Banaras Hindu University are old trees of the famous Langda variety. On an overcast day, pre-monsoon showers and windy storms were brewing. As I spoke to the young man who had bought the contract to harvest the mangoes from those trees, a gust of wind caused fruits to drop from the branches; Langda drops easily. It was a working day, and the place was overrun with students, teachers, staff, and families. The moment the fruits fell, people lost all civility and ran like monkeys after it. Some got out of their cars or jumped off their scooters to help themselves. Even the security guards!

The contractor stood by, laughing at people behaving like kids in a candy store. His nonchalance surprised me. Was he not bothered by his income disappearing? 'Unlike in an orchard, fallen fruit is free for all here. I cannot stop anybody, so why sweat it. Each person is equivalent to the vice-chancellor when she or he is gathering fallen fruit,' he said. I watched the faces of middle-aged people dressed for respectable jobs, carrying briefcases; they beamed in embarrassed delight. They knew that it was indecorous to run after fruit in this manner, yet they broke into peals of delirious laughter. They were helpless. Something inside drove them to gather the fruit. Looking for the mango's origin takes us to our own natural history.

Angiosperms have an uncanny grip over primates. When they needed fresh talent for seed dispersal, they experimented with their remuneration packages. We have been shaped by the advertising bling of flowering plants. The language of desire works in the grammar of colour, fragrance, and sweetness. Angiosperms' generosity has earned our loyalty.

Yes, we have come a long way from our arboreal origins. While all primates have some ability to stand upright, the straightening of the human back is a dramatic adaptation. It has freed up our hands for hunting and farming plants and livestock, turning us into bipedal omnivores (at the cost of backaches). Well before civilization, our ancestors had walked out of Africa to all continents except Antarctica. We have the most diverse diet ever known in the animal kingdom. To grow grain-producing grasses such as wheat, rice, and maize, we have deforested large swathes. We have exterminated a large number of animal species, some by hunting but most by destroying their habitat. We have replaced wildlife with domesticated animals for dairy and meat. We do not need to jump from branch to branch to avoid predators. Now, we sit in office chairs, our stereoscopic trichromatic eyes staring at computer screens and mobile phones, using our large brains to meet deadlines and surviving predatory bosses.

We do not like to see ourselves as primates shaped by fruiting trees. We like to believe we are in control, civilized, and rational and scientific—a view not supported by science. Over the past ten millennia, we have colonized the entire planet. Our ancestors have created diverse fields of enquiry, allowing us to gather and record knowledge, generation after generation. We worship our ancestors, displaying them in paintings and sculptures in public spaces. Emperors and explorers who founded empires. Philosophers who conjured classical cultures. Prophets who created religions. Founding fathers who created nations. Law-givers and makers of constitutions. Ideologues who gave birth to political movements. Scientists who laid down the secrets of the universe in magically pithy formulae. Inventors who gave us the printing press and the internet. Hindus reserve an auspicious fortnight before the festive season for the fathers (Pitrapaksha) followed by a nine-day celebration for the mothers (Matrapaksha). Ancestor worship permeates each and every human field. All except our natural history.

Evidence showing our descent from tree-dwelling primates mounts day-by-day. But far from acknowledging it, we keep ourselves firewalled, embarrassed by our humble arboreal origin. Words like 'ape', 'simian', and 'monkey' are derogatory, like 'villager', 'bumpkin', and 'wild'. Desperate to keep the natural world at a civilized arm's length, we look for ancestors who suit our aspirations. But wilderness does not leave us.

The sign of corporate success is a corner office atop a skyscraper or a penthouse, gazing down at downtown bustle. Exactly how primates see the world from treetops! For a vacation, we need to escape to the forest. Joy is still cooking outdoors in large groups. To seduce a mate, we use flowers. Vegetative metaphor decorates our romantic verse. For a touch of excitement, we compare our sex organs with fruits and flowers. Our desire for fleshy fruits has taken them across the world, far beyond their natural habitats. The apple of Kazakhstan is grown in Australia and the Americas. The mango has spread across the tropics and the sub-tropics. Fruit trees are planted wherever conditions permit.

When we present a basket of mangoes to grease a business relationship, we use a very old partnership. Even as we cut down the forest, we convert fields into mango orchards. We plant mango seeds in the plains, on mountains, plateaus, and in deserts. We graft mango branches and grow them in gardens, looking after them like our children. We set up university departments to research them. We write books on the mango, even though it remains a wild and untamed partner. I cannot find a more successful example of human resource management.

Chapter 10

Genetics
Randomized Field Trials

*'...the mango is a congregation of dissimilar genes.
It's crazy, you do not know what will get.'*

—M. R. DINESH

A pretty actress is discussing eugenics with a man who is a genius but looks plain. She tells him they should have a child together: 'Imagine a child with my beauty and your brains.' The man declines, saying: 'Imagine a child with my looks and your brains.' This anecdote has jump-started many a discussion on eugenics and genetic inheritance. There is no record of it actually happening. Often, the man is said to be the British playwright George Bernard Shaw, with any one of a number of glamorous actresses as the woman. It draws from stereotypes and feeds them. For example, it is never a beautiful-but-dull man propositioning a genius-but-plain woman. Yet even the stale joke tells a greater truth: we cannot foresee the nature of our progeny.

Emperors with the world at their feet have suffered at the hands of their children. Spare a thought for Shah Jahan, eating his everyday dal, staring at the Taj Mahal from a prison window. Conversely, children of unknown people have gone on to raise dynasties; the colossal Mauryan Empire began with Chandragupta who, some accounts say, was raised as a cowherd. There is no way to tell the future; yet we struggle to live with uncertainty. We crave to know what is in store. Astrologers and soothsayers were once the only way to quench that demand, reading the signs in everything from the position of the stars to the twitching of an eyelid. Now we have statistical analysts and economists, who provide paid advice on everything from our investment portfolio to climate change. For every piece of fruitful projection and advice, there is a mountain of barren soothsaying. We are better at reading the past, with the benefit of hindsight.

What about modern science and its empirical method? Not so simple! The natural sciences look for laws. Physics, for instance, studies the natural laws that govern all matter in the universe; it can describe such laws in mathematical formulae. Chemistry studies how and why matter transforms; chemists work inside labs, where they can control reactions. Biology, the study of life, is messy and complicated beyond imagination; it's all about exceptions to the rules. Its bewildering uncertainties are close to the creative arts.

The title of a 1970 paper asked: 'Are There Laws of Biology?'.[1] This was also the title of a 1996 symposium of the Philosophy of Science Association.[2] Titles of papers presented there included: 'Why Do Biologists Argue Like They Do?' and 'Two Outbreaks of Lawlessness in Recent Philosophy of Biology'. In a 2010 paper called 'Laws of biology: why so few?' two biologists listed three laws and then said: 'However, even these laws are not absolute—they come with exceptions...there is no "standard trajectory" in biology—every biological decision is optimal in a given environmental context.'[3]

Biologists cannot agree even on what makes a species. Here's one definition from 1926: '... (a species is) whatever a competent taxonomist chooses to call a species.'[4] It goes back, way back. In his 1859 book *On the Origin of Species*, Charles Darwin wrote: 'Nor shall I here discuss the various definitions...of the term species. No one definition has satisfied all naturalists; yet every naturalist knows vaguely what he means when he speaks of a species.' Darwin proposed the idea that evolution does not have a direction or a director. On this, he differed from naturalist Alfred Russel Wallace, his correspondent and contemporary, whose own theory was founded in progress, in natural selection following a purpose ('teleology').

Genetics has been found in favour of Darwin: that there is 'a design without a designer'. (This being biology, there are numerous interpretations and theories.) 'Natural selection sorting out spontaneously arising mutations is a creative process...(It) does not have foresight,' wrote an evolutionary biologist who stood by Darwin's approach.[5] 'The theory of evolution conveys chance and necessity jointly enmeshed in the stuff of life; randomness and determinism interlocked in a natural process that has spurted the most complex, diverse, and beautiful entities.... This is Darwin's fundamental discovery, that there is a process that is creative although not conscious.'

Let us see how this works in a species we know well: *Homo sapiens*. Each one of us is a reproduction of a mother and a father. Yet nobody is an exact copy. We inherit some traits from each parent—and some from distant ancestors, too. At the time of fertilization, it is impossible to say how the parents' genetic traits will line up. Siblings who get the same genetic material from the same parents are different. Most identical twins, born of the same fertilized egg that splits into two, do not have identical DNA sequences. Each person is a unique genetic arrangement. Life is biased towards randomness.

Turns out each mango fruit is unique, too. It is often said that India has about 1,000 varieties of mangoes. It's a nice round figure, helping us wrap our heads around what is otherwise unimaginable. At the last count in 2010, scientists had documented 1,682 cultivars or varieties across the world.[6] Each count, however rigorous, is incomplete. 'Such is the diversity of the mango that each fruit you touch is possibly a new variety,' said A. K. Mishra, a retired horticulturist in Lucknow who has researched the fruit over a long career.

Each mango seed is a roll of dice. When you plant one, you play blind: the nature of its fruit is uncertain. Yet generations of Indians have played that game. For most of India's history, the mangoes cultivated and eaten have been grown from seeds. It's not just the invisible Kamadeva or blind Cupid who shoot their arrows randomly. This variability of reproduction is depicted and celebrated in many cultures. British folklore has a similar character, a wild spirit or fairy called Puck aka Robin Goodfellow. A memorable depiction can be found in William Shakespeare's *A Midsummer Night's Dream*. A breakaway comedy, it marks a shift in the playwright's career.

The play is set in Athens, but the characters escape to the forest, where the unexpected is routine. Puck prepares a love potion. When applied to the eyelids of sleeping people, it causes them to fall in love with the first person they see after waking up. In a cascading riot of happenstance, several characters fall in love with somebody who is in love with somebody else. In the concluding speech of the opening scene itself, Helena—she's in love with Demetrius who is in love with Hermia who is in love with Lysander—declares:

> Things base and vile, holding no quantity,
> Love can transpose to form and dignity.
> Love looks not with the eyes, but with the mind;
> And therefore is wing'd Cupid painted blind.

Like Kamadeva, Puck and Cupid are nature's agents in our cultured world. The randomness they bring to our ideas and stories is the randomness genes bring to biology. The arrow of desire blinds people to their conditioning, making them creative but not conscious in the Darwinian view.

Our Jane Austen-powered popular culture calls it 'falling in love'. There's no saying who might fall in love with whom. No saying what nature of progeny will be borne of which union. Life's diversity thrives on such uncertainty, on the free-spiritedness of desire. The fruits and flowers of angiosperms are the evolutionary tools of Kamadeva, conducting blinded randomized uncontrolled trials!

Almost all mango varieties result from 'chance seedlings'—consequences of unintended acts. When people found a mango they liked, they planted its seed. The improvements we see in prize mango cultivars happened over decades and centuries, through the best methods known at that time. Each act of planting a mango seed is a small act of natural selection. What growers did over generations, scientists speed up for quick results. In the research institutions of the Indian Council of Agricultural Research (ICAR), long-standing breeding programmes have tried to expand desirable traits and remove undesirable ones. Yet these efforts to 'improve' the mango have not been very successful, a few exceptions notwithstanding.

It is a notoriously difficult plant for the breeder; there are many limitations. The fruit of most popular varieties produces only one seed. Each sapling takes between three to ten years to fruit, although five to seven years is common. Scientists call it a 'long juvenile phase', a description fit for humans. It takes years to obtain results of breeding experiments. Large tracts of land are needed to grow cross-fertilized trees and assess them over the years. Such long-term experimentation consumes a lot of time, land, and money. Even when the crosses are successful, the rate of fruit drop is high.

Mango growers in Uttar Pradesh have a saying: 'The mango is king. It does as it pleases. It flowers and bears fruits according to its whims.' Most prize mango varieties, such as the Alphonso and Langda, produce flowers and fruits every other year. They are 'alternate bearers' or 'biennial bearers'. 'Hapus is a big-time alternate bearer, so much so that in some trees, one side flowers in one year and the other side flowers in the next year,' said Anand Desai in

Ratnagiri, Maharashtra. 'In a good year, up to 50–60 per cent of our trees flower. This season, temperature variation and untimely rains led to a glut of male flowers. A wave of cool weather hit a second flush of flowering, too. So only about 30–40 per cent of our trees are in fruit this season,' he said. Farmers in the mango-growing region of Chittoor in Andhra Pradesh use neem trees as inverted indicators of the mango. When neem trees flower profusely, the mango's flowering is poor, and vice versa.

When it does flower, though, the mango is a sight for sore eyes. The inflorescence appears like a layer on the edge of the canopy, like sugar dust on a bulging cake. It starts as a suggestion of pale green on the end of reddish twigs, then explodes into a conical Christmas-tree shape—botanists call it cyme—that makes the panicle appear like Madana's arrow. Each panicle contains between 1,000 and 6,000 little flowers. An average tree may have hundreds of thousands of flowers, some perhaps beyond a million.

Such exuberance is not without reason. Pollination is a high-risk, low-percentage game for the mango. Vidyadhar Joshi, a grower in Devgad, gave me my first description of the mango's quirky flowering and pollination. I asked him about his yield that season. He said it was poor because of dull flowering: 'The trees had too many male flowers and not enough hermaphrodites.' How do you tell the difference? 'Hermaphrodite flowers have yellow spots in the petals and they are the ones that get fertilized to produce fruit,' he replied. The same panicle contains both 'perfect' flowers (hermaphrodite) and 'staminate' (male) ones. There are far fewer perfect flowers than male ones; their ratio can drop to below one per cent in certain cultivars and rise to 40 per cent in some others. But too many perfect flowers are not good as well; they result in an excess of small-sized fruits.

This quirky flowering is a result of 'self-incompatibility'. The perfect hermaphrodite flower cannot fertilize itself. It has both pollen-producing male parts and female parts with the ovule. But they mature at different times. Hence, they are self-incompatible. Since the late 1960s, scientists have attempted self-pollination and failed. 'From the 15th day after pollination, the selfed fruitlets were invariably the smaller, the majority dropping within about four weeks of pollination and none attaining even half full size,' wrote two scientists in 1970.[7] 'These differences were caused by degeneration of the endosperm and nucellus in fruits resulting from self-pollination.' This goes right down to its 'recalcitrant' seed; it 'cannot survive for more than a few days or weeks at ambient temperatures'[8]. This means the mango seed could not have travelled long distances without dispersers like humans.

Flowering is cued by environmental factors like seasonal change. It could be a change in the length of the day or the onset of dry conditions. Cool temperatures stimulate the mango's leaves, from where the signal moves to the ends of branches, initiating the emergence of panicles or new shoots at the end of branches. On the panicle emerge the blossoms.

'The temperature should ideally hover between 15°C and 18°C for at least a fortnight for the mango to flower,' said A. Kiran Kumar, a senior scientist at the Fruit Research Station in Sangareddy, which is a reputed plant breeding centre in Telangana. 'This temperature band helps the budding of the panicle. Once the flower emerges, though, it prefers hot and dry conditions; heat stress helps. If the temperature remains low, the panicle does not grow.' In the northern Gangetic plains, the mango flowers at the fag end of winter and in early spring.

Biology, though, is all about exceptions. Some varieties flower multiple times during the year and are called barahmasi or twelve-monthly in North India; they are quite ordinary to taste. Murugan Sankaran, a mango breeder at the Indian Institute of Horticultural Research in Bengaluru, has seen mango species native to the Andaman and Nicobar Islands flowering throughout the year. But their fruits are small, fibrous, and acidic in taste. In experiments he led, they managed to induce off-seasonal flowering even in the common mango; they had *Mangifera indica* trees flowering twice or thrice in a year. 'It is due to the micro-climate there,' he said. It was obvious even in the 1970s that north Indian varieties do poorly in the south; Langda and Dashehri do not flower well there.[9] On the other hand, south Indian varieties do well in the north; but the size and shape of the tree changes, along with the timing of the flowering and the sex ratio of flowers. The fruit's size and taste, too.

The mango tree has to balance the urge to flower with that to build its body. This conflict is hormonal. Among the five major types of plant hormones are 'gibberellins', found in more than 100 forms. In several woody flowering plants, including the mango, gibberellins promote vegetative growth in the stem, branches, and leaves. When gibberellin levels are high, it either avoids flowering entirely or flowers poorly. That's because it invests its energy in strengthening its body.[10] This suppresses the initiation of panicles that bear flowers and fruit.

The mango tree cannot have both vegetative growth and flowering, explains a review of its reproductive physiology. If it produces flowers and fruits, the exhaustion of the aftermath prevents vegetative growth.[11] Another factor that controls flowering is the amount of nitrogen present in the leaves. Slightly high nitrogen levels induce growth of vegetative tissue and slightly low nitrogen levels prevent flowering; a delicate balance of nitrogen in a narrow band is required for flowering.[12] In the Philippines, some scientists devised a technique in the 1970s to induce out-of-season flowering: they began spraying potassium nitrate on its leaves.[13] Flowering also slows as the tree ages and becomes taller; with the increase in the distance between the roots and the branches, the journey that various hormones and nutrients have to travel increases.[14]

Then there's the complex role of water, which I learned from Ashok Bang, a mango scientist in Wardha, Maharashtra. 'Water stress' causes the tree to halt or retard vegetative growth and initiate flowering, he said. Along with lower levels of nitrogen, it needs an increase in carbon—carbohydrates or sugars—for reproduction. 'The source of the plant's food is the photosynthesis in the leaves; but it cannot be stored there. The mango 'sinks' this nutrition in its roots and its stem. When there is extra energy stored as carbohydrates in the sink, the tree begins to reproduce. Once flowering sets in, however, the tree needs adequate moisture,' Bang told me. He explained a phrase that occurs in all botanical literature: the mango needs well-drained soils. He said: 'The mango's roots do not tolerate submergence in water. They also need the soil pH between neutral and slightly acidic. When the soil has excessive water, it turns alkaline.'

Scientists from across the country have been experimenting with pruning of the trees after the harvest. 'In most mango varieties, shoots and stalks that have produced fruit take about two months to die and fall off. Before that, new shoots do not emerge,' said Kiran Kumar. 'This does not leave enough time for the new shoots to emerge; inflorescence appears only on new shoots.' This is why several scientists have been recommending pruning of trees right after the harvest to prevent 'apical dominance'. This means the tree prioritizes its stem and thick branches for vegetative growth, rather than initiating new shoots for a flowering flush. But pruning a tree depends on reach; tall trees are not readily accessed. Old trees in most of India's mango orchards are already very tall. Pruning is not feasible, and reaching the canopy of each tree is labour-intensive and costly.

This string of fine balances limits production. One year the trees can have excessive flowering and fruit set. The tree then has to spread thin its resources across the canopy, reducing the average size of the fruit. The next year the tree either does not flower at all or produces very few flowers, lowering yield. Most of India's prize varieties are alternate bearers. Kiran Kumar said he had not seen one north Indian variety in all his years that was a regular bearer, even when grown in South India. In fact, the more prized the variety, the greater the likelihood it fruits in alternate years. The tree exhausts its carbohydrate reserves in a reproductive cycle, as Bang explained: 'So it abstains the following year or undertakes limited flowering and fruiting.'

Several other fruit and nut producing trees are also alternate bearers. They include pistachios, oranges, apples, pears, apricots, hazelnuts, and olives. Growers and orchard owners have known this forever. If they have several trees, they can compensate because when some trees are in an 'off' year, others are 'on'. It averages out! For the mango grower, this is a problem; each tree has to earn enough in each season for two years. The investment of two years, however, can disappear in an afternoon storm that causes largescale fruit drop.

How do you induce a tree to flower regularly? A solution came from an unexpected entity facing an unrelated problem. The power lines of US electric utilities were getting tangled in trees that had grown very tall. Mechanical trimming was expensive. In the 1950s, they funded research to chemically control tree growth.[15] Researchers produced a synthetic plant hormone that retarded tree growth by preventing cell division. This chemical, however, had to be painted on the wounds of pruned branches. That was as costly as mechanical pruning. Then, in the 1970s, further research produced a class of chemicals that could be injected into the trunk or the roots. These prevented plant cells from elongating, which also retarded vegetative growth. Only one chemical from that research is used today as a plant growth retardant: paclobutrazol (PBZ). Beginning in the 1980s with the apple, it is now used across the world to stunt the growth of a range of plants. It has become a controversy all its own, but that's for the ensuing chapter.

In 1991, Murad Burondkar, a plant physiologist, was the first to use PBZ to induce an Alphonso mango tree to flower in the 'off' year. I found his number from a grower and called him. A gentle, pleasant voice greeted me. He invited me over to his office. So I turned towards Dr Balasaheb Sawant Konkan Krishi Vidyapeeth in Dapoli, Maharashtra. Burondkar had assisted the famous scientist R. T. Gunjate in some mango-breeding experiments in the 1980s and 1990s. They began to suspect that gibberellins were impeding flowering. They set up an experiment to check the hormonal levels of two varieties: Alphonso and Neelum.

'Everything else was comparable. But the Alphonso's gibberellin level was thrice that of the Neelum. That nailed it,' he said. I asked him why the Neelum mango from Tamil Nadu was used in so many breeding experiments, even though it does not figure among the most desirable varieties. 'Neelum and Totapuri are the only proven regular bearers. And Totapuri is not useful because its undesirable shape persists in all breeding lines. Neelum is more malleable. It is possible to keep its better features and breed out the undesirable ones,' he answered. 'Besides, Neelum is naturally a dwarf variety. Its trees do not get too tall.' India's mango trees are so tall they cannot be reached easily for management interventions, especially in North India. Scientists say they take up too much space in exchange for meagre returns. Dwarfing is much needed.

The humble Neelum—its fruit is slightly starchy-chalky to taste—has been bred with big-ticket varieties in several long-standing breeding programmes across the country. More than forty-five new varieties have resulted from them.[16] Scientists have looked for crosses that have the taste of the treasured varieties along with the Neelum's regular flowering and size. Neelum-Alphonso crosses have been used at the Fruit Research Station in Vengule, Maharashtra, to create two varieties that have acquired a niche presence: Ratna and Sindhu. But the two most successful varieties created by scientists in India are Amrapali and Mallika. Both are crosses of Neelum with Malihabad's

Dashehri. Both taste like the Dashehri but are regularly bearing dwarfs like the Neelum. The Amrapali is quite popular in Odisha; Mallika in Karnataka.

Scientist Ram Nath Singh led the research that yielded these two. In 1978, he listed the six features that make 'an ideal mango variety'.[17] One, dwarf size. Two, medium-sized fruit. Three, regular bearing. Four, tolerance to bacterial and fungal diseases. Five, a stable pleasant flavour. And, six, a long shelf life. Singh was a rigorous and celebrated scientist, admired by his peers as well as by farmers. This laundry list, however, reminded me of the George Bernard Shaw joke, with the perfect match of beauty and brains and the conscious search for stereotypes. Of classified matrimonial adverts and websites, where 'traditional with a modern outlook' or 'convent educated but homely' are common terms.

Sure enough, Amrapali and Mallika have not acquired the cult following of varieties like Langda or Alphonso or Dashehri or Mankurad or Banganapalle or Imam Pasand—all irregular bearers. I've heard growers in Uttar Pradesh complain that their Amrapali trees have grown quite high and are not regular bearers. Both varieties have another problem: like Neelum, they fruit very late in the season, which is a big disadvantage in the mango market where the early bird makes a killing. In the markets of Konkan and Mumbai, traders have asked me to spot a single Sindhu or Ratna in the mandi. Even when the best of scientists put in a lifetime of work in making the desirability of mango predictable, there's no saying how that cultivar will perform in varying conditions of orchards spread across a large country.

Everything points to a deep instability. In its evolutionary journey, the mango has not yet hit upon a stable, repeatable reproductive formula. Scientists hope to find it through genetics and bioinformatics.

Why is the mango so difficult to breed, I asked M. R. Dinesh, the former director of the Indian Institute of Horticultural Research in Bengaluru, the country's premier institution of its kind. 'The mango is a congregation of dissimilar genes. It's crazy, you do not know what you will get,' he said. I went to see him during the mango season a few years ago. His lab was overrun with fruits from the research plots. The sweet and musky odour of the hormone ethylene flooded the laboratory while he dealt with research scholars and their work, lifting his head over the computer screen to answer questions. He said research teams across the world have been working since the early 2010s on deciphering the genome of *Mangifera indica*, exchanging their findings and data online. But it has not been cracked yet.

This search is not new. In 1950, S. K. Mukherjee, one of the most cited scientists in mango studies, wrote that the mango seems at least partially an 'allopolyploid'—that it possesses genes from dissimilar species.[18] Because of the stability of its chromosomes, the mango 'out-breeds' or 'out-crosses' readily. Hence its genetic variety. His investigations revealed two sets of twenty chromosomes, totalling forty. Now scientists have settled on this finding and agree that the mango is a diploid. But its genetic ancestry shows ancient polyploidy.

We know this from an Indian team that reported the draft genome of the Amrapali variety in 2018; it found at least three instances of ancient whole genome duplication (WGD) events.[19] Two years later, a Chinese team reported the genome of the Alphonso; it also found evidence of ancient WGD events; it found that the mango diverged from the litchi and the sweet orange about 70 million years ago. Moreover, it found a more recent genome doubling from about 33 MYA.[20] Such factors contribute to the difficulties in mapping the mango's genome. Several research papers regret that 'studies of these phenomena in the mango fruit are limited by the lack of genome-scale data.' The draft genomes of a few varieties have been available for several years; it is only recently that reference genomes have become available to scientists across the world.

Forget the difficulties of genome mapping, it is not easy to imagine. If our chromosomes are maps, each gene is like an instruction on it. This map has several route options (alleles) for each point (locus/loci) on the map. At the time of fertilization, two sets of chromosomes have to align on the genetic map. At each locus, there are several alleles that can slot in. An organism with similar alleles on both sides for the same loci is called homozygous. This indicates that the organism has acquired stability through inbreeding. But when an organism is heterozygous, it's not predictable which allele will slot into which locus. You cannot say which parent will provide the beauty/ugliness trait and which one will provide the brain!

'It is highly heterozygous. There's much variation in alleles and loci in the mango DNA,' Dinesh said. A 2013 paper described the mango's genes thus: '...the allele size ranges from 11 to 39 with an average of 24.21 alleles per locus.'[21] In other words, at each intersection of its genetic map, the mango studied had an average of twenty-four route options. Two dozen! With so many choices arranged so randomly, each act of pollination is a new deal from multiple decks of shuffled cards. Other studies have thrown up different numbers and such matters will take a while to resolve.

With such genetic variability, how is it that we have any defined cultivars at all? How is it that we can tell a Langda mango from an Imam Pasand by its appearance and taste? In part, that's down to whether a particular allele is self-imposing (dominant) or a shrinking violet (recessive). Certain traits are dominant, so they steer the embryo in that direction. This does not mean the recessive genes go away; they are also passed on, without expressing themselves. One copying error, one mutation can tip the scale between dominant and recessive genes.

Any sought-after variety has certain dominant genes that account for its character. In the common mango varieties, the gene responsible for biennial bearing is a dominant one. Some mango species have multiple (poly) seeds (embryos) inside the stone (polyembryony). Only one seed is a result of pollination. The others are clones of the mother plant from the nucellus tissue of the ovule. They are 'true to type' and the trees that result from them are exactly like the mother tree. Most mango species in Southeast Asia are polyembryonic. They are also found on India's west coast, from Kerala to Goa. Scientists who studied these have surmised that the Portuguese brought these from their colonies in Southeast Asia.[22]

Among the multiple seeds, the sexually produced seed is the runt of the pack. It lacks the vigour of the cloned seeds. Either it does not survive germination or is much smaller in size.[23] Such mangoes do not have the varietal diversity of 'monoembryonic' mangoes, which have only one seed produced sexually. Researchers have established that a single dominant gene present in certain mango species is responsible for polyembryony.[24] If that gene is present, the mango stone will have multiple asexual seeds along with one sexually produced one.

In 1988, two scientists analysed observations on more than 1,000 mango crosses. They found that recessive genes control dwarfness, regular bearing, and precocity (the tree fruiting at a young age).[25] All three are critical for the mango grower's profitability. Dominant genes are responsible for alternated bearing. This is why it is so difficult to get the kind of tree growers need: manageable dwarves that bear flowers regularly and begin to produce fruit at a young age. Regular bearers are also precocious.

It is the same story all over when it comes to the colour of the flesh and skin. Breeding experiments have revealed that the gene responsible for colouring the flesh light-yellow is dominant over orange-yellow. A number of locations on the mango genetic map decide the colour of the fruit's skin. This is a great concern for those interested in expanding India's export market. Red-coloured mangoes are preferred in international markets.

'We eat the mango with our tongue but in the West, they eat the mango with their eyes. They have not tasted our fine mangoes,' said B. M. Chandrasekhara Reddy, the former vice chancellor of the Y. S. R. Horticultural University in Venkataramannagudem in Andhra Pradesh's West Godavari district. Consequently, mango breeders have been trying to bring red to the superior tasting desi varieties. Again, the experiments are long-winded and the success rate discouraging. Likewise, breeders have been trying out ways to create new varieties that have in-built resistance to pests and diseases and to vagaries of weather. It might help to have the mango genome mapped. Then again, it might not.

Genetic inheritance does not account for all of a variety's characteristics. Some traits result during a tree's lifetime in the form of genetic mutations. These mutations are 'somatic', which means bodily in Greek. They occur in bodily cells due to environmental stresses. They are not transferred to offspring via sex cells.

Somatic mutations cause cancer. Not all are harmful, however; some are quite desirable. Since they are non-transferable through pollination, the only way to propagate them is to replant the vegetative tissue from that particular tree. The graft is a clone with the parent's somatic characteristics. Farmers and gardeners have been propagating plants by grafting since ancient times. All the well-known mango varieties are results of grafts from chance seedlings that had desirable characteristics.

Grafting techniques have evolved over millennia. Everything from the stem to buds to roots to branches is used in about eight methods known currently. Their several variations are also reported. Apart from retaining the genetic character and somatic mutations of the parent tree, grafting has some other advantages. Grafted trees are precocious; they tend to flower within three to four years of plantation. Seed-grown trees take five to ten years. That is because the grafts overcome the juvenile stage of the plant much quicker. For the same reason, they do not grow up to be too tall and unwieldy.

A graft or a 'scion' can grow its own roots; but its vitality is compromised. A tree grown from seed has a more robust root system. To get the best of both worlds, horticulturists graft the desirable tree's 'scion' on a seed-grown sapling called 'rootstock'. Usually, it is a local and hardy variety; in recent years scientists have found that the polyembryonic varieties, not popular in India, make for good rootstock. This helps surmount the mango's crazy reproductive variability.

Yet, even this method is not infallible. A grafted tree's performance depends on the conditions. The taste of fruit from the same graft grown in different soil and weather conditions can vary. For example, the agricultural university at Dapoli took grafts of the most sought-after Alphonso trees and created planting material for northern Karnataka. These were planted in the region surrounding Dharwad and Bengaluru. The best horticultural practices were used and the yields reported were far greater than the Konkan region, home to the Hapus. In some cases, the yield was three to four times higher. And yet Karnataka's best Alphonso does not compare in quality with the best from Maharashtra. The shape of the fruit is oblong in some cases; the skin is tough in others; the hallmark colour and aroma are absent.

Nature's wild child refuses to be tamed by grafting also. '(The) Alphonso of Ratnagiri cannot be duplicated away from the coastal region in regard to its fruit quality. Thus commercial varieties of mango, although having a wide range of adaptability, are specific to different regions of the country,' wrote Ram Nath Singh in 1978.[26] Even the Alphonso, which in Singh's study was the only variety with 'a stable pleasant flavour'. In 2011, a team of scientists published the results of their comparison of the volatile compounds in Hapus grown in three different locations.[27] 'Despite having so many virtues, cultivation of Alphonso in different localities in India does not result in the same quality of fruits. Because of this fact, Alphonso cultivation is concentrated in a 700-km long, narrow coastal belt of western India, the Konkan region. Even within this region, the fruits show conspicuous variation in their taste and flavour,' said the authors.

The trees were genetically identical and the agronomic practices followed across the Konkan region were uniform. '...the difference observed in the volatiles at different cultivation localities is likely to be regulated by external environmental factors,' the authors noted. The difference in taste and flavour 'is reflected 15 days after the mature fruit has been detached from the plant'. This means it is impossible to tell the difference in quality till the fruit is ready to be eaten. You cannot identify the finest fruit by how it looks, no matter how much money you paid for it in the market.

Even as the desirable traits fade out, the genetic defects remain. For instance, alternate bearing. The parent tree's vulnerability to pests and diseases is also transmitted via the graft. While grafts yield more predictable fruits, they narrow the genetic range of trees in a plantation. With entire orchards planted with grafts of the same variety, the trees offer controlled, laboratory-like conditions for pests and diseases to better train themselves. In the complex games of biology, they become sitting targets.

Scientists have been talking about looking for answers within the mango's wild character. Specifically, in wild species, though actual experiments have been few and far between. Such enquiries focus on Indonesia and Malaysia, since they have the greatest diversity of wild mangoes. There are species unaffected by fungal diseases (*Mangifera laurina*) that can grow in wet and flooded areas; those that flower and fruit at high altitudes and in relatively cold climes (*M. dongnaiensis*); species that produce fruit with no fibre at all (*M. magnifica*); ones that bear flower and fruit in the off-season (*M.*

rufocostata and *M. swintonioides*); those that bear prolifically and have a high proportion of perfect flowers (*M. pentandra*); prolific species that bear small, black, sweet fruits (*M. casturi*); ones with in-built resistance to a range of serious insect pests (*M. altissima*); and species that produce fruit with exotic tastes (*M. pajang, M. foetida,* and *M. caesia*).[28]

India may not have the species diversity of Southeast Asia but it does not lack wild mangoes. They have been found across the country, especially in Odisha, the northeastern states and the Andaman and Nicobar Islands.[29] 'Tribal villages and surrounding forests have several unique mango types,' said Ravi Rebbapragada, who has worked for several years among the forest-dwelling communities of Andhra Pradesh and Odisha. 'These mangoes do not taste very sweet, with some exceptions. The fruit of each tree tastes different,' he told me. Forest-dwellers and their knowledge is ignored and underrated; they themselves have no interest in educating others. As Yale social scientist James C. Scott has shown over five decades of research, forest-dwellers stay away from the state and its coercive ways.

Hunter-gatherers are 'generalists and opportunists ever alert to take advantage of the scattered and episodic bounty nature may bring their way. Botanists and naturalists have been continually amazed by the degree and breadth of knowledge hunters-gatherers have of the natural world around them,' Scott wrote in his 2017 book.[30] 'Their taxonomies of plants are not classified in Linnaean categories, but they are both more practical...and quite as elaborate....What is perhaps just as astonishing is that this veritable encyclopaedia of knowledge, including its historical depth of past experience, is preserved entirely in the collective memory and oral tradition of the band.'

Such observations draw support from mango genomics. Two biotechnologists at the Utkal University in Bhubaneswar in Odisha used multiple DNA markers to determine the Indian mango's diversity.[31] Based on the origin of the mango germplasm they were analysing, they split them into four groups: northern, southern, eastern, and western. The results followed an anti-clockwise pattern: northern mangoes showed the least genetic diversity, followed by western ones. Southern mangoes were a lot more diverse. But the eastern mangoes had the highest genetic variation. More than genomics, perhaps the future of mango research lies in looking for desirable traits in these wild mangoes. That, however, requires their preservation and conservation.

Seven regions of India are listed as hotspots of mango diversity.[32] One: the Humid Tropical Region (Manipur, Tripura, Mizoram, Meghalaya, and southern Assam). Two: the Chota Nagpur Plateau (the Bihar-Jharkhand-Odisha tristate region). Three: the Santhal Pargana (spread across Bihar, Jharkhand, and West Bengal). Four: the Chhattisgarh-Odisha-Andhra Pradesh tristate region. Five: the Dhar Plateau (Madhya Pradesh-Maharashtra-Gujarat). Six: the Humid Tropical Peninsular Forests (southern Kerala). And seven: the Andaman and Nicobar Islands.

'No other fruit does possess this amazing glory,' said Bang. 'One just gets fascinated at the varietal wealth and diversity. The mango is unique!' Scientific literature says wild mangoes offer great possibilities for breeding programmes. This will widen the genetic base of the mango, instead of narrowing it down. This approach does not try to fight the naturally wild character of the mango; it tries to go along. It also bears the cultural imprint of how the mango has been traditionally seen in India. But such research is more complicated and messy than doing genetic analysis in labs. It involves finding wild species in association with forest-dwelling peoples.

These random riches, however, offer no ready solution. The modern horticulturist needs predictability of quality and quantity, along with flowering and pollination on schedule. For the grower, it is all about return on investment to keep the business afloat, especially with tastes changing along with changes in markets and industrial systems. That is the paradox of our relationship with the mango: we desire the king, but we expect it to obey our orders like a loyal servant.

The mango gets its open-ended nature from the invisible Kamadeva. From a state that is creative but not conscious. It does not acquiesce to all our expectations. All the same, it can surprise us with what we cannot imagine. That is not good enough for those who cannot accept this in-built uncertainty. They consciously search for stereotypes, for the perfect combination of beauty and brains.

Chapter 11

Ecology
Biocasino

*'Few insects have a greater impact on...(the) world trade
in agricultural produce than tephritid fruit flies.'*

—J. E. PENA

Agriculture is a venture in ecology. In many ways, it is the opposite of industrial production, which frames how we view production and consumption today. Industry has its own set of variables: fluctuating markets and raw material supplies, changes in oil prices, the availability and training of labour and management, among other things. Nevertheless, factories are controlled environs, and conditions can be managed to a great extent. Agriculture, though, contains all of those variables plus the inherent risks of dealing with nature and its uncertainties.

The mango's internal genetic volatility encounters the external world's variability. The tree is shaped on the inside by how it deals with the outside. We cannot hope to know the mango without understanding its ecology and its physical environment. Science divides the external pressures into two. When a living organism, like a pathogen or a predator, affects another organism, it is a 'biotic' stress. A strain from the elements, like rain or drought, is an 'abiotic' stress. The two are often connected, as we shall see.

It is worth repeating that the mango is a flowering tree of the tropics. A quick way to understand its setting is to look away from the equator towards the poles. In colder temperate regions, biotic and abiotic conditions are predictable; there are few pests and diseases, and the weather forecasts are reliable. Along such latitudes lies the industrialized North—the rich countries of Eurasia and North America. They have incomparable modern infrastructure of science and technology, including the horticultural sciences. Gymnosperm conifers dominate several temperate forests. While they have unusual and rare organisms, one does not go to temperate regions for biodiversity.

The tropics and the subtropics are a different world. Direct sunlight generates heat from the ground that makes the weather unpredictable, tormenting even meteorologists. Monsoonal weather is common across the tropics and the subtropics. For example, India gets almost all its rainfall during the three-month southwest monsoon, perhaps the world's most intensively studied weather system that still remains a mystery. Former union minister Chaturanan Mishra was known for saying that the monsoon—not he—was the real agriculture minister of India.

Life thrives closer to the equator. 'The genetic "homes" of the 30 crop plants that, in aggregate, give humanity 95 per cent of its nutritional requirements are all to be found in Asia, Africa, and Latin America,' wrote authors Cary Fowler and Pat Mooney in a 1990 book.[1] 'Were one to list the top five crop species for every country, only 130 crop species would be named, virtually all originating in the Third World.' Called the global South or the 'developing world', the area is almost entirely in the tropics and the subtropics. But all their biological wealth counts for nothing in economic comparison to the industrialized North; the global South contains the world's poorest.

If the mango originated in tropical rainforests, so did its innumerable friends and enemies. From pollinating insects to pathogenic bacteria, from frenemy fungi to seed dispersing primates. All non-human primate species live in the tropics or the subtropics, and our ancestral hominins evolved in tropical Africa. Everything is more diverse here, be it flowering plants, animals, insects, microbes, or fungi. In tropical ecology, there is always a lot going on everywhere. Naturally, a lot is always going on in mango trees. Each mango tree is a planet unto itself. Many critters big and small live on it. One review of major pests of the mango listed 322 species of insects and mites.[2] Of these, four are serious pests found across the world that require regular control measures. The big four are fruit flies, seed weevils, tree borers, and mango hoppers.

'More than 400 pests of the mango have been recorded and twelve of these are serious pests in India,' said Ghanshyam Pandey, the former head of the department of crop production at the Central Institute for Subtropical Horticulture (CISH) near Lucknow.[3] The most serious pests of India's mango crop are the sucking hoppers. These minute insects from a variety of species suck the plant's sap and they particularly relish the tender tissue that emerges during inflorescence. They lay their eggs in the panicle, leaf, and leaf-stem. Their bodies secrete honeydew, a sweet nectar-like fluid that they leave on the tree; this attracts sooty mould disease. Hoppers are vectors of several plant pathogens.

In the rich mango-growing regions of coastal Andhra Pradesh, the mango hopper has become a raging problem in recent years, said B. M. Chandrasekhara Reddy, the retired vice-chancellor of a horticultural university. G. V. Ranga Rao knows this well; he has managed about 100 acres of ancestral orchards on behalf of his joint family near Nuzividu town, about 40 kilometres north of Vijayawada. 'My neighbour there has some 30 acres under the mango. He has received awards for organic farming. He decided to not use any synthetic insecticides on his mango. I warned him the hopper will devour his entire crop,' Rao told me several years ago. 'Sure enough, the next season, he lost everything to the hopper. He came to me to get mangoes for his family to eat.'

Rao is also a highly regarded entomologist who spent a major part of his career researching a crop protection approach called integrated pest management (IPM). Till 2017 he was working at the UN-established International Crops Research Institute for the Semi-Arid Tropics (ICRISAT) in Hyderabad. When I first met him at his office, though, it was not the hopper that had him worried. His orchards were under severe attack from a fruit borer relatively new to India. Called the red-banded mango caterpillar (RBMC), it has come all the way from Indonesia, spreading along the rim of the Indian Ocean to Thailand, Myanmar, Bangladesh, and then into India. From West Bengal and Odisha, it reached Andhra Pradesh.

With his international connections in scientific circles, Rao enquired how other countries dealt with the RBMC. 'It is a minor pest in Southeast Asia, from where it has come. There, it hits about 5 per cent of the crop. But here in India, it has deadly consequences. I know quite a few growers who have lost 100 per cent of their mango crop to this caterpillar.' Rao was concerned about its further spread: 'This region sends raw fruit across the country. How long before the RBMC spreads and becomes a countrywide problem? Nobody can say.' After he tipped me off, I began noticing reports of this insect inflicting damage to mango crops in Telangana and Karnataka. In 2022, I visited mango growers near Bhagalpur in Bihar, who had lost their entire crop to the RBMC.

'In East and West Godavari districts, mango cultivation is fast disappearing due to this pest.' In and around Nuzividu, about 75 per cent of mango growers have shifted to seed cultivation or feed crops for dairy and poultry, which are more manageable and profitable. Rao has begun to shift away from mangoes himself. With a palm oil processing factory coming up nearby, he dedicated 25 per cent of the land he manages to oil palm and another chunk to eucalyptus. 'I advise all farmers in irrigated areas to get rid of mango trees. It saddens me, but everything from guava, citrus, oil palm, and eucalyptus is more profitable.' Once the borer gets inside the fruit, it heads straight to the seed; no amount of insecticide spraying can reach it. 'The only way is to fumigate. But how much can you fumigate each fruit?' Rao said.

How did it get so bad? Rao had a one-word answer: irrigation. The mango trees of coastal Andhra Pradesh were not irrigated earlier. A spurt in irrigation projects let to an increase in mango production. 'That also increased pests. Insects thrive in wet and moist conditions. Besides, as it is in all fruit crops, irrigation compromises the quality and taste of the fruit.' He said his mother had always asked him to not irrigate the trees that produced pickling mangoes; the water reduces the acid content and sourness of the fruit.

He has encouraged young scientists to work on this threat. One study recommended the use of light traps. Said Rao: 'Light traps do not work very well in the thick foliage of mango orchards.' He kept underlining the need for more research. But science struggles to keep up with the dynamic world of subtropical insects. To give an idea, the mango leaf webber (*Orthaga euadrusalis*) was not a serious pest of the mango earlier. But it has become so recently in the two biggest mango producing states of Andhra Pradesh and Uttar Pradesh. Rao says we need an all-new approach to plant protection. And he had not even mentioned fungi.

'Anthrax!' said a trader, screwing his face, lowering the corners of his mouth, as if recoiling from a contagion. 'Anthrax has spoiled the mango crop this season. All the produce is blighted,' he said, showing me black spots on Alphonso mangoes. Several traders had complained of anthrax. This was a few years ago while I was visiting the Vashi fruit market in Navi Mumbai. I knew anthrax is an animal disease caused by bacteria. How could it affect plants? Then, somebody explained: 'It's anthracnose, a fungal blight. They call it anthrax because they don't know any better.'

When it comes to fungi, such ignorance is common. Fungi are a separate taxonomic kingdom. Their mysteries confound scientists, too. Their study comes under botany, although they are closer to animals in many ways. A single individual colony of the honey mushroom is among the world's largest organisms, spread across twenty-seven acres in Michigan, USA (look up humongous fungus). Fungi can live off both living plant tissue as well as dead tissue; it is called hemibiotrophy.

The fungus behind anthracnose is found across the world. It kills parts of the mango leaf tissue, causing black 'necrotic' spots, and these can join up to afflict the margins of the leaves. It can also attack twigs and kill them off. It can infest flowers and panicles, turning them brown-black. It can infest and abort small fruits, mummify them to later spread its spores via splashes of raindrops. A severe case of anthracnose is rare but it can kill a tree. Then there are the more serious forms of damage for the grower. It affects harvested fruit, leaving brown-black spots on the skin. The fruit turns unappealing. The fungus can go deep into the pulp.

Scientists believe this fungus has co-evolved with the mango. Why? Because the mango fruit contains elaborate chemicals at high concentrations to repel it.[4] The fungus just lies low, masking its presence with its own chemicals, waiting for the fruit to ripen. Ripening emits the hormone ethylene, which signals the conversion of starch into sugar, making the skin tender and colourful. (We recognize the sweet, musky odour of ethylene as the fragrance of ripening fruit.) The hormone lowers the fruit's chemical defences. The fungus can detect ethylene and blight the fruit when it is in transit or when it reaches the market.

Anthracnose is the biggest reason mango growers fear untimely rains. Vidyadhar Joshi of Devgad described his travails from the 2014-15 season. The flowering was poor with a low proportion of 'perfect' flowers. By the end of February, his trees had fruited. Then came unseasonal rains. 'I observe the rain like some other people pray to their gods. They light lamps and incense, I observe the rain gauge with religious attention,' he said, his hands joining in a gesture of reverence. 'On 28 February 2015, it rained 5.8 millimetres. The next day, 1 March, another 5.4mm came down over three hours.' It was a warning; humidity causes fungal spores to spread rapidly. He immediately set about spraying the fungicide difenoconazole on his crop. 'Farmers from all over were calling for advice on how to handle the effects of untimely rains. Communication among farmers is quite poor, they are desperate for leadership,' he said.

With all his knowledge, Joshi couldn't do enough. 'The trees I managed to cover in three days did not have the fungal infestation. But I need thirteen days to cover all my trees because I take on contract several orchards spread across a wide area. The trees I reached late produced blighted fruit. The larger the fruit, the greater the infestation because there's more surface area.' This despite the fact that Joshi hires more full-time workers on average than anybody I have met; despite staying in touch with a range of scientists.

With operations spread over more than 500 acres, Desai Bandhu of Ratnagiri had it even worse. 'It was already a poor year. Then came the rains,' said Anand Desai Pawas. About 80–100 workers are employed full-time throughout the year, with an additional 450 being hired for the two-month harvest period. 'It still takes me eight days to cover all my trees,' said Desai. The large-scale operations of the Desai family are organized, well-funded, and informed with the best of science and technology. The family is proud of its heritage and takes its mango business seriously, enjoying immense brand value and profitability. Yet the Desai Bandhu struggle with untimely rains and 'anthrax'.

It is merely one of many fungi that live on the mango, along with a host of other agricultural crops. *Fusarium moniliforme* is even more dangerous. It causes 'mango malformation', a complex of three types of deformation of the tree. One affects younger plants and their vegetative growth, the other mangles the inflorescence, and in the third the two combine. The fungus does not kill the tree, just mangles its canopy and lowers the flowering. This is more common in the northwestern parts of India where a high proportion of the trees have this disease. It is not so prevalent in eastern and southern India. Western Uttar Pradesh and Punjab have seen cases of malformation destroying up to 80 per cent of their mango crop. Growers near Lucknow complained that Amrapali, the IARI variety, is especially vulnerable to malformation. In severe cases, there is no option but to destroy the trees and choose new, uninfected planting material. That amounts to loss of time, money, and labour.

Then there are the black spots of alternaria rot caused by another fungus that lives on the tree throughout the year. It hits leaves and flowers, retards pollination and fruit setting. If successful, it blights the fruit. Another serious disease is the 'stem-end rot', indicating the part of the fruit from where it begins. A range of fungi cause this disease. They also ruin fruits after the harvest, particularly those from old trees. The mango has a few other fungal enemies. For example, black mould is caused by a fungus called *Aspergillus* and 'transit rot' is caused by one called *Rhizopus*.

'In the month of March, hot and dry conditions are ideal for powdery mildew, a disease caused by a clutch of fungi. It appears in the form of clearly visible white powder on the leaves and stems,' said a scientist from Lucknow. 'But if it is cold and humid, then you get blossom blight, caused by another set of fungi.' What can growers do to save their crops? One, closely observe their trees and the weather conditions to anticipate fungal infestation. Two, spray fungicides in a timely manner. Three, treat the harvested fruit with expensive vapour heat, which can easily ruin the fruit. Four, pray for good luck.

Just on the disease front, the mango has a formidable list of afflictions. In one review, I counted twenty-one diseases of the foliage and flowers. From rot to spots to blights to necroses to blotches to moulds to disorders to malformations to wilts to mildew to scabs to sudden decline.... You can blame everything like fungi, algae, bacteria, viruses, parasites, and midges. Then there is the range of soil diseases. There's black root rot, nematode damage and white root disease, to name a few. 'Diseases affect all tissues and developmental phases of the mango,' says a review.[5] And that's without counting insect pests.

I feel a weird tension when I visit entomology departments of universities, like walking into the climax of the movie *The Silence of the Lambs* (1988). The world of insects is a messy sound-and-light affair. Viewing insect specimens arranged and labelled neatly in display panels makes me uneasy. The mind drifts to diabolical plots, imagining a cold-faced scientist walking up to me in a white lab coat, wielding a clipboard with the details of a sinister plot!

Instead, at CISH, Rehmankheda, I met Balaji Rajkumar, an animated young entomologist in jeans, t-shirt, and sports shoes. He walked on his toes, keeping his weight forward, ready to pounce, in the manner of natural athletes. His friend and colleague Gundappa told me Balaji's nickname is 'Jonty', after the former South African cricketer and fielding phenomenon Jonty Rhodes. Balaji, Gundappa, and I walked out to the institute's mango orchards where they carried out their entomological experiments on dedicated plots.

The trees were numbered. Balaji approached a tree that had white powdery clusters. 'This is the giant mealybug, so called because it is the largest of such bugs,' he said. It sucks on the plant's juices, sticking to crevices and small juicy twigs. It secretes and covers itself with powdery wax to protect itself. The male does not feed as an adult, dying after copulation; the female of this species is more dangerous. 'It can do without food for months altogether, waiting for hot and moist conditions,' said Gundappa. When such conditions arrive, a frenzy of feeding and reproduction occurs, visible only as white wax powder. It secretes honeydew, inviting infestation of sooty mould and other fungi. After it has had its fill by June, the female just drops off the tree to lay eggs in the soil—down to a depth of half a foot. Come December, and nymphs emerge from the eggs and climb the trees. It thrives if there is rainfall while laying eggs and a dry spell when hatching.

So how do you control them? 'Insecticide sprays do not help because they simply shed their protective powder. The only way is to spray soapy water that penetrates the wax,' said Balaji. Horticultural recommendations include hoeing the soil at the time of the laying of eggs; even sprinkling chlorine-containing pesticides directly on the soil does not help. Another suggestion is to band the tree trunk with chemicals that prevent the nymphs from climbing up the tree. 'The mango mealybug has more than sixty known host trees. How many do you band?' Balaji quipped. He picked up a dead pink mealybug and placed it on his handkerchief to carry out a postmortem: 'If it is killed by a fungus, the body hardens; a bacterial infection leaves the body soft, smelly, and putrid; a virus leaves the body undisturbed. This one was killed by the insecticides we are trying out on the tree.' He carried it like a trophy to be put in some glass panel.

A tiny insect came and settled on my notepad. I decided to test my guides. 'That is a thrips nymph,' Gundappa said, barely glancing at it. A large moth suddenly caught Balaji's attention. He handed over the dead mealybug to Gundappa and whipped out his phone to photograph it; he was also the institute's photographer. 'The mango hosts many kinds of spiders,' he said. We walked past a tree trunk teeming with hoppers. They are reared here to check the efficacy of new pesticides.

A large fly began to hover over my hand, moving with it in synchronous alacrity. 'That is the mango's greatest friend, the syrphid or hover fly,' Balaji lit up. Its larvae kill the larvae of aphids, a class of serious sucking pests. After the syrphid larvae become fully-grown hoverflies, their diets change to nectar and pollen; then they become even more valuable to the mango tree. 'They are by far the most important pollinators of the mango here. Then come honeybees and houseflies,' Ghanshyam Pandey had told me earlier inside the CISH building. Similarly, the rufous treepie is frequently found in mango orchards. The bird eats the black hairy caterpillar, another mango pest, in the higher reaches of trees that are otherwise difficult to reach. (Not all birds are friendly; a pandemonium of parakeet can savage a tree laden with fruit.) The aphids don't take it lying down; they pay for protection by secreting honeydew that draws red ants, another major pest. But the ants also feed on some weevils that harm the tree in other ways. The entomologists just went on and on and on....

Then Gundappa switched to ethnobotany, pointing to a shrub growing on a farmer's orchard close by: 'This is the hill glory bower (*Clerodendrum infortunatum*). Villagers call it bhattar and use it to heal wounds. It is also supposed to be helpful for pregnant women.' Balaji said the plant is a good indicator for the emergence of the mango mealybug: 'The mealybug nymphs congregate on this shrub. If you spot it, you can expect a mealybug attack on your mango trees from January.'

A half-day spent with Balaji and Gundappa left me feeling that the fruit isn't the most interesting part of the mango ecosystem. The world of insects, invisible to the casual eye, has all the drama of a mobster movie! There's violent crime, sex, retribution, love. Attacks and counterattacks. Informants and counter intelligence. There is nature's rule of law with its own enforcement agents.

Seeing the mango's ecology through the eyes of two young entomologists was a novel experience. I could not help but compare it with how mango growers see their orchards. They have their own terms for these pests and diseases, their own ways of dealing with them. They do not have the interest, training, or the resources to seriously study the pests and the friendly insects. Their only concern is to have fruits at the end of the season to sell in the market. But that is merely one aspect of the mango tree. Agriculture is not industrial production. Perhaps it is in temperate regions, but surely not in the tropics and the subtropics.

Biotic stress comes in many shapes and sizes. Each year during harvest season, mango growers in Maharashtra's Konkan region employ more than 80,000 migrant workers to protect their crops from monkeys. Most guards come from Nepal, where the paddy crop is harvested by the time the mango tree is full of fruit in Konkan. The monkeys have become a serious threat. 'Earlier, there were buffer forests between the Sahayadris (Western Ghats) and Konkan's coastal regions,' said Vivek Bhide, a physician in Ganpatipule, Ratnagiri, one of the better-known cultivators of the Alphonso mango. 'Now those forests are so degraded that they cannot feed the wild animals. So they come down to the farms and orchards of Konkan.' Under natural, forest-like conditions, monkeys are likely seed dispersal agents for the mango tree. But when they are driven to orchards out of a desperate lack of food in the forest, they cannot wait for the fruit to ripen. They attack trees, going for unripe fruit.

Andhra Pradesh is no different. G. V. Ranga Rao said monkey depredation is one of the main reasons mango cultivation is losing viability. 'Our village has about 5,000 monkeys. Because of religious reasons, monkeys cannot be culled,' he said. When the menace gets out of hand, the forest department catches some monkeys and releases them near another village. 'So now each village has hordes of mango-marauding monkeys,' said Rao. He was forced to hire two guards and several dogs just to keep monkeys out. This increased the cost of plant protection.

Then, a researcher had an idea. He developed an electrified fence running on battery; it shocks and stuns the monkeys with a small charge. Rao tried it on his farm. 'In order to not kill the animals, the charge is low. The monkeys figured out that it is only the first marauder who will get a shock. Thereafter, the fence takes five minutes to charge from the battery. In that duration, the entire horde is on your mango trees. So they willingly risk the first shock. What are they going to do, they have nothing to eat!'

Rao worked with the inventor, asking him if the fence could get recharged a little quickly with, perhaps, an additional battery. It just didn't work. To each technological solution that a grower finds, nature's research establishment has a workaround. The cost of protecting the mango crop from monkeys has been steadily increasing across the country. Rao said it was an important consideration in his switch to palm oil.

Monkeys were a major reason farmers switched from growing mangoes to oranges in Maharashtra's dry Vidarbha region. While the mango is a monkey favourite, the orange peel contains an oily chemical that stings their eyes, repelling them from the fruit. 'As orange cultivation became more profitable, farmers cut down their mango trees and used the timber for making boxes for carting oranges,' said Ashok Bang in Wardha. In recent years, farmers have observed monkeys peeling oranges behind their backs, so that the irritant in the peel cannot squirt into their eyes. That is how quickly organisms learn new tricks without setting up research and development establishments.

'Fruit fly' is a wide term often encountered in biology textbooks. The common fruit fly (*Drosophila melanogaster*, remember?) is the most studied organism among geneticists. It is easy to study in laboratories; it has a simple genome and breeds profusely. Several studies have bred on it to win Nobel prizes, sustaining many a career in genetics. Hundreds of species of the family *Drosophilidae* have been recorded; they lay their eggs in rotting fruit and are a nuisance outside of the lab.

The other family of the fruit fly is *Tephritidae*; it makes *Drosophilidae* seem pleasant. It includes thousands of species in hundreds of genera that, put together, is the world's most destructive all-round pest in agriculture. 'Few insects have a greater impact on international marketing and world trade in agricultural produce than tephritid fruit flies,' said an entomological review.[6] Each of the world's mango-growing regions has fruit flies of some kind or other.

The review listed about sixty species that attack the mango. Among them, the Oriental fruit fly is the most dangerous mango pest in India. 'It accounts for about 27 per cent of the harvesting loss,' said the ICAR.[7] In northern India, it disappears in December due to the harsh winter, only to reappear in February. The warmer conditions of southern India make it a year-round pest of not just the mango but several fruit and vegetable crops. When an infected fruit is cut open, the pulp is gnarled by maggots; it can be a revolting sight if you are not a committed entomologist. Apart from directly destroying the harvest and opening the path for a range of infections and pests, the fruit fly is the major reason Indian mangoes do not get exported to international markets.

The female pierces the skin of the mango fruit just before it is fully mature and lays its eggs inside the pulp; the incision is no bigger than a pinprick. One female can lay more than 1,000–1,500 eggs in a lifetime. It takes an average of thirty-seven days for an egg to become an adult; under ideal circumstances, this fruit fly can get through nine, even ten generations in one year. It is highly adaptable, feeding off a wide range of plants. It attacks the banana, citrus, and papaya plants with equal gusto. Even if one orchard does well in controlling it, the fruit fly can find sustenance elsewhere. Once the maggots drop out of the fruit, they bury themselves in the soil and pupate, emerging as adult fruit flies ready to attack more fruits.

There are only two ways to control fruit flies. The first is aerosol spraying of synthetic insecticides. Groups of organophosphates and organochlorides used earlier were found to be severely toxic to humans. They are cheap and still on the market. Those who have the money to buy synthetic pyrethroids, which are not as toxic. With pesticide residues becoming a major concern, scientists do not recommend spraying as the main line of protection.

They suggest fly traps based on the Male Annihilation Technique (MAT). These contain chemicals mimicking the hormonal scent of the female; the males cannot resist it and walk into the trap containing an insecticide. For such devices to be effective, however, the fruit fly must be prevented from breeding. That requires a range of 'biocontrol' measures, which need regular monitoring of the orchards by the growers.

Every approach to plant protection, in fact, requires the grower to watch over not just the trees—and their branches and leaves and flowers and fruits—but also the soil, surrounding plants, and foliage, temperature changes, weather forecast, and the rain gauge. Not to mention keeping themselves abreast of the latest from research stations and laboratories. This is easier said than done anywhere in the subtropics, especially in India.

∫

Several growers and traders use the same analogy for Alphonso: 'It is the king of mangoes but it is blind in one eye!' They refer to the second serious problem with this variety, after irregular bearing: a high proportion of the fruits have spongy tissue. When you cut open such a mango, you do not find the flawless, orange-tinged fibreless pulp; instead, you find yellowish corky tissue that looks like sponge, often with gaps of air. In bad cases, the tissue turns an ugly grey and has a stench. It is not an infection, hence non-transferable.

For some inexplicable reason, something goes wrong just before the harvest. The starch of the unripe fruit does not convert into sugary pulp upon ripening. What makes it worse is that there is no way to tell if the fruit has spongy tissue. (Except to dunk all the fruits in water; the healthy ones sink but the ones with spongy air pockets float.)

It is a form of physiological disorder called 'internal fruit breakdown' (IFB). It has some other forms with dodgy names, too, that ruin the fruit: 'tip pulp'; 'soft nose'; and 'stem end breakdown'. The most common form of IFB in North India is 'jelly seed', noticed in Dashehri, Chausa, Langda, and Amrapali. It causes the flesh close to the stone to turn soft and jelly-like.

There are a few theories. One says it has to do with unusual heat around harvest time[8]; a few growers in Konkan told me the side of the tree facing the Sun produced more fruits with spongy tissue. A grower in UP said calcium deficiency is behind the jelly seed problem; some studies support this view.[9] A third hypothesis, first reported in 2006, is perhaps most likely.[10] It attributes IFB to a premature onset of seed germination.[11] This gets triggered by a lack of certain fats inside the fruit and seed. The role of fats is critical to all seeds; the Alphonso fruit is reported to contain the highest percentage of fats in both pulp and seed compared to other mango varieties.[12]

There are no proven measures to prevent IFB.[13] Spongy tissue and jelly seed really affect the mango sector. Some estimates say an average of 30 per cent of the total Alphonso crop is lost to spongy tissue, in bad cases increasing to 35–55 per cent.[14] More recent estimates put the losses in the 20–60 per cent range.[15]

In some experiments to prevent spongy tissue in the Alphonso, researchers sprayed some ordinary nutrients mixed in seawater. It reduced the incidence of spongy tissue from 54.3 per cent to less than 5 per cent. While the results are promising, the practices they suggest have not caught on yet. Nevertheless, it is a spark in an otherwise gloomy field: an elegant solution for a serious problem.

IFB is a genetic trait. It afflicts at least sixty-five mango varieties across twenty-three countries.[16] Indian mango varieties and cultivars derived from them show a higher incidence of IFB, particularly Alphonso and Dashehri. Yet some popular varieties rarely demonstrate these disorders; they include Rajapuri, Banganapalle, Pairi, Kesar, and Neelum. This is why ICAR's mango breeders have created and released crosses with desirable traits.

In the Konkan region's Sindhudurg district lies the famous Regional Fruit Research Station of Vengurle. It has been the centre of the most important experiments on the Alphonso, providing a bulk of the planting material used along the Konkan coast. Its Alphonso-Neelum cross called Ratna has a large fruit free of spongy tissue. Balaji had told me that a variety called Elaichi—it has aroma like cardamom—is resistant to mango malformation. Murugan Sankaran, mango breeder at IIHR, said two mango species from the Andaman and Nicobar Islands, *M. captosperma* and *M. andamanesis*, seem resistant to the anthracnose fungus.

The vulnerability of varieties to biotic and abiotic stresses varies wildly. Several researchers have documented parts of this bewildering story. A study in Punjab tested five cultivars. Their findings, presented in 2005, showed the Mallika was most vulnerable to the hopper and Dashehri the least; Mallika and Langda were the most prone to malformations. The same Mallika turned out to be strong against the powdery mildew.[17]

Two Pakistani horticulturists measured varietal susceptibility to the hopper in 2003. The Langda and Saroli (another name for Bombay Green) were the most vulnerable, while Neelum, Dashehri, and Zafran least so.[18] Horticulturists have told me that hoppers avoid trees that have an open canopy with free flow of wind and sunlight. Scientific literature also attests to this characteristic.[19] However, no variety has substantially shown any resistance to the mango hopper.

The humble-yet-sought after Neelum has a peculiar weakness against black spots caused by a bacterium (*Xanthomonas campestris pv. Mangiferaeindicae*). This is passed on to all progeny when Neelum is used as the female parent but not when it is a male parent. A commercially unviable variety from Florida, called Sensation, has proved to be highly resistant to bacterial black spots in Australia; although, it is beset by IFB.[20]

No mango is entirely immune to the fruit fly. A variety's vulnerability is determined by the taste of the fruit and the nature of the skin. B. M. Chandrasekhara Reddy said the juicy-fibrous varieties of coastal Andhra Pradesh are never infested by fruit flies. Two scientists at IIHR reviewed existing studies to show that Banganapalle was the most susceptible variety to the fruit fly and the Langda was the least vulnerable; on one trial plot the Landga tree had zero fruit fly maggots.[21] Just as the immature fruit has an in-built arsenal of chemicals to keep anthracnose at bay, some varieties have irritants that the fruit fly cannot surmount.

'Irritants like turpines and turpinols repel fungi and microbes and slow down the onset of rotting in the fruit,' Ashok Bang told me. He was missing the Langda, not readily available in Wardha. The irritant also constitutes the signature Langda taste that hits the throat. These irritants are concentrated in the fruit peel and not in the pulp.[22] But the Langda drops off readily. In UP and Bihar, when people joke about somebody with poor nerves, they summon this variety: 'He shakes like a Langda mango in the face of a slight breeze!'

Plant breeders face impossible odds. There is the mango's confounding genetic variability. Then, each cultivar responds unpredictably to different conditions. Each variety comes with its own contradictions. The most underrated wild mangoes that command no market or conversation have enviable traits. Each choice is a mixed bag. The best-informed growers have to deal with vast uncertainties. Each grower must make a range of make-or-break choices without fully understanding them. It takes a few hours for conditions to turn from ideal to disastrous. Each abiotic stress brings its own biotic stresses.

Mango scientists have it slightly easy, holding down permanent jobs. But the mango frustrates them to no end; the simplest experiment can take months or years to test out—and the odds are poor even then. As attractive as the mango tree is for commercial and cultural reasons, it is a nightmare for a horticulturist looking for control. Countless microbes and insects have complex relationships with the mango. The tree cannot do without them but they can make its life difficult.

Each tree is a casino of variable abiotic conditions. Innumerable biotic players try their hand at it. Each victory results in several losses to other players. Every player has a sporting chance. Modern science and horticulture give the impression that we run this casino. But the truth is we are also one of the numerous players gambling on the tree. All our technological advances cannot fix the play in our favour; not with any consistency, anyway.

In this casino, as in all others, the House always wins, regardless of the outcome for each gambler. The House has built this casino tree over millions of years. Each prize mango fruit that we devour is a small victory against pests and diseases. The fruit fly or anthracnose collect their winnings from each mango that they infest.

In the biological casino of the mango tree, the House is not us. It is the invisible hand of ecology that plays us! Our odds improve when we go along with it. When we try to control the House, it has ways to sucker us into playing impossible odds. Desire readily turns into a gambling addiction.

Chapter 12

Cultivation
Who Grows Your Mango?

'The main reason for low productivity...is neglected orchards as owners are absentee land lords.'

—B. S. KONKAN KRISHI VIDYAPEETH

We are interested in only one organism betting on the mango: ourselves. The Indian obsession with mangoes results in their cultivation on more land by more growers than any other fruit. However, growing the mango is very different from not just other agricultural crops but also from other fruits. 'A mango orchard is a long-term investment. It is not comparable to a five-year commitment required for, say, grapes or pomegranate,' said Vivek Bhide of Ganpatipule in Ratnagiri. It takes a different kind of grower to cultivate mangoes. And a very distinct kind of grower, if you want quality. 'I earn for three months: March, April, and May. But I have expenses all through the year, both personal and in the orchards,' said Vidyadhar Joshi of Devgad in Sindhudurg.

You can tell an engaged grower from an indifferent one by how they reply to questions about their work. If they reply with only frustrated rants, the conversation is fruitless. Bhide and Joshi are two of the most engaged and articulate mango cultivators I have met. They responded to every question in detail. They engaged with follow-up questions. They are immersed in each aspect of their work; they talk with conviction, authority, and passion. It is from such growers, along with a few scientists, that I could piece together the requirements for growing quality mangoes.

The work begins right after the mango season has ended. For the Alphonso on the Konkan coast, that is May–June. In large parts of South India, it is June–July. In North India, the last major variety is the Langda in July–August. Then the cycle begins again. The first task is pruning the trees. This prevents 'apical dominance' that suppresses flowering. This requires reaching the canopy; tall trees make pruning difficult and expensive. Pruning also exposes the leaves to sunlight and wind, which controls pests like the mango leaf webber; its female lays eggs in July on leaves. If inspections reveal its presence on any tree, the grower needs to target pesticide sprays, tear down infested branches, burn them, and then rake the soil underneath to remove fallen larvae.

The rains cause weeds to grow. The rainy season requires weeding operations and the application of both organic and inorganic fertilizers. Well-kept trees have a stone periphery around them to press down on the roots. It is called gudhmi in Maharashtra and thanwla in Uttar Pradesh. This forces the emergence of white-coloured superficial roots good for tree health. It also corrals the water, fertilizer, and leaf litter, enriching the soil. The rains inevitably break the water channels on the orchard floor; these need repair. If water logging is not prevented, the soil turns alkaline. An alert grower must watch the floor as keenly as the canopy. Each biotic and abiotic stress requires a customized response. No one insecticide prevents all pests and fungi. Besides, diagnosis is not enough. An attentive grower must keep pace with the dynamic world of scientific research—all of it in scientific language.

Still, the monsoon and its aftermath are not hectic. A watchful grower can detect early signs of pests and diseases, and prevent it before it blows up into, say, a fungal infestation in January. It helps to keep an eye on the surrounding vegetation and neighbouring lands for insects that survive on other trees and can then migrate to the mango. A stitch in time and all that. By November–December, trees begin to flower along the southwestern coast. In North India, this happens two months later. This depends on the region and the variety.

With flowering begins a rigorous round of plant protection. New shoots and flowers are tender and juicy. Plant predators love tender and juicy. The grower has to keep an eye out for a list of fungal infections, as also minor shifts in weather. As we have seen already, the slightest variation in weather invites its own set of pests. 'The winter fog is essential for healthy flowering,' said Kaleemuddin Siddiqui of Amroha, a rich mango-growing region 125 kilometres east of Delhi in Uttar Pradesh. 'The fog covers the trees like a warm blanket and coaxes a prolific flowering. If the fog disappears or the winter rains fall at a bad time, the flowers suffer paala (frost).' The grower hopes for a narrow band of ideal weather—it arrives once in several years.

A successful flowering ends with fruit set within a month, which is January–February in the south and March–April in the north. The level of alert can come down at this stage. Raw fruits have their own chemical defence to tackle their natural enemies. But as the fruit matures, these defences begin to get withdrawn. This is when the fruit fly appears; its population peaks in May–July. The most potent of toxic pesticides, along with fly traps, are not enough to prevent this formidable foe. The grower needs to adopt 'biocontrol'—to prevent it from breeding. For this growers need to scan both fallen fruits and those dangling from the branches, and infested fruit must be destroyed completely. Again, the orchard floor needs constant scanning and raking, along with a keen eye on surrounding lands and vegetation.

The maturing fruit responds in complex ways to water. All irrigation and watering must stop twenty to twenty-five days prior to the harvest for the sugar levels to shoot up inside the fruit; or acid levels for sour, pickling mangoes. A shower around this time can also cause sap to gather near the twig, making the fruit bitter.

After the harvest, the fruit needs a round of treatment to keep away anthracnose or fruit fly. One way is 'vapour heat treatment'. A slight change in temperature settings here can ruin the fruit.

'If you wish to get high prices for your produce, especially in the export market, you have to take care of each and every fruit from the tree to the market shelf,' said A. Kiran Kumar in Sangareddy. There are just too many variables. In fact, in rural UP, a term for growing an orchard is jungle paalna, that is, to culture a forest. 'The mango grower needs a big heart. The sum of a season's hard work can disappear in a matter of minutes in a shower or a storm or a pest attack,' Siddiqui said. If a season is hit or ruined entirely, the mango grower does not give up. Gamblers do not leave the table when they are down. They have to find the resources—beg, borrow, steal—to have another go at the mango table, hoping for a better hand in the next deal.

'When his labours come to nothing, you can often hear a grower quip with nonchalance, 'Bagh paala hai!' Siddiqui added. It was a pun. Bagh from the Farsi means an orchard or a garden; bagh in Hindi/Urdu means a tiger, from the Sanskrit vyaghra. Tending an orchard is like taming a tiger! Even growers like Bhide and Joshi cannot control all variables. The only example of successfully following all horticultural guidelines to grow the perfect fruit that I have seen was on the internet. It was about Miyazaki mangoes; the name comes from a prefecture on the Japanese island of Kyushu.

One famous brand grown here is the 'Taiyo no Tamago' or the egg of the sun. It is a red mango weighing about half a kilogram. The plants are kept dwarfed and reared inside greenhouses. Each fruit is supported by a string and has its own net. It ripens on the plant itself and falls into a net hanging underneath; reports say it is extraordinarily sweet. (I have met only one person who has eaten it in Japan and he was emphatic about the taste.) Each mango is tended to by trained professionals like a high-end artefact; it is packaged like an item of jewellery. Each mango sells for more than

US $25 (more than ₹1,800). The first produce that hits the central fresh market of Miyazaki prefecture is auctioned. In April 2019, two mangoes were auctioned for half a million yen (about $4,500 or about ₹333,000).

In 2017, a farmer from Jabalpur bought a mango sapling from a stranger on a train. A few years later, when it produced fruits, he realized it was the Miyazaki mango.[1] Word got around and thieves broke into his orchard to steal the mangoes. He employed a team of guards to protect the tree. Gujarati traders from Surat and Mumbai began offering thousands of rupees for his mangoes, though he does not sell yet. The Gujarati passion for the mango has, meanwhile, created a surprising orchard. While it is not like the greenhouses in Japan's Miyazaki prefecture, in some ways it is even more remarkable.

Consider the ideal horticultural conditions for growing mangoes. Jamnagar is the exact opposite. The district is the northern edge of the Saurashtra peninsula, diving into the Gulf of Kachchh. In this semi-arid region, sand grains remain loose, unable to hold water and nutrients, making it difficult to grow plants. The soil is full of salt, measuring 9 on the pH scale (7–14 is alkaline and 0–7 is acidic). While the annual average rainfall is 500–600 mm, it is highly erratic; there are parts that get very little rain and very infrequently.

'In 1996–97 when Dhirubhai Ambani and Parimal Nathwani of Reliance were looking for land for a refinery, everything from Jamnagar to Dwarka was a wasteland,' said Ashok Bargaje, head of horticultural operations across the sprawling campus of the world's largest oil refinery that Reliance Industries Ltd (RIL) built here in 1998–2000. Here, the company has orchards with more than one lakh mango trees.

How did the mango arrive here like a glittery casino in the Mojave Desert? It has to do with adaptation. One of the statutory requirements for setting up a refinery is creating a green belt around it. If Reliance was going to invest in raising a green belt, it wanted one that could earn its keep with mango trees. The company was looking to expand into the markets for agricultural produce, particularly fresh fruits and vegetables; Reliance Fresh was formally launched in 2006. The horticultural operations around the refinery began to pick up steam.

Scientists were summoned from Junagadh, 140 kilometres south of Jamnagar, known for its horticulture and its Kesar mango variety. They threw up their hands; the soil was too saline and barren. RIL enquired about the top names in horticulture. R. T. Gunjate, a horticulturist in Maharashtra renowned for his expertise with the mango, was contacted. He took up the challenge, beginning with preparing the land marked with gullies and ravines. Instead of trying to level it, they left it undulating. 'The large holes were filled with the dirt excavated during the refinery's construction. It was black cotton soil,' said Bargaje. Better soils were spread in rows to create raised banks on which were planted the saplings. Even this soil was too saline for mango trees and needed treatment. 'Inorganic fertilizers increase the soil's salinity further. Instead, we added copious amounts of organic compost and added to it gypsum brought from Rajasthan,' Bargaje added.

Carbon-rich soils harbour microbes that release organic acids, bringing the soil alive. Slowly, the salinity was rectified and the soil pH is now between 7.5 and 8. More importantly, soils with carbon retain moisture, a critically important quality in a water-stressed region; it rains here for an average of fifteen days in a year. This region gets water from the Narmada River canals, but that water is too precious. Even the rainwater collected is saved for fruit trees. All other plants are irrigated with treated sewage. On the periphery of the orchards are planted casuarina trees, tolerant of salinity, as windbreakers.

RIL's founder had the Gujarati passion for mangoes and had heard of the Lakhi Bagh set up in Darbhanga at Akbar's initiative.

After his death in 2002, the orchards were dedicated to him: Dhirubhai Ambani Lakhibag Amrai. Actually, it is more a bagh than an amrai. Whatever the name, it is a marvel of ingenuity. Its stats are comparable in scale to the refinery. Horticultural operations here extend over 2,500 acres with more than seven million trees and shrubs. Of this, 600 acres is under mangroves and 650 acres under fruit trees. That includes about 140,000 mango trees of 138 varieties. The biggest prize of this orchard is a variation of the traditional variety: Reliance Kesar. It is larger than the Kesar of Junagadh; the largest fruit has been weighed at 910 grammes.

Kesar is a regular bearer; Bargaje said they've noticed only a 10 per cent variation in flowering and fruiting from one season to another. The rootstock used here is a polyembryonic variety from Israel called 'M13-1'. It is very tolerant of saline soils. While its own fruit is bland with leathery skin and a heavy turpentine taste, it provides a robust base for finer varieties. More than 330 full-time workers attend to the horticultural operations here, supervised by thirteen horticulture officers. There is an extensive network of drip irrigation to save water and deliver it precisely to the plants.

The priority here is to produce mangoes round the year. They have pursued two varieties from Maharashtra: Bajrang and Niranjan. Both produce large fruits, both have a variable flowering character—water stress causes the trees to flower readily. 'So we do the opposite,' said Bargaje, 'we irrigate the trees and prune the branches to delay the flowering'. It is not too difficult to delay it by up to two months. Here, they put it off by four months. When the Kesar mangoes are being harvested, Bajrang and Niranjan are in flower.

I saw large, mature fruits hanging from the same trees that had much smaller fruit from a second round of flowering, while some branches had flowers. The same tree reflected three different seasonal conditions! A bulk of the mango is harvested in May but some fruit is harvested as late as December. The Ambani household in Mumbai gets its Alphonso mangoes from Maharashtra in March;

by May, when the Alphonso season ends, this orchard begins to supply Kesar. Bajrang and Niranjan keep coming right down to December. Their off-season is three months, a gap that increasingly looks under threat.

It seems unimaginable but there are some advantages, too, of cultivating mango in such punishing conditions. Despite it being a coastal area, the humidity is not very high. Hence, there are no major insect pests. And there's no need to spray synthetic pesticides. The mangoes from this human-made oasis are organic. The horticulturists here could do all this because they were encouraged to take risks that government research stations cannot. They had no fear of failure. RIL was willing to back them with every investment. On account of the scale and intensity of the problems overcome, as also the sheer audacity, the Lakhibag is a demonstration of how very far we can go in thrall of the mango.

Vivek Bhide was growing Alphonso mangoes in the late 1990s. Like most growers, he used a synthetic pesticide called chlorpyrifos to deal with hoppers. 'In 2000 I read that Canada had banned it after realizing it causes skin problems when sprayed on the lawns. I stopped using it right then and there.' It seemed logical. Bhide is a doctor of medicine, a man of science with deep environmental concerns. Besides, his medical practice gives him another line of income, increasing his ability to take risks. His use of all agro-chemicals is based on study, experimentation, and ethics. Yet chlorpyrifos came back to haunt him six years after he had stopped using it.

In 2006, he decided to tap the high-value foreign markets. After a tie-up with a famous agri-business, he shipped a load of his mangoes to Japan. His consignment came back. 'They had rejected it after the skin of the fruit was found to have high levels of chlorpyrifos,' he said. When he told them that he did not use that pesticide, they informally asked him to check the processes followed in India's paper industry. The problem lay in the boxes. The factory that made the cardboard boxes added chlorpyrifos to keep out termites. 'The chemical vaporized from the cartons and got on the skin of the fruit,' he said. Japan has since changed the protocol; now the mangoes have to be packed in thermocol boxes.

The Desai Bandhu Ambewale had a similar experience. 'Our mango consignment went into the vapour heat treatment plant at the Vashi market. Right after that, we checked for pesticide residue and it was 0.02 parts per million (PPM). Fumes from the mandi had laced our produce with pesticide residues,' Anand Desai told me. When the consignment landed in Japan, the pesticide residue level was 0.04 PPM. 'It was still under the limit of 0.06 PPM required under Japanese regulations. But our Japanese partner did not wish to risk their brand value. So the exports stopped,' he said.

Chlorpyrifos is a neurotoxin; it attacks the nervous system of insects. It is also dangerous for humans, causing reproductive and developmental disorders, especially among children. Several countries have banned or restricted its use. It was patented in 1966 by the Dow Chemical Company of the US. Even the US, which tends to be lax when it comes to regulating industry, fined Dow $732,000 in 1995 for failing to report 249 poisoning incidents related to its chlorpyrifos-containing pesticide sold under the brand Dursban.[2]

Then, in 2007, the US Securities and Exchange Commission fined Dow $325,000 on charges of bribing an Indian official of the Central Insecticide Board in Faridabad to the tune of $200,000 for clearing three pesticides between 1996 and 2001. One of these insecticides was Dursban. 'Without admitting or denying the allegations in the Commission's complaint, Dow consented to pay a $325,000 civil penalty,' the commission's website says.[3] India's Central Bureau of Investigation filed an FIR and investigated Dow's Indian subsidiary; the case was dismissed in 2014 when the bureau failed to provide adequate evidence.[4] Dow's unpopularity in

India also has to do with its purchase, in 2001, of Union Carbide Corporation, responsible for the world's biggest industrial accident, the Bhopal gas disaster of 1984.

Regardless, chlorpyrifos continues to be used and sold in India, especially for controlling termites and mosquitoes. You can go to a hardware store and ask for 'termite oil'; it will likely be a chlorpyrifos formulation. It is also approved for use in fourteen crops by the Central Insecticide Board of India, as of October 2019.[5] It remains one of the most widely used pesticides in India. Several studies and reports have shown how its indiscriminate use endangers public health, with its residues getting detected in fruits and vegetables frequently.[6]

Bhide is an exception. I have met very few mango growers who refuse to spray a synthetic insecticide because of health and safety concerns. Almost all mango growers are willing to risk poisoning from insecticides rather than tolerating pest attacks. They end up spending vast amounts of money on concoctions of toxic chemicals, available readily on the market, even if they harm human health and the environment.

I have interviewed numerous farmers and mango growers over the years about their plant protection measures. Most buy pesticides based on advice from their pesticide dealer, not scientists. They know these are poisons but they are clueless about the long-term harm they cause. 'I will spray any pesticide I can lay my hands on because all my labour and investment rides on that crop of mango. If I do not get that profit, then there is no point to what I do,' said a grower near Malihabad.

Antsy mango growers are always desperate to recover their investment each season. Most cannot even name the pesticides they use. It is common knowledge that a lot of spurious pesticide is sold and used. There is no way to control this. A. Bhagwan, a senior scientist at Sangareddy in Telangana, said even when the pesticide is not spurious its efficacy depends on judicious use: 'The timing of pesticide sprays is critical. Farmers do not follow the bulletins issued by scientists. So the sprays become ineffective.'

Indiscriminate use of insecticides, spurious or not, has three disastrous effects that are never reported. One, major pests develop resistance to the synthetic poisons. Two, minor pests gradually become serious threats. Three, pesticides kill friendly insects and pollinators, like the hover fly, which not only pollinate the mango but also kill and eat insect pests, naturally checking pest populations.

'The leaf webber, thrips, and fruit borer were not major pests earlier. They have become more dangerous because of pesticide abuse,' said Pandey of CISH. 'During the mango season in West Bengal, blue flies used to appear, attracted to the mangoes. In fact their name in Bengali is aamer maachhi or mango fly,' said Swati Bhattacharjee, the Kolkata journalist who coordinates a network of reporters working on agriculture and the environment. 'They have disappeared entirely because of pesticides. I hear so many people complain that they miss them now,' she said. Joshi of Devgad said the loss of pollinators from pesticide abuse has begun to hit mango yields.

Entomologist G. V. Ranga Rao said there is more than enough evidence to show that the war against insect pests cannot be won, not with synthetic pesticides anyway. Indiscriminate pesticide use also has a grave economic cost—the law of diminishing returns sets in. The increasing cost of cultivation makes the grower ever more insecure, pushing him deeper into a vicious cycle of desperate decisions. Horticultural sciences are also in a myopic trap. Because they have to provide advice that produces immediate results, they are biased towards recommending pesticides. There is little fundamental research on insect pests and their natural predators.

Vivek Bhide, otherwise understated and soft-spoken, became animated and his voice acquired passion: 'We need to seriously look back at what we have done over the last fifty years in the name of agricultural development.'

It is not merely mango varieties that divide North and South India. There are deeper cultural differences. In South India, when I asked growers some questions to which they did not have an answer, several had names and phone numbers of horticulture scientists in their area to offer. A few called them to field the questions. In Konkan or Andhra Pradesh, many growers had a direct communication with their state horticulture departments and agriculture universities, making demands on both to resolve their problems.

Not so in North India. Vijay Singh, a major mango grower from Mall village close to Malihabad, gave an example: 'CISH scientists developed a harvester. It is essentially a net with a blade at the end of a pole. It reduces wastage of produce at the time of the harvest. It is available for ₹200 but many growers here do not buy it.' He said two-fifths of all mangoes grown in the region are lost due to shoddy handling practices. Outdated horticultural practices are quite common in North India.

The most stinging indictment came from Vivek Bhide of Ratnagiri, who had visited Malihabad and some other mango belts of UP. 'The growers there are idiots. They are sitting on a goldmine but they have not got a clue,' he said. 'Not just of the genetic wealth of their wonderful varieties but the excellent natural conditions. They continue with horticultural practices that were already outdated in the last century. Their marketing is poor. They do not spare a thought for presentation.' Why is this so, I asked. 'Illiteracy,' Bhide answered angrily. He was not being literal. Many growers in and around Malihabad are very well educated. But I have found out more about growing mangoes in twenty minutes with growers such as Bhide and Joshi than in hours and days I spent with growers in North India.

In Malihabad, I visited the state horticulture department's office during the mango season and otherwise. I never found anybody there. Mango growers could not name anybody who worked there. In fact, they do not even mention CISH, one of the country's top two horticulture research institutes. It is located 8 kilometres away in Rehmankheda, across the railway track.

It is a lot more than the railway track that divides Malihabad's mango growers and the research centre. The growers have a litany of complaints about how the scientists are apathetic to their problems. 'We went to CISH a few years ago,' said Ram Asrey Maurya, a mango grower from Malihabad's Nai Basti locality. 'They spoke a language we just could not understand. We are more willing to take our horticultural advice from shopkeepers who sell us pesticides; at least they talk our lingo.' Scientists at CISH had a different story. They say the growers there are not open to what modern science has to offer. Or that they are suspicious of scientists.

Then there are some natural problems. 'Mango trees in North India tend to be naturally taller to capture the more horizontal sunlight, especially in the winter,' said G. V. Ramanjaneyulu, an agricultural scientist and the executive director of the non-profit Centre for Sustainable Agriculture in Hyderabad. Trees in the south have an advantage in staying short and spreading a wide canopy. Most orchards were planted at a time when horticulturists recommended keeping a wide distance between trees; this gives them room to develop heavy canopies that are unmanageable and inaccessible. 'Hailstorms and wind storms damage tall trees a lot more than shorter ones,' said Kiran Kumar. 'The idea is to not have a heavy out-of-reach canopy. A large number of leaves under the shade are useless. That's called the "shading effect" and it must be avoided. Each leaf must earn its place by maximizing photosynthesis,' said Murad M. Burondkar of Dapoli.

Kiran Kumar took me to look at fifty-year-old trees they were rejuvenating. The old branches had all been cut down and the trunks stood only five feet from the ground. New branches were emerging. 'We have reduced their age by twenty years. Now they will fruit like younger trees. Some have begun to fruit already, one year after they were downsized,' he said. I have seen farmers conducting such rejuvenation experiments in Andhra Pradesh, Telangana, Tamil Nadu, Karnataka, and Maharashtra. I saw only one farmer try this method in UP.

High-density cropping is now in fashion. Traditionally, scientists recommended planting fifty to forty mango trees in an acre, leaving 10–17 metres in every direction between trees. In the 2010s, this was revised to seventy trees per acre under high-intensity cropping. Now, scientists recommend 160 trees per acre under 'ultra high-density cropping'.

The idea is not very new. 'Farmers have traditionally known that it is better to grow a large number of plants in a small area if the conditions do allow high yields, perhaps because of poor or rocky soil,' said Ramanjaneyulu. 'That is how tribal farmers have farmed their lands over hundreds of years. But scientists began recommending large spaces between plants in order to push the high-yielding varieties they had developed. Now, they are recommending exactly what ordinary farmers did earlier, without so much as an acknowledgement.'

For all their market savvy and progressive ways, Konkan's Alphonso growers have created a new problem for themselves. The quality of their mangoes is declining due to the indiscriminate use of paclobutrazol (PBZ), the growth-retardant that induces trees to flower. Balasaheb Bende, a mango trader in the Vashi market, said: 'If I serve bad tea to somebody, he will ask me whether it is made of PBZ!'

Growers think differently. 'It is a useful tool if you know how to use it, like a doctor uses a sensitive drug,' said Bhushan Nabar, a structural engineer who manages his family's orchards in Vengurle. 'I use it only on robust trees and only after an interval of three years or so. I have about 1,400 trees and use PBZ on 300–400 each year.' Abuse of the chemical lowers tree branches and, sometimes, splits the bark, he had noticed. 'I can recognize trees on which PBZ has been used; they look droopy and exhausted.'

Scientists stand firmly behind PBZ. Burondkar, the first scientist to prove its efficacy on the mango, said PBZ is a very useful chemical, provided farmers know how to use it. Bharat R. Salvi of Vengurle agreed. It is a rare case, however, that a grower has read the literature on PBZ and actually knows the dos and don'ts.

Vivek Bhide has. He recounted his trials with PBZ. 'From 1992 to 1997, I tried it on trees of varying ages and varieties, planted in different locations. The results were unsatisfactory. The fruit got smaller after the first two to three years.' He noticed the trees began to look deformed and the branches began to droop. 'When you use this growth regulator, the plant requires more fertilizer; if you do not apply additional fertilizers the rate of fruit drop increases. Fertilizer brings a flush of vegetative growth that leads to intensification of pest attacks on the tree.' He said the tree does not handle it well, perhaps because this tinkering with its hormones compromises its immune system, making it vulnerable not only to auto-immune problems but also to certain fungal diseases.

'PBZ is the reason Devgad lost out on its quality,' said Yadnyesh, Bhide's son. 'They had the largest mangoes. Excessive flowering, however, reduces fruit size. You get a larger number of fruits but of smaller size.' Bhide stopped using PBZ as soon as he noticed the side effects; the 'Fruit King' brand under which he sells his produce is prominently advertised as being free of PBZ on each box. This earns him a premium. He said he can tolerate alternate bearing but he cannot afford to lose the most desirable traits of the produce he sells. In neighbouring Pawas, Anand Desai said that orchard owners do not abuse PBZ because the health of the trees is important to them. He is careful in using PBZ. However, contractors who lease orchards for one season abuse it widely because they are desperate to get returns on their investment.

'My neighbouring farmers use PBZ on their mango trees year

after year. They are not bothered about the effect on the trees, and they say they will not be growing mangoes after about twenty years,' said Noshirwan Mistry of Mumbai. Mistry is an economics graduate who bought 15 acres of mango orchards in Dapoli and has a reputation in Mumbai for selling organic mangoes. 'I get sixty to eighty fruits per tree; if I use PBZ, that might increase tenfold.' In Vashi, Bende summed it up: 'The quantity has increased. The quality has declined.'

Having wasted many hours of many days meeting indifferent mango growers, I remember each engaged grower I met, and the insights they shared. Like Kaleemuddin Siddiqui of Amroha. He described how his family's orchards are laid out; the varieties that yield early are closest to the road and the ones that yield at the end of the season at the farthest. This makes horticultural operations easier to manage, from monitoring to harvesting. It was a memorable moment because I had read about this same scheme in an 1897 book; it attributed this practice to a British botanist.[7] Perhaps it is much older. But it is still thriving in the Siddiqui orchards. The owners are proud of their heritage.

Mujeeb Khan was a great surprise. He has rejuvenated the old trees of his ancestral orchard in Mohammadpur village near Malihabad. He used the best of modern science to get the most out of old trees. He had the energy and the resolve to try the best of new ideas. Vidyadhar Joshi is a passionate man whose enthusiasm is infectious. 'The mango is not just my livelihood. It is my career, my profession. I have no off-season.' It is a business and he runs a tight ship. He holds a part of the profit of a good year as a buffer to tide over the bad years. He buys health and accident insurance for his workers each year. 'If you love your trees, you end up loving your workers who look after them. If you can insure cars and motorcycles, why not buy insurance for your workers.' He keeps a bulk of his work force employed through the year to look after the orchards, which is why his mangoes have a reputation.

It took lots of luck and much time to find growers like Siddiqui, Khan, and Joshi. They are outliers, just like Bhide and Mistry. Most mango growers are so uninterested in their work that it is difficult to have a conversation. Initially, when I visited an orchard, I walked up to the workers and asked: 'Where can I find the owner?' They appeared baffled and evasive. It took me a long while to figure out the mango business. Eventually, I found the right question: 'Where is the thekedaar (contractor)?' The response was always swift.

This is the biggest reality of India's mango sector. And nobody mentions it. Most owners do not visit their orchards; they are simply not interested. Land is a fixed asset; landowners seek the surety of a regular, predictable income, as if from a fixed deposit in a bank. They do not wish to sully their feet or minds with the earthy complications of growing mangoes. They 'sell the crop' to a contractor for a bulk payment in advance. How many orchard owners do this? It is impossible to say because there is no data. Almost all reliable sources say a significant majority of orchard owners across the country sell their crop on contract.

A major part of this sale, if not all, is settled in advance. The remainder is settled after the contractor sells the harvest. Some landowners forward sell the crop when the fruit has set; some do it at the stage of flowering. Both ways, the contractor has an estimate of the produce. Then there are those who sell the mango crop in advance right at the beginning of the season; this is the most common deal in my observation. Most such deals are for two years to account for biennial bearing. Some sell the crop for five years.

Contractors pay at least half the agreed amount upfront. But they cannot get a realistic estimate of the quantity and quality of the produce. Almost all the risk rests with the contractor. He brings the workers; attends to all horticultural operations from spraying pesticides to harvesting the fruits. It is the contractor who sells the mango in the market, not the owner of the orchard. I have seen brawls in orchards! I have seen contractors with injuries on their faces and bodies from fights, blood dripping from their arms. I have seen them arrive with a weapon for a conversation; they never know what to expect of a stranger. The contractor is the arch villain of India's mango badlands; he is also the arch victim.

The roots of this contractual arrangement can be traced all the way back to the Mughal times when orchards were exempted from land revenue. Landowners employed gardeners on contract for raising orchards. The gardeners doubled up as enforcers to collect land revenue; Pathans were popular choices given that they came from Afghanistan as mercenaries. The arrangement between the owner and the gardener depended on their relations. Whether the gardeners were paid separately or not, the fruit harvest was a part of the payment. (Provided, of course, that the best produce was sent to the owner; if it was of high quality, the owner could use it as a gift to please the subedar or nawab or sultan.)

This arrangement continues in varying forms. Most landowners sell the crop to contractors they know or have dealt with previously. The landowners of a bulk of orchards I have visited live in a city or in a foreign country, not having set foot on their land in decades. But the incentive has changed from Mughal times. Then, it was about having productive land without paying land revenue; it was about producing refined fruit that could be gifted. In our times, it has become a means to subvert land ceiling laws brought in after India's independence, abolishing the zamindari system of the British Raj.

I asked some former colleagues who research land matters. Their preliminary research showed that horticulture and plantations were useful in undermining land ceiling in at least nine states: Maharashtra, Tamil Nadu, Andhra Pradesh, Telangana, Karnataka, Uttar Pradesh, West Bengal, Bihar, and Kerala—all are major players in horticulture.

Land being a state subject, each state drew up its own rules on land ownership after independence. Land ceiling laws took away land parcels from zamindars to distribute them to tenant farmers. Each state also had its own exceptions to land ceilings. Orchards and plantations were a common exception because, it was argued, production from trees requires large and consolidated land holdings to remain viable. Land ceilings were tighter for irrigated lands growing crops like paddy and wheat. For unirrigated parcels of land, the ceiling was lax. Orchards were labelled unirrigated. This allowed zamindars to keep more land within the family.

The researchers pointed to political scientist Shalendar D. Sharma. In his 1999 book *Development and Democracy in India*, he compiled the many loopholes used to defeat land reforms: '...the ceiling laws exempted several categories of farms, especially those organized along capitalist lines, including... "farms under orchards"... if one could get good agricultural lands, with occasional fruit-bearing trees planted here and there, recorded as an orchard, the entire area went out of the ceiling provisions.... Not only was there a veritable rage in the planting of "orchards", numerous dummy institutions such as religious "trusts" and "educational institutions" were created overnight to evade the ceiling laws.'[8] Sharma lists examples of cheating ceiling laws from UP, Andhra Pradesh, Maharashtra, Bihar, Gujarat, Mysore, Rajasthan, Madhya Pradesh, and Haryana.

Take the example of UP. The state's 1960 law allowed for land holdings of 40 acres; in case the family had five members or more, this could go up to 64 acres. Several exemptions—loopholes, really—were applicable and were fully utilized. For example, landholders planted a few fruit-bearing trees to capitalize on exemptions for orchards. Then, in 1972, the state amended its land ceiling laws, bringing down the limit to 7.3 hectares (18 acres); about 13.3 hectares (33 acres) for a family of five or above. This reduction also came with loopholes. The law now said: '...one-and-one-half hectares (3.7 acres) of unirrigated land or two-and-a-half hectares (6.1 acres) of groveland or two-and-a-half hectares of usar [salt-affected] land shall count as one hectare of irrigated land...'

Old people in the state remember how this helped evade land ceilings. In Lucknow, Farsi scholar Khan Mohammed Atif recalled the rush to set up mango orchards in his native Malihabad: 'Hamid Khan was the first landholder to sell his crop in bulk in Malihabad in 1906 for ₹6,000. He rallied people to plant orchards. He told those with large landholdings that the Congress will abolish the zamindari system after independence, and that only the land with mango orchards will be spared.'

Ramesh Dixit, a retired professor of political science at Lucknow University, hails from Sandila, a rich mango-growing area west of Malihabad. He acknowledged that people set up mango orchards in the 1950s and 1960s in that area as a means of keeping as much land away from the ceiling as possible. But the incentive for growing orchards has changed. 'Now it is to have a fixed annual income. Cultivating land for various crops has become a Herculean task. We do not have big land holdings in our area. With the division of land among members of joint families, it is difficult to survive on agricultural land.'

After independence, orchards became a means of saving family properties from land ceiling, all the while getting some regular income without the hassles of actual cultivation. Mango being India's most popular fruit and well suited to this region, mango orchards became very common. These tie together with the emergence of new mango markets after independence. Such owners are called 'absentee landlords'! Their involvement in the horticultural operations is minimal, if at all. They have no interest in their orchards or trees. Only their contractors know a little bit.

Even those opposed to land reform and ceiling laws have to admit that absentee landlords have a terrible effect on agriculture in general, and the mango in particular. It is the single biggest problem with India's mango sector. All other problems can be addressed and resolved, or bypassed and tolerated, when the owner has a hands-on interest in the land's productivity. But there is no solution to indifference.

'Whatever suitable land is available for cultivation is owned by several members of the family. Most of them are absentee land lords,' said a 2009 booklet of the agriculture university at Dapoli.[9] 'The main reason for low productivity of Alphonso mango from Konkan region is neglected orchards as owners are absentee land lords.'[10] Joshi said 75 per cent of the land under orchards in the Devgad region is under contracts. He takes hundreds of acres under contract each year; among other things, it helps to keep his workers gainfully employed. 'I take on contract orchards and lands inherited by people who are not interested in farming or mangoes. Often, they cannot sell the land because of family disputes over property; some have a sentimental attachment to the land, even though they have moved to cities,' Joshi said.

The signs of this apathy can be seen in orchards all across the country. 'Absentee landllordism is rife in UP. Nobody looks after the trees from July to December. Most such orchards are the feasting grounds of insect pests,' said Pandey of CISH. One sure sign of this is branches destroyed by the leaf webber, he added. They are very common in neglected orchards. The contractors take every shortcut available. India's mango sector is built on fly by night operations of desperate contractors. The fruit over which the shauqeen hold forth is actually grown and traded in an environment of indifference.

Chapter 13

Markets
Caution to the Wind

'Everything's a huge risk in the mango business and one has to play blindfolded!'

—SUBHASH HANDE

Each summer a makeshift mango market appears along the highway running past Malihabad. Not for drive-by sales; traders leave those to assistants. They focus on wholesale deals. Sitting by the roadside, sipping tea from small paper cups, they negotiate prices, and estimate demand and supply. The sights and sounds are those of a rural haat or a seasonal fair. But the players arrive in expensive cars, looking to speculate, wearing an attitude that is more casino than mofussil north India.

This is where I met Shariq Khan also known as Shanne Miyan, a mango trader from Lucknow, a few years ago. We sat on plastic chairs by the highway. How goes the season, I asked. He reacted as if I had lanced a boil: 'Yeh kaccha kaam hai (This business is unreliable/infirm).' He threw up his hands, suggesting he wanted to be rid of it, his face creased into lines of self-loathing.

Which business is pukkah (firm), I asked. 'Fast-moving consumer goods (FMCG) like toothpaste!' he said. Why? Both supply and demand are predictable. Payments are cleared on schedule. A minor disruption does not pile up the losses. If a deal is unfavourable, one can always say no, choosing to take a long position on his inventory. Most of all, the return on investment—time, money, and labour—is predictable through the year! Not so with the mango. 'Bahut kachcha kaam hai yeh (This is very unreliable),' said Naseem Beg, another trader in Malihabad.

The mango fruit is a three-month affair. Those who trade in mangoes need to find other work for the remaining nine months of the year. The big wholesalers primarily deal in other fruits, switching to mangoes in the summer, and then back to apples or guavas or oranges or pomegranates or whatever else is in season. They are few in number, operating from their godowns or offices in agricultural markets or mandis. A trader who buys from growers to sell either to another trader or retailers is called a commission agent because he charges a transaction fee.

Such big wholesalers buy their stock from mid-level or small-time traders who could be shopkeepers or journalists or bank clerks. They have the connections to make a quick profit because nothing matches the hysteria of the mango. This puts them in the same category as the contractors who are the actual mango growers. Often, though not always, the mango trader is also the mango contractor. Shanne Miyan is both, which means his risk exposure is the greatest, both from the weather (as a grower) and from the market (as a seller).

Just like the contractor, your average mango trader must have a short-term view and be willing to react quickly to dynamically changing markets. He should be able to casually withstand a series of setbacks. He should seize each opportunity ruthlessly! Like the contractor who does not hesitate from using pesticides excessively, the trader must be willing to slip in low-grade produce along with premium products. Get the job done, by hook or by crook! Like contractors, traders are habitually suspicious and evasive. Revealing one's hand in this game assures losses. Like the contractor, the trader is seen as a villain.

Shanne Miyan is not one; he is a lawyer by training. He was dressed in an impeccable white kurta-pajama when I met him, and his leather sandals were polished. He spoke with a rare clarity and intensity. (As we shall see later, the most important things in conversations among traders remain unstated.) 'Each step is a high-risk gamble. A light shower at the wrong time can blight a flowering. In 1996, storms damaged the entire produce in this area. In 2008, it rained incessantly.' Then there are local superstitions. He remembered one year when a rumour convinced everybody that 10 June was an auspicious day to harvest the fruit. Suddenly labour was in short supply. And the market was flooded with immature fruit. A price crash was inevitable.

Nobody actually understands agri-markets. Even when backed with accurate data and elaborate research, each analysis applies only to a limited area, for a limited time, for a limited crop. It is a maze of operators, governed by complicated rules and regulators that change from place to place, from time to time. Khan said whatever comes under the purview of 'gormint' (north Indian for government) gets destroyed: 'If there is a government scheme to promote the mango, it stays in their files. Growers and traders have no information.' The Mandi Parishad in Uttar Pradesh imposes a tax of 3–4 per cent on items traded in its market; but Bihar has abolished the mandi tax. Its operations are opaque, including its elections.

Trucks ferrying mangoes have to deal with the vicissitudes (read bribes) of the local transport department. Traders are dealt with differently from farmers. 'The courts have ruled that a consignment can be considered farmers' produce only when the farmer is sitting in the truck, along with his land documents. But mangoes do not mature in one go and growers don't sell the produce altogether; it comes out in batches. Several growers and traders use the same truck for efficiency. Which farmer is going to sit in the truck all the way from Malihabad to, say, Delhi?' Khan said. Besides, most growers are contractors without land papers.

Everything is ripe for a transport official; he has only to insist on inspecting every scrap of paper to find a loophole. Even when the paperwork checks out and the vehicle meets all norms, truck drivers are routinely harassed and extorted at borders and checkpoints. A truckload of Vivek Bhide's fine Ratnagiri Hapus mangoes left for Indore in 2013. 'All the produce on the top of the consignment was destroyed by the heat,' he said. Poor roads shake and rattle the vehicle, damaging the texture of delicate fruit, he added. India's agricultural value chain is designed like an obstacle course—or the board game Snakes and Ladders. It handles a delicate, perishable commodity like the mango in the punishing summer heat and monsoonal downpours.

At the Vashi fruit market in Navi Mumbai, trader Subhash Hande took a call from a grower in Ratnagiri. He told him not to send the produce affected by anthracnose: 'A Ratnagiri grower sent me thirty boxes. I had two major buyers, so I quoted him a liberal price of ₹2,500 per box. Once I quote a price to a grower, there's no going back—it is tantamount to signing a bearer's cheque! But the truck that brought his produce had a tarpaulin cover. It trapped heat, ripening the entire consignment. The retailers who took the produce from me brought it back because they were not able to sell it.' The price, meanwhile dropped by ₹500 per box after they picked up the produce from Hande. 'It gave them an excuse to dump the thirty boxes back on me. I could not say no to those retailers because they are my regular clients.'

'Farmers are eager to sell everything they grow. But each customer wants the best produce; they never pay for damaged produce. Which means the traders have to find a way to build in the cost of damaged fruits in their deals,' said Sanjay Pansare, former chairperson of Vashi's Agricultural Produce Marketing Committee (APMC). Vasant Chaskar, a trader, recounted how an export company went bust: 'They owed me ₹6 crore when they went under and disappeared. I had to sell my ancestral land to clear the payments to the growers.'

One time, when I met Hande at the mandi, his workers were

segregating the produce that could be redeemed. Hande was calling vendors, trying to convince them. 'Even if I find retailers to lift what can be saved, I will lose ₹1,200 per box on this deal,' he said. He had fronted an advance to the grower at the start of the season. 'If rain or anthrax (anthracnose) spoil the produce, my investment stagnates,' he said. The retailers pick up the produce from him on credit, settling after selling. 'My money is stuck everywhere. But returns are not guaranteed. Everything is a huge risk in the mango business and one has to play blindfolded!'

Then there are traders who are quite happy with their business. In Murshidabad, West Bengal, I met Ramakrishna Das. He had been buying the contract for all the trees at the Katgola Palace for four years. He made a profit each year, and his capital investment and labour were always viable. He said this work is well worth doing. Because he seemed so happy, I asked: 'Will you let your son get into this business?' The reply was immediate: 'No.' Why, I asked. 'It is so risky! A minor storm can dent the return on your investment and all your labour. This work is good for someone illiterate like me. My son goes to school. I want him to get a proper job.'

∽

Vivek Bhide answered a call from an unknown number. It was somebody who wanted to buy his mangoes. Bhide said he sells only through his agent in the Vashi market in Navi Mumbai, and then gave his name and number. Surprised, I asked him why he declined the sale. He said three generations of his family have dealt with traders in Mumbai. Was he not losing a share of his profits? 'We earn off each other. They provide me service. Through them, my produce gets into markets that I cannot tap directly.'

Such trust sounds odd. The agri-trader has a negative image in India, going back more than a century. The trader was seen as an ally of the British Raj. During the great famine of 1876–1878, the colonial government was more concerned about free trade than famine relief. At its behest, Indian grain traders made a killing exporting a record amount of food grain, even as millions died.[1] After Independence, this image deteriorated further, both in popular culture and academia. We heard of the 'evil middleman' creaming off the poor farmer's profits. Imagine the character of Sukhi Lala in Mehboob Khan's hit film *Aurat* (1940). The usurious and lecherous trader-moneylender, played by actor Kanhaiyalal, eyes the helpless widow Radha! When the director remade the film seventeen years later as *Mother India* (1957), Kanhaiyalal was the only actor to reprise his role of Sukhi Lala.

The evil moneylender peaked in the 1960s. In this period, India depended on shipments of US food grain in aid; some commentators have described this phase as India's 'ship-to-mouth' existence. It was about then that the Bollywood actor Manoj Kumar transformed into the 'progressive farmer', the celluloid personification of the slogan 'Jai Jawan, Jai Kisan'. Such maudlin idealism could not have worked without an equally generalized foil: the 'regressive' trader. When Manoj Kumar first appeared as Bharat Kumar in his own production *Upkar* (1967), the role of the villainous moneylender Lala Dhaniram fell to—of course—Kanhaiyalal. The wicked Lala meets his comeuppance when farmers, under Bharat Kumar's progressive stewardship, refuse to sell him their produce because not only does he underpay them, he also plays the market through hoarding and black marketeering.

This image is reflected in academic writing. Policy documents have repeatedly proposed reforms for liberating the farmer from the clutches of intermediaries and moneylenders. But the well-being of farmers has not been the actual focus of the agriculture sector in about two centuries. Government policy has centred on attaining a surplus of food grain. Which is why in the 1990s, with the godowns stocked with food grain, the government deprioritized agriculture in its pursuit of structural reforms of the economy. Farmers were reduced to recipients of 'sops' to solicit their votes in elections. Free-market economists began parroting about 'moving people out of agriculture' to sectors that exist—even today—only in their policy fantasies. Farmers and traders were left to fight it out.

The second priority in the policy to regulate agricultural markets is control over food prices. Such is the power of even mild discomfort to the urban consumer that in 1998, the ruling political party in Delhi was said to have lost the assembly elections over high onion prices. This has limited farm income, even as input costs began to spike in the aftermath of the Green Revolution. The farmer was already dependent on the local trader for credit and sale of the produce. Now he came to depend on him for expensive inputs like high-yielding seeds and pesticides.

Exploitative or not, the farmer's relationship with the trader is a direct one, tangled up in self-interest and mutualism. Even the richest mango growers need advance payments now and then; the advances traders offer are difficult to turn down in an hour of need. The farmer-trader relationship is definitely more direct than the government's relationship with the agriculture sector. In the case of a high-end seasonal commodity like the mango, the markets work very differently from food grain or vegetables. I wanted to meet these traders Bhide trusted. I took their numbers and reached Navi Mumbai.

✧

Brothers Sudhir and Ganesh Bhor, Vivek Bhide's partners-in-trade, worked out of shop number 59 in the fruit market of the Vashi APMC. I found them dealing with regular retailers. As I waited and watched the negotiations, I folded my arms across my chest. Sudhir Bhor came and whispered in my ear: 'You cannot cross your arms like that here. It is bad for business.' I remembered the first time I had gone to a court as a young crime reporter, the clerk there had rudely ticked me off for sitting with my legs crossed. When you are a new guy, such matters are par for the course. I straightened my arms.

When money is at stake in such deals, gestures matter more than words; it is more Desmond Morris than Deepak Chopra. Amid a lot of staring and mumbling, the retailers shifted their gaze from box to box, examining the produce, asking about other stock. Sudhir Bhor asked his workers to open the other cartons; the grower's name was painted on each box. Bhor's eyes were fixed on the retailers, looking for a tell to make his move. It's poker.

Bhor extended his arm, covered under a gamchha, a thin cotton towel that most people here have dangling around the neck. The retailer extended his arm, putting his hand on Bhor's hand, concealed under the cloth. Their eyes focused on each other's. The contours of the towel indicated moving fingers. The retailer shook his head in disapproval. More finger movement occurred. Not copacetic still! After more such movements, the retailer's eyes softened and he nodded his agreement. Bhor drew out his hand from under the cloth and slapped it against the retailer's palm. The soft, desi low-five signals approval; it is used frequently to appreciate a joke or a quip!

The deal sealed, Bhor asked his worker to load up the boxes on a small transport vehicle some retailers had hired collectively. He conducted the same covert negotiations with two other retailers. After they left, he offered me a cup of tea and a makeshift chair. What was the idea behind the hidden hands, I asked. It helps customize each deal to the demands of each retailer, he replied. He laid out the nuance of such digital transactions: the fingers and their phalanges function like price counters. After the retailer declined the initial price, Bhor reduced it by hitting the right phalange.

This digital counter is called a hattha (the hand). I asked him about the lack of transparency in such negotiations of price. 'When the government issues a tender, why are the quotes closed? Why do not they keep the quotes open?' he asked. 'The hattha is exactly like that. It helps me get the best price, just like the closed tender helps the government. If I can keep the retailers guessing, they will offer me their best prices.' With more retailers arriving, the day was turning busy. I left to go see someone who turned out to be one of the most colourful people I have met in the entire mangoverse.

Ashok Seth Hande is a force of nature. Among Vashi's biggest mango traders, he has the corner on perhaps the highest-rated mango: the Alphonso of Devgad; he is called the 'Devgad King'. 'When Prime Minister Jawaharlal Nehru died in 1964, I was a young boy in the Crawford Market (in downtown Mumbai). I was the first graduate working that market; that, too, wearing a pant and shirt (amidst traders in dhoti-kurta). I started working on my own in 1984.' An actor in Marathi vaudeville and a famed innovator of theme-based musical programmes, he breaks into dramatic song and poetry at the slightest excuse. A master of Marathi double entendre, he can conjure up a fruity sexual metaphor for everything. He shouted at workers unloading a truck to be careful. Then, he turned to me and said: 'Is dhandhe mein nazar gul to aanda gul!' (In this business, avert your eyes for a moment and you lose your testicles!).

When a bald retailer turned down the boxes offered to him, Hande broke out: 'If you do not want it, go oil yourself! But how will you oil your bald head!' Then he turned to me and said: 'If you put very good quality produce in front of an unlikely candidate, he gets surprised and rejects it. What will you do if a beautiful woman walks up to you in an empty train coach, sits next to you, and wants to kiss you?' He stopped, sized me up, and then said: 'You will probably freeze!'

Hande can switch instantly from playing a knave to the intentness of an air traffic controller. He sized up people and produce while attending to his two mobile phones, answered queries, instructed his bookkeeper and staff, and still had time to dish out a few quotable quotes in my direction. 'There is a big difference in perspective. The grower looks up to the mango, standing underneath the tree. We look down upon it, lying in a box. We do not determine the price. It is the quality of the produce. If the grower gives me quality, I will always find a willing buyer,' he said.

When I asked him to name his best growers, he also mentioned Vidyadhar Joshi of Devgad. When I met him, Joshi had told me he did not like to sell to traders. 'I never take advances from traders; I would rather take a bank loan,' he said. 'Agents will tell you they lend money based on relationships. Yet if I do not sell to the trader who gave me an advance, he will come and sit at my house or send his people to collect,' Joshi said, adding that the traders play with the prices. The covert price negotiations favour the wholesaler and retailer, but they short-change the grower. 'That is why Vashi traders come here in cars that cost more than ₹50 lakh, while I go from orchard to orchard on my rickety motorcycle.'

Back in Vashi, an old man arrived, and Hande suddenly went quiet; his bearing indicated an important buyer. I was introduced to mango exporter Mohan Dongre. 'He buys nothing but the best,' said Hande. Two boxes were opened for him; he asked the workers to show him the produce in the second layer and his expert eyes rejected both. ('There's no theory to this business. Everything has to be measured right then and there,' Dongre later told me at his office in the mandi, from where he runs Raj Impex, a mango export business set up in 1980.) As he exited, Hande told me: 'Even our best growers send low-quality fruit these days. This lot is from a lovely old man I have known over the years. Yet this lot is damaged by anthracnose. I'll find a retailer and sell it at a lower price,' he said. He remembered a name, called the number, and proceeded to hustle.

After another bulk buyer turned down the best he had to offer, Hande said: 'He is smart. This lot has a high proportion of fruits with spongy tissue. He is like an old nawab looking for ever newer, ever younger women.' He broke into old film songs, describing the lust of old men for young women. Almost on cue, the buyer found a suitable batch. Hande laughed and replied: 'It is but nature's law!' Another time, while haggling over rates with a buyer, Hande looked at me and said: 'If a young man quickly selects a woman as his bride, her dowry goes down. If he is not that keen on her and she has to wait, the dowry goes up.'

All this while, his son Sujay stood by at attention, observing quietly, dressed in upmarket jeans and Nike trainers. He was shy and unwilling to talk, as if he were inside a classroom. He has a master's degree in economics and another in business administration. Yet he hangs around here. 'This is the greatest university to learn business,' he told me when I finally managed to drag him aside for a chat. Where business is good, the younger generation feels encouraged to join the family trade. Hande used to be a keen table tennis player and a state champion. 'A fellow player called Sujay Ghorapade had a very attractive game. I named my son after him.'

There is a generational shift in the mango business, amplified by the physical shift from Crawford Market in downtown Mumbai to suburban Vashi in Navi Mumbai on 12 March 1996. Ganesh Bhor explained this on a subsequent visit: 'Earlier, Muslim traders controlled the entire fruit trade in Mumbai. But their younger generation did not follow up on the early mover's advantage. They lost out. Now, Muslims make up only 15 per cent of the trade volume; another 5 per cent is Sindhi traders. The rest is all Marathi traders now.'

Pansare was more effusive: 'You will find Gujaratis in most businesses in Mumbai but not in the mango trade. They are well-educated hence they are risk-averse. The mango business is a kachha dhandha (unreliable business), it requires the raw mentality of a warrior who is willing to jump into the unknown. Which is why only Muslims and Marathas get into it.' A well-informed person had told me that more than 70 per cent of the fruit traders in Vashi today are from the Khed-Rajgurunagar area between Pune and Ahmednagar; most had come to Mumbai for work as hammals (porters) for the established Muslim fruit traders at Crawford Market and Vashi. Gradually, they took over the business.

'We have worked with families of growers and retailers over generations,' Ganesh Bhor said, pointing to his thick, well-thumbed diary. 'I'm in a pickle this year because the produce is affected by anthracnose. I have already paid an advance to my best growers. Now I have to sell poor produce to reluctant retailers on credit,' Bhor said. His workers were opening boxes and re-sorting produce for retailers.

He can still run a profitable business because the rates at which he buys from the grower and the rates at which he sells to the retailer are determined by current prices. The advances and credits are settled on the price of the sale on that day. (Unless the price is already agreed, as Hande had mentioned.) 'But even that is becoming increasingly difficult. The younger generations are intolerant of losses. They will risk an old business relationship to avoid even a minor loss,' he said, overseeing the unloading of a truck from Andhra Pradesh carrying Totapuri, Banganapalle, and Dashehri, and another from coastal Karnataka carrying the Lalbagh variety. Working with known growers ensures quality and timeliness of supply. Working with known retailers helps avoid petty losses. Each new retailer is carefully screened over a test period.

In other cities, retail sales of mangoes are handled by regular fruit vendors; at the most, some of their associates join them to handle the additional demand. But there is nothing regular about how crazy Mumbai gets about Alphonso. 'Up to 80 per cent of the retailers come from UP only for the mango season. They rent out shops and street corners. They deal with wholesalers they can trust year after year,' Bhor said. Even the retailers prefer the covert negotiation of price, although open bidding in an auction is more transparent. Given the temporary nature of the mango season, retailers do not have the time for auctions and a level-playing field; they look for quick returns for their labour, for a wholesaler sympathetic to their circumstances. 'Each retailer has his limitations. Each one prefers a personalized deal. They are looking for quick work in, quick money out,' Bhor said.

A retailer named Jagdish turned up in the middle of our conversation. He complained of poor sales because the municipality had dug up the road in front of his shop, obstructing walk-in customers. He said he wanted only the best produce that he could sell over the phone to the best customers, then deliver it personally. But most of the mangoes lying there had anthracnose. Jagdish got up to leave. Bhor was aggressive at first, then he softened. He reminded Jagdish of their twenty-five-year association. The retailer heard him out and said he would come back after scanning the market.

Crawford Market is a colonial-era heritage building with corroding iron awnings and a 30-foot-high ceiling. Fruit retailer Ram Morde operates out of his 48-square foot, family-run shop here. A customer wearing a toupee turned up and asked for a couple of boxes of Alphonso mangoes. Morde assessed him and then cut open a box selling at ₹600 per dozen. After he had paid and walked away, Morde explained it to me: 'He wants a bargain but does not want to run the risk. I offered him a box that sells at ₹1,200 per box, but it is more likely to have spongy tissue inside. At the full rate, I guarantee quality; not one mango will have spongy tissue. The person who pays double is paying for my eye.'

An experienced retailer has done enough people-watching to qualify as an amateur sociologist. A young man stopped by and made enquiries. Morde gave him the short shrift. After he left, Morde told me: 'He knows nothing. He would not know a good fruit if somebody hit him in his face with it.' Young people today are not used to buying fresh produce in the market, he said, because their tastes and shopping habits are formed in glitzy malls. Perishable items do not do well in malls and upmarket spots, he said: 'I got nothing but grief from working with Big Bazaar, Hypermart, and D-Mart. The wealthy rich who own shops in malls come to us to get their mangoes. They have large shops of 50,000 square feet or more. But I handle greater volumes out of this small shop and my 200-square-foot godown.'

He was in the mood to dish out. 'Workers in malls who wear ties and speak in English are paid about ₹12,000 per month. My workers...Mama here (Bajirao Bhide) gets ₹25,000 a month. The most skilful workers get up to ₹30,000. Otherwise you cannot expect them to sort and grade a high-value commodity with care,' he said, dressed in a crisp, starched half-sleeved linen shirt and brown polyester trousers. He checked for updates on the day's rates and volumes from the Vashi market on his phone. 'We are audited. I pay income tax and deal in crores of rupees. I have twenty workers handling operations here.' Anybody who calculates the margins closely cannot do this business, he said, because this is gambling in its entirety; produce worth ₹500 sometimes sells for double the price and the opposite also happens. 'All perishable items are risky. But other fruits carry much smaller risks. The mango? That is big-time gambling.'

Does one have to be an old-world Maratha strongman like Morde to sit at the retail end of this gambling table? Not really, thousands of migrant retailers come to Mumbai for the mango season. But, if it is difficult to find an articulate grower, it is several times more difficult to find a retailer willing to open up. Small-scale retailers survive by camouflage. Tormented by policemen and municipal officers, plying their trade in uncertain spaces, bullied by anybody looking to make a point, losing their earnings to petty criminals from the public and private sectors.... They must keep their heads down and their noses clean!

That modus operandi does not include opening up to a reporter. After several failed attempts, I mentioned this problem to food writer Rushina Munshaw Gildiyal. 'Oh!' she exclaimed, 'You must meet Hanuman Mangowala. He has sold us mangoes for years and will talk to you if I ask him to. He has a new mobile phone number each year. Let me try this one.' A few minutes later, I had his number and location near Girgaum.

Hanuman Prasad Gupta is from Babhnan village in eastern UP,

between Faizabad and Basti. There, he ran a scrap metal shop that he lost to family disputes. In 1986 he started coming to Mumbai in the mango season along with somebody he knew. He suffered some losses initially; by his fourth trip, he had learned the business. He has never lost money in the city's mango season since. 'Thousands of people from my village and the surrounding region come to Mumbai during the mango season,' he said. His mango calendar begins in mid-March when he arrives in the city. There are agents who charge about ₹2,000 to arrange a small room on rent; he pays ₹45,000 for three months. This tiny room is the ripening chamber and sorting area. At night, Gupta and his three associates hang their hammocks over the ripening mangoes and sleep in the intoxicating smell of ethylene. During the day, they handle the operations. By the end of June, the season is over. In the first week of July, Gupta is back in his village.

This is a four-man operation. Gupta brings his son with him now, along with a cousin and his son. The cousin handles the purchase of fruit from the Vashi commission agents. The four of them work together to ripen the mangoes and then go door-to-door to sell them. Given his experience and the loyalty he has earned from clients like Rushina over the years, Gupta is the salesman. He knows how to segregate produce based on the customer's purchasing power; or how to stay in the good books of his patrons.

He is invited to family celebrations; gifts are earmarked for him during weddings. At times, even his advice and opinion are sought. There are no disputes over rates; he sells to rich families in south Mumbai. They do not mind paying a premium for the guarantee of quality fruit that he brings. Besides, it is a seasonal connection. His regular clients wait for him. When he calls out to them, it triggers the kind of emotions that the cuckoo bird's call in mango orchards has stirred in poets for centuries. He is the bearer of the most prized gift in the city. Regular vendors enjoyed a kind of proximity with clients that is difficult to imagine in our times.

In Tagore's short story 'Kabuliwala' and the poignant 1961 film of the same title, the little girl Mini had her Rehmat 'Kabuliwala'. Rushina has her Hanuman 'Mangowala'.

All the mangoes traders buy and sell in bulk are unripe. Only unripe fruit has the hard texture to survive transport over long distances; ripe mangoes of most varieties are too delicate to travel. Traditionally, mangoes have been ripened in dark, closed places. Perhaps a dark corner of the kitchen where the fruit is covered under leaves—the medicinal plant adusa or vasaka (*Justicia adhatoda*) was popular. Or a dark and closed room deep inside the house.

Like bananas, papayas, apples, pears, and peaches, the mango is a 'climacteric' fruit: it can be ripened after it has been plucked off the tree. 'Climacteric' is the final stage of the fruit's life when it undergoes terminal hormonal changes, releasing lots of the hormone ethylene for ripening. (Non-climacteric fruits—like citruses, pineapples, berries, and grapes—ripen on the plant and cannot be ripened artificially.) Out in the open, a climacteric fruit in the final stage is ripe for attack from insects, fungi, and microbes. Artificial ripening in closed chambers is good for both quality and control. It helps conserve the fruit's flavour and taste; the ripening can be monitored and the fruit can be eaten just after fully ripening—the ideal time to have a mango. That was the traditional practice.

Customers today, however, don't have such patience. They are used to buying fresh produce ready to be consumed. This leaves the business of ripening the mangoes to the last node in the mango marketing chain: the retailer. Some of them still follow old practices, and they buy unripe fruit from wholesalers and ripen it in their storage facilities. Those in a hurry need airtight ripening chambers flooded with synthetic ethylene gas; these ripen the fruit evenly and quickly. Most retailers today lack the time to ripen the fruit slowly and traditionally; nor do they have the space or their own ripening chambers. The mango is merely a seasonal source of income for them. 'Retailers typically lose about 20 per cent of their stock while ripening the fruit. Nobody pays for that,' said Dnyanoba G. Makode, deputy secretary of the Vashi APMC.

In their desperation, many retailers resort to a quick fix to ripen mangoes in one to three days: calcium carbide. In contact with moisture, this powder produces acetylene gas that mimics the natural hormonal effect of ethylene. There are critical differences, however. Acetylene acts quickly to change the colour of the skin, but it does not ripen the fruit evenly. Its fast-acting effect is stronger on the outside of the fruit. This makes the skin appear attractive and ripe. But the fruit does not undergo the changes, like the generation of enzymes, which cause the starchy tissue to transform into sugary flesh.

This is why a lot of the mangoes in the market appear luscious but are mediocre to taste. Carbide adversely affects vitamins and micronutrients inside the fruit. A more serious concern is the presence of impurities like arsenic and other carcinogens in carbide. These poisons can reach the fruit from the powder and then the consumer, putting them at risk of arsenic poisoning. Its symptoms include vomiting, diarrhoea, and irritation of the skin and the eyes. It can be especially harmful to pregnant women and the workers who use these chemicals in makeshift conditions. Several recent studies on animals have shown the harm that can come from calcium carbide.

It was banned for ripening of fruits in 2011 by the Food Safety and Standards Authority of India (FSSAI) under the Union Ministry of Health and Family Welfare. Enforcement of the law, however, is another story. It is common knowledge that retailers continue to use calcium carbide, not just for mangoes but for a variety of fruits like papaya. In recent years, the insecticide Ethephon, a plant growth regulator, has been used to ripen mangoes; when it comes in contact with moisture, it decomposes to synthetic ethylene and other chemicals. Food safety regulators keep playing catch up with the innovations of dodgy operators.

At the lower end of the mango market, the worst practices have to do with retailers minimizing losses. Calcium carbide is to retailers what PBZ and pesticides are to growers: desperate measures. Their insecurity is matched by the consumer's impatience and willingness to pay exorbitant rates to get mangoes earlier and earlier in the season. Growers and traders know the best prices for their produce are realized at the beginning of the season, with the consumers ripe with anticipation. In March 2022, the first five crates of Devgad Alphonso were auctioned for ₹18,000 to ₹31,000 each in Pune.

There is general agreement that the size and quality of Devgad's fruit have declined, even as its quantity and temporal availability have increased. A trader described to me the shift in mango availability in terms of traditional festivals: 'Earlier, on Gudi Padwa, we saw one or two boxes of mangoes. The markets peaked from Ram Navami to Akshaya Tritiya. The festival of Vat Poornima marked the end of the season. Now, it begins so early that half the season is over by Gudi Padwa!' Earlier, the growers and traders followed the festivals for their seasonal calendar, from Makar Sakranti to Shivaratri (in February–March). That calendar has gone for a toss. If it is not the insane pressures of the market, then the effects of climate change are making everything unpredictable.

The increase in quantity owes to the increase in total cropped area. The longer duration is due to PBZ. The attractive appearance is due to calcium carbide. They are examples of short-term measures to control this fruit of the wilderness. And of growers and traders trying their best to score a profit in the short mango season. The variability, the kachha nature of this business, the numerous obstacles, the threat of losing all investment...nothing deters the players in the mango market. Come next season and they are ready to play the game, again. They know people go to great lengths to satisfy this shauq. The high rollers go for the riskiest side of this business: the export market.

Chapter 14

Exports

Source: 'Exposed'

'Our businesses have only a domestic perspective. They do not possess an exporter's imagination.'

—ABHIMANYU MANE

I was headed for UP's mango heartland at 80 kilometres per hour when the 400-kilogram motorcycle hit a pothole on the highway. The impact sent it flying into the air. It landed upright; I was unhurt. But the rim of the front wheel was damaged. Harley-Davidson (H-D) sent a flatbed truck to retrieve the vehicle. I found alternative means to press on.

I had the idea to explore how mangoes and H-D motorcycles brought Indo-US trade relations back on track in 2007, after a nine-year freeze. The US had imposed trade sanctions against India after the nuclear tests in Pokaran in 1998. Trade negotiations resumed in 2005 in Washington, DC, when Indian prime minister Manmohan Singh and US president George W. Bush issued a joint statement. This was followed by an agreement in the following year in New Delhi. While there was bonhomie and informal agreement on both sides, it was still difficult to overcome the legal barriers and narrow national interests to finalize the deal.

Trade associations of both countries joined the cause. The negotiators zeroed in on two items to give dour diplomacy some colour and popular appeal. US trade officials in India had been lobbying on behalf of H-D to lower vehicular emission standards, allowing their import. H-D motorcycles have infamously inefficient and overpowered engines. In 1983, the company was close to shutting down due to competition from better engineered and more powerful Japanese motorcycles. The then US president, Ronald Reagan, who was otherwise a champion of free trade and small government, protected H-D by imposing US duties on foreign motorcycles. All countries protect their own businesses by restricting foreign competitors. This violates the principles of free trade but companies have to face similar restrictions in foreign markets.

The US protection of H-D led to its revival, supported by the booming US economy in the 1990s and by Hollywood; remember Arnold Schwarzenegger in *Terminator 2: Judgement Day* (1991)? But this socialist gift from Reagan led to H-D facing protective duties and high tariffs in foreign markets. Barriers in international trade are not restricted to high tariffs. They can also be in the form of measures that are taken apparently for other reasons, like public health or biosafety. These are called non-tariff barriers or NTBs, a way to keep out another country's products without changing import tariffs. H-D could not export to India because of vehicular emission norms. India had begun to tighten its emission standards in the late 1990s, following a major case in the Supreme Court. H-D engines did not meet the revised norms. For US trade interests, this comprised an NTB. They sought a relaxation in India's emission standards.

Indian mangoes also faced non-tariff barriers in the US. Their import into America was banned in 1989. Reason: food and agricultural safety, as defined in 'phytosanitary standards'. The premise was that Indian mangoes had either high pesticide residues, likely to harm the health of US consumers, or insect pests that could spread in the country. There was evidence of both. In particular, they feared the fruit fly and the stone weevil. But anybody who knows anything about international trade in farm produce will tell you this is not actually about consumer health or farming ecosystems. It's about protecting domestic business. American business interests are aligned with mango producers in Mexico and other Latin American countries. The US was protecting them with an NTB.

This was a great annoyance for the large number of Indian migrants in the US. They had grown up in India's unique mango culture, eating superior mangoes. Latin American mangoes, bred from ordinary Indian varieties, cannot compare in taste or flavour with Indian mangoes. One such migrant is Bhaskar Savani, a dentist in Pennsylvania. He has described to interviewers how his father used to smuggle a few mangoes in his bags each time he travelled from India, as many Indians do.[1] In 2000, he brought along an entire bag. Officials at the JFK Airport asked him to throw it away. Instead, his father tried to eat as many as he could, emerging three hours later smelling of mangoes. Savani has told this story many times over.

He had inroads into trade associations. Along with some others, he lobbied Ron Somers, president of the US-India Business Council within the US Chamber of Commerce, to persuade the US government to lift the ban on Indian mangoes.[2] Somers was called 'mango king' because he knew the taste from his twelve-year stint in India as a corporate executive. Having tasted some quality mangoes in New Delhi during his 2006 visit, George W. Bush was sympathetic to their cause. Exactly how the negotiators worked out the trade deal is known only to them because such negotiations are confidential. But in April 2007 both governments agreed on a 'mango-for-H-D' deal, bringing bilateral trade back on track. The newspapers lapped it up in their headlines. Two years later, H-D arrived in India amid much fanfare.

Several years later, I thought it might be a good idea to take a top-of-the-line H-D motorcycle to Malihabad for a Dashehri-H-D hang-out. The company was happy to issue me one of their trial vehicles from the Grand American Touring line. It was a white and chrome Electra Glide, an ocean liner on two wheels. I hit the highway directly. Some way before Kanpur, in the middle of a smooth stretch, suddenly potholes appeared on the road, sending me flying. The highways of north India's mango country had not taken into account the Indo-US trade agreement.

I had another go, this time in Mumbai. I went small with the Street 750, a motorcycle H-D designed for Indian tastes, put together in their Indian factory. I was to take it to Nashik and Ratnagiri in the middle of the mango season. But two visits to the Vashi mandi in Navi Mumbai filled me with doubt. The heat from the engine was intolerable, despite it being a liquid-cooled variant. I returned the trial vehicle before the trial.

Ratnagiri produces famous mangoes. But why go to Nashik, the district which is Maharashtra's highest onion producer? Because mango export is tied up with onion exports. (The state is India's largest producer of onions, while India is the world's second-largest producer, after China, and fourth highest exporter). Exporting onions is a tricky affair because they germinate in transit. To prevent this, onions are exposed to nuclear radiation in controlled conditions; in this case, it is gamma rays. This prevents germination, making it possible to transport onions over long distances by ships. When nuclear energy is emitted, it is called radiation. When this energy falls on a surface, it is called irradiation. It can be compared to radiotherapy to treat cancer.

To irradiate onions, the Bhabha Atomic Research Centre (BARC) set up an irradiation plant in 2002 at Lasalgaon, Nashik. Called the BARC Krushak Kendra, it was built here to irradiate onions right where they are produced, instead of building expensive storage at the Vashi mandi or other points of exports. Besides Mumbai's expensive real estate, the humid weather there makes it difficult to prevent the germination of onions. The plant can irradiate about ten tonnes of produce in an eight-hour operation.

For US food regulators, irradiation is an essential part of both food safety and extending the shelf life of foods, because it kills insects and microbes. The Indo-US trade deal required irradiation of mangoes for export. The mango season does not clash with the onion season. So the underutilized plant was pressed into the mango's service in 2007. To see this plant, I ditched the H-D motorcycle and hitched a ride in an air-conditioned car with Harshad Doshi, the managing-director of Agrosurg Irradiators, the company that runs the plant. In the four-hour drive, Doshi recalled that his father used to buy mangoes in bulk and gift them to friends, neighbours, and business relations. 'We had all sorts of mangoes, not just Hapus,' he assured me. An electrical engineer trained at IIT-Mumbai, he worked in power corporations before branching out on his own.

We arrived at a large, white building at Lasalgaon, 56 kilometres northeast of Nashik city. We climbed the cast-iron steps of the side entrance into a large hall with a high ceiling. It had assorted cartons of mangoes lying around a narrow, circular assembly line that disappeared into a concealed chamber at the far end. Doshi took me to a small laboratory-cum-office next to the hall. Here, the effect of irradiation became clear in two six-inch tall beakers with onions soaked in water. One was labelled 'Irradiated'; the treated onions inside looked ordinary. From the same batch, they had taken out some that were not treated; labelled 'Control', this had onions germinating shoots from one end and roots from the other.

From there we went to Nashik city to collect a couriered package from the US. It contained 'dosimeters'. These are hand-held devices that measure the amount of radiation. They allow the operators to optimize radiation exposure to the 'transfer standard'. (Again, it can be compared to radiotherapy.) Costing US $8,000, the dosimeter was made by a company called Far West Technology in Santa Barbara, California, in accordance with norms laid down by the US National Institute of Standards and Technology (NIST).

Each 'optichromic dosimeter' costs ₹200; the plant needs sixty for a day's operations. That's ₹12,000 a day on the dosimeter. This is used along with NIST's kit of dosimeters (cost: $2,035 for the dosimeters, ₹3,000 for the courier). It records exposure to radiation in a hidden metric that only their sensors can read. Readings from the two sensors are tallied and then plotted on a graph. After checking these, NIST sends a certificate indicating the precise dose of radiation that has been applied.

That is not enough for US food safety norms. The certificate has to be examined by an inspector of the US Department of Agriculture's Animal and Plant Health Inspection Service (APHIS). The inspector comes to India and must stay for the length of the mango export season. That used to cost ₹1 crore per year. No exporter can bear that cost, so it is borne by the central government's Agricultural and Processed Food Products Export Development Authority (APEDA). All this adds to the cost of export. Latin American mangoes do not have to undergo such rigorous and punitive regulation to be exported and sold in the US. Did Harley-Davidson's motorcycles also face such obstacles to export to India?

One estimate puts the average cost per tonne of imported mangoes at $318, which is about 12 per cent of the cost of the fruit and its transport.[3] Call it a 12 per cent tax! No wonder the total volume of India's mango export to the US was only about 1,300 tonnes in 2019.[4] This despite another irradiation plant coming up at Vashi, this one from the Maharashtra State Agricultural Marketing Board (MSAMB), which handled 620 tonnes along with the 680 tonnes processed at Lasalgaon. During the Covid-19 pandemic, the volume of exports fell to 813 tonnes in 2022, according to APEDA. But it increased to 2,043 tonnes the next year.[5] (Databases tracking international trade report that Mexico exported about 470,000 tonnes of mangoes in 2020, the largest proportion of it to the lucrative US market and most of the remaining to Canada.[6] Of all the mangoes consumed in the US, Mexico supplies about 60–65 per cent, followed by Peru's 12 per cent and Ecuador's 9 per cent.[7])

Doshi took me for lunch at a guest house close to the plant. In the afternoon, the day's operations began, after attending to each regulation. Empty boxes were lined up first on the conveyor belt for trial. Boxes of mangoes with labels of a Mumbai exporter were ready to be loaded onto the assembly line. There were three varieties: Alphonso, Kesar, and Badami. We could not stand there after the belt began to move. So I had a peek inside the six-metre-high irradiation chamber before the operations began. The walls were more than a metre-and-a-half thick in the front and the back and one-metre thick on the sides.

Inside, it smelled of stale onions. On the wall was a signalling system with three lights indicating the state of the 'source' of radiation: 'Exposed', 'In Transit', and 'In Shield'. I hopped over the rails of the conveyor belt for a peek. The 'shield' was a deep tank with water to restrict the radiation. It had a pulley to draw up the 'source'. Assorted onion debris floated on the water. Near the bottom of the tank was the 'source': a cube of radioactive cobalt glowing electric blue! It looked like a wormhole into an unknown dimension! I retreated to safety. The chamber was closed. The light signal changed from 'In Shield' to 'In Transit' as the pulley drew up the cube. Then it stopped and the signal changed to 'Exposed'. The conveyor belt began to roll, first taking in the dummies and then boxes of mangoes around the cube of radioactive cobalt.

In September 2020, Harley-Davidson announced that it was leaving India, the world's largest motorcycle market, because of poor sales and a lack of demand for its products. It sold off the factory it had opened in 2011 in Bawal, Haryana; a scooter company owns it now. But then it did not quit India entirely. It got into an agreement with Hero Motocorp, India's largest manufacturer of small-capacity two-wheelers. It runs a scaled-down operation now. The export business is risky. Much riskier than a 400-kilogram motorcycle bouncing off potholes.

'We operate blindfolded in this business,' Tanveer Hussain told me. 'Fog, sunlight, dew, the slightest variation in weather...everything affects us. There's no saying what might leave us blindsided.' He works year-round to make everything happen in the ninety-day mango season. I had met India's largest exporter of mango pulp at the Kaggalahalli mango market, 35-odd kilometres south of Bengaluru. He looked impressively urbane and out of place in the makeshift mandi—glowing skin, expressive face and the tall, muscular build of an athlete or a model.

His family got into the mango business in 1968; he learned from his father, the late R. M. Muneer—Tanveer's company RMM, incorporated in 1999, bears his initials. 'He was the first trader to realize the potential in Totapuri,' said Javed Hussain, Tanveer's cousin, who learned the ropes from his uncle Muneer after losing his father at the age of seven. 'In the 1970s there were only two pulp factories in this city. One was ours and the other was owned by the Kissan brand.' Earlier, mangoes were sold in batches of 100 in this region. 'My father and uncle brought in weighing scales, along with grading, sorting, and other kinds of quality control.' In the 1980s, mango drinks like Maaza and Frooti flooded Indian markets. The pulp came from here.

Tanveer and his family have expanded operations dramatically. Their other company is TASA Foods, bearing the initials of the three brothers. Tanveer, the oldest, handles the business and the export markets; Arshad, the one in the middle, handles the operations of the four pulping factories; and Shadab, the youngest, has specialized in handling the supply chain and mandis. The region east and north of Bengaluru has extensive commercial mango orchards spread across Karnataka, the Rayalseema region of Andhra Pradesh and Tamil Nadu. The Chittoor district of Andhra Pradesh, in particular, is a major producer of the Totapuri variety, the mainstay of the mango pulp business. The district has more than 100 plants producing mango pulp, with another sixty-odd pulping plants in Krishnagiri district.

While the orchards are mostly in Andhra Pradesh and Karnataka, the pulp is exported from the port city of Chennai, Tamil Nadu.

RMM and TASA are the biggest operators here with a turnover of more than ₹150 crore. 'We compete with Coca-Cola and other giants in this business,' Tanveer said. The mango pulp business is built around just two varieties: the bulk of it is Totapuri, providing texture and body; the flavour comes from a small amount of Alphonso. The small, low-quality Hapus that do not make it to the table go to the pulping sector.

Tanveer is a bundle of nervous energy, pacing and bouncing about like a tiny gymnast; it's comical because of his large frame. 'Each season, I struggle to meet the demand. We need to buy at least 40,000 tonnes of Totapuri and 10,000 tonnes of Alphonso to meet our commitments with export partners. I just got a call that Totapuri's price went up from ₹12 per kg to ₹14 per kg. This ₹2 difference will make a dent of ₹40–50 lakh in my accounts.' A slight drop in prices can also bring about a windfall.

Requiring mangoes in such large quantities, they cannot deal with the growers. Shadab works the mandis in the manner of an uber wholesaler, buying from big traders who aggregate supplies from smaller traders. A lot depends on Shadab's connections in the mandis and his market intelligence. I went to Chittoor and met him as he was getting ready for his mandi rounds. He has his own market calendar. 'The season's first produce lands at the Bengaluru mandi. Then it moves to Krishnagiri district in Tamil Nadu. After that the Andhra Pradesh mandis get busy. First Rayachoti, then the Kodur mandi in Kadapa district, and the last of the season's bulk produce lands here in Chittoor,' he told me.

In the oppressive heat of Rayalseema, I travelled to some of these mandis with Shadab in his SUV. My lack of Telugu and Kannada meant I couldn't understand the negotiations, or how Shadab gathered intelligence. It featured many phone calls, a lot of sipping of tea with random people in mandis, many random conversations with pointed questions dropped at the right moment. His trade network is immense; he has built on his father's work. 'Totapuri is called 'Bangalora' or 'Collector' here because fruit traders from 'Bangalore' used to come collecting this variety. Old traders tell me this name is owed to my father, who was the first trader to regularly buy mangoes in bulk at the mandis here,' Shadab said.

After his mandi rounds, we landed up at the TASA factory. I met Arshad who manages the factory, where operations follow a specific order. First the unripe fruit that Shadab brings in from the mandis has to be ripened. Then it is loaded on to the conveyor belt for sorting. Then a chlorine wash for eliminating microbes, followed by a wash to clean the chlorine residue. Another round of sorting begins, in which the tip of each fruit has to be cut manually. From here on, the process is entirely mechanical. After that, the mangoes go into a machine that removes the stones and skin. Then the pulp goes into a 'finishing machine'. The pulp is heated to 65°C, decanted and taken to balancing tanks that ensure the sugar-acid ratio is appropriate. Thereafter, the pulp is sterilized at 108°C, and cooled to 35°C. It is then packed into aseptic bags that are ready to be transported. The produce of the soil turns into produce untouched by human hand!

'The first priority is labour. We need 500–600 workers who come from Tamil Nadu, Andhra Pradesh, Bihar, Maharashtra, and UP. We have permanent dormitories for them with water and toilets; we give them weekly food allowances and seasonal festival allowances. We pay for their tickets. The contractors who organize the workers are paid in advance,' he said. Over a period of time, workers from one region acquire a certain speciality. 'Workers from Nagpur are swift and nimble with knives, so we put them at the stage where the tip of the fruit has to be cut.'

From January, he has to accumulate firewood for the boilers because it is not possible to get it in bulk during the season: 'This plant needs about 2,500 tonnes of fuelwood and the older RMM factory needs another 1,000 tonnes. The rates are around ₹4,800 per tonne.' A lot of this wood comes from chopped down mango trees, another use for the tree. The next priority is water. The TASA plant requires 250,000 litres of water for each day of operation. Every third day the plant has to be cleaned; it takes 50 million litres of water. Their three bore-wells can meet only half the demand. This is a water-stressed region; the water table hovers at a depth of 600 feet. The shortfall has to be met with tankers of 4,000 litres that supply groundwater from nearby regions.

Arshad took me on a factory tour. The sewage treatment plant, built at the cost of ₹35 lakh, puts treated water into a pond. Farmers from nearby areas come to take this nutrient rich water for irrigating their paddy. (I asked if rice from Chittoor's paddies has a mango flavour; didn't get an answer!) At the balancing tank, I watched workers measure the pulp's sugar content through a large glucometer called the Brix measuring machine; it measures the sugar content of a liquid on the Brix scale of degrees. The required Brix value is above 14.2, up to 15. Another worker was measuring the pH value for acidity.

We sat for lunch and a thundering storm rolled in. Shadab rushed out; I followed. He looked worried: 'It is a bad omen. This storm will damage trees. A lot of immature fruit will fall. The growers will gather them and head to the markets to minimize their losses. The glut might bring down prices, which is good for us. But we need mature fruits of good quality that we can ripen properly.' He began to check his phone for market updates and was distracted right through lunch, which was a joint family affair.

During the rounds, I'd missed the large godowns on one side of the factory, from where the ripened mangoes were coming into the plant. I walked in there on my own, closing the door behind me. Eight rows of crates were stacked, one on top of the other, going up to about eight feet. The ceiling was lined with a silver-coloured heat-reflecting material. At the far end, industrial-grade fans ventilated the enclosure, running on pump-set engines. There was barely room for one person to walk in the passage between the rows, like in old libraries. In the dark, I stumbled on a little contraption throwing up a mist; it was releasing synthetic ethylene, the ripening gas permitted under regulations.

The heat was oppressive, it felt like a sauna. The respiration rate of mangoes increases when they are ripening, leading to the accumulation of carbon dioxide. Along with the ethylene, it left me gasping for breath. The darkness! The industrial clamour of the fans! A multimedia nightmare descended upon me! My overripe imagination ginned up the plot for a horror film. As I panted my way towards the door, a title for the movie occurred to me: *The Ripening Chamber*!

⁓

For mango growers and traders, foreign markets hold the promise of premium returns in forex. Yet these markets flatter to deceive. According to APEDA, the largest importers of Indian mangoes over the past few years are the UAE and other countries of the Gulf region, followed by the UK and USA. The consumers of these mangoes tend to be South Asian expatriates.

India produced close to half of the world's mangoes or about 44 per cent in 2021, according to international market intelligence platform Tridge. China was second with less than 6.67 per cent, followed by Indonesia (6.27 per cent), Pakistan (4.71 per cent), Mexico (4.3 per cent), and Brazil (3.6 per cent).

Yet India was only the third largest exporter of the fruit in 2022, accounting for 11 per cent of the total export value of the mango. Top exporter Mexico accounts for about 28 per cent and the Netherlands is second with under 17 per cent. Brazil is fourth with over 10 per cent; it is followed by Spain (5.8 per cent), Egypt (5.3 per cent), Hong Kong (4 per cent), and China (3.3 per cent).

Exporters from Hong Kong, Thailand, and Vietnam control the Chinese mango markets. The premium markets in the US are controlled by Mexico and Peru. The Netherlands obviously does not produce mangoes; it is a major importer from Brazil, re-exporting a bulk of the produce to Europe. The three largest volumes of mango trade from one country to another are: Mexico to the US; the Netherlands to Germany; and Brazil to the Netherlands. India's exports to the UAE come tenth on this list.

It is a peculiar story. The fruit gets its name from India. The country is the world's largest producer. The quality of India's mangoes has no comparison outside of South Asia (exceptions include the Carabao mangoes of the Philippines and the Miyazaki mangoes of Japan). And yet its mango exports trail behind Mexico and the Netherlands, comparable to Brazil.

The reasons are far too many and too complex. Five stand out. One, the mango is a tropical fruit that does not travel; Indian varieties are delicate, unlike Latin America's leathery but colourful mangoes. Two, shoddy methods of plant protection leave Indian mangoes with either insect pests or pesticide residues—or, worse, both. Three, India's government agencies and private businesses fail to coordinate their efforts to follow up on market opportunities in a concerted way. Four, countries protect their markets with NTBs for political reasons. Five, the phytosanitary protocols required by importing countries are a powerful deterrent; the premium markets in the USA, the EU, and Japan all have numerous requirements that call for varied and expensive phytosanitary infrastructure.

Not all importing countries insist on irradiation. The European Union (EU) keeps banning the import of agricultural items from India on varying grounds, from the presence of fruit flies and other insect pests in fruits and vegetables, to high residues of banned pesticides. The EU banned the import of Indian mangoes in 2014; the stated reason was the prevention of fruit flies.[8] But the real reason, a person familiar with that incident told me, was the EU inspector's horror at the shabby phytosanitary standards and the complete lack of quality control, both at the Vashi mandi and APEDA's set-up at the airport. Apparently, they inspected the registers; they noticed that no export consignment had been rejected on grounds of quality. It was common knowledge that exporters could bypass the quality control system with small bribes. 'The ban over fruit flies was a way to convey their disappointment with handling of produce meant for their markets,' the source said.

Access to European markets depends on earning the trust of retail chains and the discerning customers of rich countries. This happens through a certification system called GlobalGAP (Global Good Agricultural Practices). Major European retail chains created this system in 1997 from norms set by the Food and Agriculture Organization (FAO) of the UN. Initially called EurepGAP, it provided assurance of quality to retail chains that were facing growing concerns over food safety among their customers. In 2007 EurepGAP was broadened to GlobalGAP as the world's most widely accepted food safety standard. It lays down rules for quality control that the growers must follow. It requires inspection and certification of orchards by inspectors authorized by the GlobalGAP programme.

Maharashtra government authorities were quick to act. 'They began with much confidence, asking growers to get their orchards certified under GlobalGAP,' said Vivek Bhide, who is often consulted on such matters by government agencies because of his knowledge and experience. 'I had told APEDA and the state's horticulture board that we should devise our own certification system that takes our conditions into account. But they did not buy it.' This requires a coordinated effort to bring various parties on to the same page. 'Each arm of our government works on a different wave length from the others. That is why we failed to capitalize on the access to US markets from the 2007 trade agreement,' Bhide said ruefully.

He gave examples of the differences between India and Europe. 'They [US] demand that growers practise monoculture, growing the same crop uniformly. Our orchards have inter-cropping of herbs and shrubs and other crops. Besides their beneficial effects on the soil, some of these plants flower at the same time as the mango trees.

This becomes a refuge for insect pests, attracting them away from the mangoes,' he said, emphasizing the difference between tropical and temperate conditions.

Another example: some countries require the small piece of stalk above the tip of the fruit to be removed because it contains acidic sap that can contain allergens. 'If you break the stalk, it secretes the sap, creating an unpleasant ring around the tip. Our scientists and regulators could not explain to them that the piece of stalk and the sap under it protects the fruit's vulnerable opening from fungal infections,' he said.

Yet another: GlobalGAP disallows the application of cow dung compost and urine. 'In their conditions, it might be unsanitary. But in the Indian heat, gobar degrades in less than ten days. Our heat-affected soils desperately need organic fertilizer,' said Bhide. An Indian certification system would have accounted for such factors and standardized the best practices; these could then be tuned with GlobalGAP requirements. This is what China did. In 2009 European regulators accepted the ChinaGAP certification system. Not so in India!

Another way to meet fruit quarantine requirements is Vapour Heat Treatment (VHT). Japan requires it, as do some European countries and Australia. It destroys insects and their larvae by keeping unripe fruit in dense, saturated water vapour at a given temperature for a specified length of time. I visited MSAMB's VHT plant at the Vashi mandi; imported from Japan in 1998, it began operations in 2000. Abhimanyu Mane, the agriculture development officer in-charge of operations there, was ticking off workers, reminding them to maintain standards of hygiene. It took me a while to convince him that I posed no threat, and that I merely wanted to have a look and understand the operations. He ticked me off, too, making me take off my shoes. 'Countries such as Mexico and the Philippines dominate the international market because they have adapted their production practices to the export market,' Mane told me later. 'Our businesses have only a domestic perspective. They do not possess an exporter's imagination.'

He opened a mango crate that had fruits lumped inside haphazardly. He took off the synthetic netting covering the fruits and started arranging them in an orderly manner: 'This is what we lack, a good presentation. Customers in industrialized countries take for granted the fruit's uniformity, hygiene, and appeal. They buy on visual appeal since they are not educated about the taste of our fine mangoes,' Mane said. He ranted about the exporters and the irregularity of their paperwork, about their failure to follow the packaging norms.

He repeatedly threatened to throw out their produce, but then always got the job done—and done right. His manner was fiercely scrupulous; he asked a worker to leave the premises because he had turned up in a lungi. I spent half a day watching workers prepare and line up boxes containing Kesar, Alphonso, and Benishan mangoes that were to go into the treatment chamber. It had temperature sensors and vapour nozzles. After the treatment, the fruit goes into a cooling chamber where the heat makes it impossible to stand.

Several traders and growers had told me that the heat of such treatment leads to premature ripening of several mangoes. 'The rejection rate of fruit in VHT plants is 30 per cent. That means a third of the produce is damaged,' said Sanjay Pansare of Vashi. 'The norm is sixty minutes of heat treatment. But this was developed for mangoes weighing 500 grams. Our mangoes are half that weight, on average. Such heat damages the fruits.'

Besides, there is so much temperature variation in transport and handling that growers and traders find it nearly impossible to move quality produce in good condition over long distances, even using refrigerated trucks. Abhimanyu Mane looked like he was fighting a futile battle.

BOOK THREE
The Fruit of the Senses

Chapter 15

South
Tradition and Individual Talent

'I hated the mango when I was growing up because my father made me taste up to forty different varieties in a day.'

—HARITA KONGARA

You can be right there and still miss it. I visited Banganapalle in 2006 to cover a farmers' training programme. It is 45 kilometres east of the Hyderabad–Bengaluru highway in the district of Nandyal, Andhra Pradesh. This semi-arid part of the Rayalseema region is known for its high-quality long-grain paddy that goes into a proper Hyderabadi biryani. On that visit, I heard nothing about mangoes.

Several years later, I wrote about the bland taste of Banganapalle mangoes. A former editor of mine hailing from the state complained; he asked me to sample the fine mangoes of coastal Andhra Pradesh. Some years later, while travelling through Vijayawada, I realized the outstanding mangoes of coastal Andhra are the fibrous, sucking type. Chinnarasalu (meaning: small juicy), Peddarasalu (big juicy), Cherukurasam (sugarcane juicy), and Panchadhara Kalasa (pot of sugar) are all prized and celebrated here but hardly known outside. If I knew more people in coastal Andhra or had some Telugu, I'd have access to the stories behind these mangoes. In a large and diverse country, one must be mindful of one's limitations.

Perhaps no single mango variety travels across India as much as the Banganapalle. In the markets of North India, it is sold as

Safeda. In West India, it is called Badam. But in Andhra Pradesh and Telangana, it is called Benishan, meaning spotless in Urdu. (Actually, here it is Baneshan.) While I did taste some fine Benishan mangoes in coastal Andhra, they were not significantly better. Scientists have explained how water affects taste. This region has many irrigation projects. Commercial growers irrigate their orchards for better yields. It compromises the taste.

Where does one find the best Benishan mangoes? I got two names. The first was 65 kilometres west of Hyderabad in Sangareddy. The Indian Agricultural Research Institute (IARI) has a reputed research farm here. Upon reaching the city, however, I could not locate it. Nobody had heard of it. Then somebody on the road told us we should ask for the astabal or the stable. Immediately, we got accurate directions. Inside, we saw an old building painted a deep maroon. It was a stud farm and a horse stable under the Hyderabad nizam's rule, along with orchards and some experimental farms. That is the name people remember, not making the effort to memorize the formal name: Sri Konda\Laxman Telangana State Horticultural University's Fruit Research Station.

Here, I met A. Kiran Kumar, the senior scientist in charge of horticulture, whom we met earlier. Before matters got too horticultural, I asked him if I could taste a good Benishan. Of course, he said, instructing a worker to get a fruit from a particular tree. I was brought a mid-sized mango that had ripened on the tree itself. As it was cut, Kumar's little room was filled with a powerful fragrance. The flesh had no fibre but it was not too firm like the Benishan mangoes I had tasted. A few drops of juice dripped from it. It was sweet with subacid hints.

I had tasted many kinds of mangoes in several places. Yet this Benishan created new sensations in my mouth. The front of the tongue was yo-yoing between the sweet and the sour. The back of the mouth and the nasal passage were flooded by a rich aroma. I did not offer anybody else a slice. I felt a rush and then checked myself to consume it slowly. Kumar sat across the table, laughing at my helpless excitement. After I was done, we walked through the orchard. He turned out to be one of the better informed scientists I had met. He described the experiments they were conducting. When we got back to his office, I pulled out my phone and texted my apology to my former editor: I confessed I had done wrong by the Banganapalle. Just that I had not washed my hands. My phone began to smell of that mango supreme and my apology.

I wanted more. By now, I had connected Banganapalle mangoes with the place I had visited in 2006. The next morning, I found myself waiting for a bus, bright and early, at the Aramghar intersection south of Hyderabad. When an air-conditioned roadways bus finally arrived, it turned out that I was the only passenger who could not get a seat. The driver offered to seat me on a narrow bench behind him; I accepted gratefully.

To make the buses appear up-market and aircraft-like, the operators had put vents above seats to blow cool air into the passenger's face. Such a vent dangled three inches above my head, having lost two of the four screws that once held it in place. I moved my head from one side to the other, dodging the blast of cold air. The legs, however, were in a different state. The only place to rest them was by the driver's seat, right next to the Leyland engine blowing hot air. For most of the 280-kilometre journey, I had a cold head and singed feet. On the average, I was a statistician's joke!

I got off at the Krishi Vigyan Kendra, where the caretakers gave me the same room I had occupied in 2006. I headed out to Rajeshwar Reddy's 10-acre orchard. His father had bought the 30 acres after independence for ₹2 lakh from the nawab of Banganapalle. At the orchard's gate, I met Abu Jani, the contractor who had bought the crop. He wasted no time in offering me mangoes to eat. He took me to his shelter, where his workers were sorting the harvest. He pulled out a few ripe ones and offered them to me. The skin colour was a saturated golden-yellow, the shape was irresistibly plump, the skin was thin, and they were mid-sized. The fragrance hit me even before cutting the fruit. It was a thing of beauty, like nothing I had seen before. The Benishan mangoes from the loamy soil of coastal Andhra are a pale yellow; their shape flattens in the middle; their skin is thicker; their size is large.

I had four mangoes within a few minutes, sitting under a tree that looked a century old. They were sweeter than the mango I had tasted at Sangareddy; much sweeter than the sweetest Benishan of Vijayawada. The smell was incomparable. Abu Jani was smiling; he asked me to try the fruits of other trees. We walked around the orchard. He said the marquee taste of the Banganapalle was due to the soil—light, porous, sandy. There is no irrigation in this semi-arid area so there are not too many insect pests either. He sprays pesticides only once a year, during flowering.

From there, I went to Kausar Bagh, the erstwhile pleasure garden of the nawab. This, it is said, is where the Benishan mango was born. Salman Hussein, the contractor, asked if I'd like to try some of their mangoes. Of course, I said, I'd come a long way just for that. He reached into a bucket of water and pulled out a mid-sized mango, with a deep yellow colour. The water had done nothing to contain its aroma. I pulled out my pocket knife and cut through the delicate skin; the juice immediately ran down my arm. The flesh was a dark shade of orange, it was as sweet as it was creamy, and there was no fibre.

The first bite sent a wave of joy down my gullet and through my being. I must have uttered sounds of delight because the contractor and his associates began laughing out loud. This mango brought me to my knees, leaving me moaning in a puddle of pleasure. The fragrance of a true Benishan is like nothing else! It is as overwhelming as it is long-lasting. My clothes, shoes, notepad, pen, phone, backpack...everything smelt of mango. (Two days later, I could still smell Banganapalle in my sweat and in my urine.)

I roamed about Kausar Bagh in a cloud of mango vapour.

Hussein said the nawabs had planted orchards across 3,000 acres here. They were also fond of horses and riding. Iqbal Basha, who drove my taxi, said 'sarkar' preferred four mangoes instead of a meal during the season. He was also very fond of a small mango variety called Shakkar Guthli meaning sugar stone. We came upon an old well, lined with cement and weeds. I leaned over to look down at its dry base; an owl flew out of it and numerous squirrels emerged. Next to the well was an old room with windows on each wall; the roof had fallen in. There were grand old trees and a very old water channel running through, lined with stones and disrepair. An old step-well had its stone wall lined with new cement, its steps buried under dirt. All signs of what must have been a pleasure garden. In another country, they might have turned this garden into a conserved monument!

Hussein said the descendants of the nawab family now lived in Hyderabad. They had sent their manager the day before to take with him 2,000 mangoes from Kausar Bagh and the same number of fruits from Pari Bagh. The produce here is sold by the number. I saw them sell 100 mangoes for ₹3,500. The grower gives an additional twenty mangoes free. This custom is called rijhavan or rungawan in Hindi/Urdu; it is a way to please the buyer. The contractors buy the crop for ₹1 lakh and spend about ₹50,000 on maintenance. They still make a profit of about ₹1 lakh per season. Mangoes from here are not taken out. The buyers come to the orchard.

Pari Bagh, divided from Kausar Bagh by a stream called Jureru, had been purchased by a lawyer named Balaram Reddy. Old trees here have been cut down and new ones planted. Iqbal Basha said his uncle had purchased the orchard from the nawab's family, but they had to sell it to get money for his uncle's treatment after he suffered a stroke and was paralysed. 'In my childhood, I've played in this orchard. We used to be scared of the large trees above and large snakes underneath. And the stream had water right through the year,' he said.

The Benishan quality has as much to do with the shauq of the nawabs of Banganapalle as it has to do with the dirt—the mineral composition of the soil. Banganapalle lies in the Nandyal valley, surrounded by hills on every side. This semi-arid dirt yielded diamonds in the past, the stones sometimes appearing on the surface after rains. This mud still yields diamonds, only they appear in the canopy of the mango trees. The mineral-rich mangoes are not forever, they do not last. But their fragrance lingers. Their memory lasts forever. On the way out of Banganapalli, I kept thinking of my 2006 visit, of what I had missed, of what I might have missed still. Of how sure people are of the mango they know.

It was a stroke of luck. I heard of a farmer living near Vishakhapatnam who breeds new plant varieties. Two of his wondrous creations had received much publicity in the Telugu press: one, a chilli variety in which the fruit grows upwards to avoid damage from dew; and the world's only mango variety that can be frozen for months. I reached Tarluwada village in Anandapuram mandal one morning.

Scores of people were waiting for Kongara Ramesh outside his charitable homoeopathy clinic. He had been called away to a nearby town to treat a six-year-old child who had fallen into the swimming pool of a resort. Such was the family's faith in Ramesh that he was asked to treat the boy inside the ICU. The doctors there distrusted homoeopathy, but they let Ramesh rub some tincture on to his skin. When he returned, I asked him about the patient. 'I just got a call that the boy is better. Doesn't matter which medicine worked, as long as he is okay,' he said.

He knew a little Hindi and it was easy enough for us to talk. He was tall and well-built, dressed in a white lungi and shirt. Between his white beard and bald head were blue eyes and the kind of peaceful smile one sees in brochures of godmen. Around here, this man is treated as one. Over weekends, he ends up seeing up to 800 people in his homoeopathy clinic. They come from far and near—and he is not even from around here. He grew up in a village in Guntur district. Already recognized as a producer of high-quality seeds, his plant breeding skills brought him fame in the late 1980s. That is when he met with an accident.

'In 1991, the wheel of the motorcycle I was riding got jammed. I got thrown off, injuring my head. I lost my memory,' he said. Among those who had come to him for his chilli seeds was Sridhar Rao, the managing director of Navayuga Engineering that operates the Krishnapatnam port in Nellore district. Sridhar had offered him treatment, inviting him to stay in Tarluwada. Ramesh leased out his ancestral land in Guntur and moved his family to the company's extensive farms here. That is when he began to study homoeopathy—to fix himself.

Fame followed him. Soon, people began to flock to him for relief and treatment. I saw him in his consultation room, speaking to each patient softly, pulling out his homoeopathy books every now and then. They felt relief by just talking to him. The patient about to meet him next held token number 292. That's how many people had consulted him or his assistants that day. By 10 in the morning, another seventy-two waited outside; the queue outside the adjoining room had seventy-five people; twenty-odd people waited outside, next to a dozen auto-rickshaws. A group of eight had arrived the previous night from Rajahmundry, 215 kilometres away. They had camped there to see Ramesh. He maintains scrupulous records that are open to anybody. It had 79,715 patients. In between two consults, he told me: 'In this area, a lot of people come with pancreas affected by tuberculosis and its medicines.'

I left him there and joined his daughter Harita and son Venugopal to taste the mangoes he had bred. She has followed her father's interest in plant breeding, studying botany and then horticulture. She described the intricacies of cross-pollination and the techniques that go into hybrids: 'Regardless of how many flowers get pollinated, only one flower out of 1,000 results in fruit-setting.' The variety that's made her father famous is called Swagatham. 'We call it No. 6,' she said. Her father had crossed several varieties with desirable traits. 'He created about thirty crosses and was not very careful with all the labels. Only twenty-four of them were numbered and tagged,' she said.

No. 6 was a cross of the Imam Pasand with Chinnarasalu, a small, fibrous, and juicy variety that's a local favourite. Likewise, No. 8 was named Amrutham; it is a cross of Amrapali and Chinnarasalu. Ramesh has created another fine-tasting variety with very low sugar content, crossing the Rumani with one of his hybrids. Harita had a cheerful smile but she chose her words carefully. She seemed to value an objective distance. She described her father's small errors in the manner of a mischievous student who had caught out the teacher.

Her brother Venugopal brought me four Swagatham mangoes on a plate with a knife. Bright yellow, ripe, and fragrant. I first picked out the one that looked most ripe and ready. Its thin skin came off like butter paper. Some thin strands of fibre held the flesh in place. It seemed quite firm but, as the knife ran through it, juice began to drip all over my hands. I drew up the sliced cheek and the volatile compounds hit my nose like a mist. The flesh was sweet but not overwhelmingly so. There were subacid hints. It was really a complex, well-rounded fruit. Its parents are very dissimilar. Imam Pasand, also called Himayat, is a large table mango with firm, fibre-free flesh and a refined sweetness. Chinnarasalu is small, fibrous, and full of sweet-sour juice. No. 6 had the best of both!

Venu told me about the time he carried twelve Swagatham mangoes in a cardboard box with him in the train. 'The entire coach was fragrant with its aroma,' he said. His favourite is No. 8 or Amrutham: 'It is much sweeter than Swagatham. But not excessively sweet. Because you can freeze it, you can eat even a small mango slowly, over twenty minutes, like a frozen dessert.' But it appears late in the season and I did not get to taste it. Harita said its skin is thicker and layered: 'The outer layer comes off easily but the inner layer of the skin is thicker and bitter. My father says that thickness protects the fruit and is a major reason it survives freezing.'

This is extraordinary because the mango does not tolerate freezing. When stored for long periods at temperatures below 13°C, the fruit of all mango varieties degrades and gets damaged. It is a physiological disorder called 'chilling injury'. The fruit loses aroma; the flesh goes brown; the porous parts of the skin get discoloured; finally, the skin collapses, like it has been scalded. But Ramesh's miracle mango Amrutham avoids all these injuries for several months at freezing temperatures. At some point, some entrepreneur will work out the vast market potential of this variety! (There's no saying somebody isn't working on it already!)

I had eaten all four Swagatham mangoes. Harita and Venu took me to the orchards. A mango tree had keeled over, knocked down by a storm; its roots hanging in the air, as were green mangoes dangling from the branches. Harita said it was a seedling mango of indeterminate parentage. Then she pointed to another tree born of a Banganapalle seed; its fruit is high in pulp with a very small stone. 'I hated the mango when I was growing up because my father made me taste up to 40 different varieties in a day,' she said.

This is when her father joined us with a bottle of cold water and apologies for his absence, even as scores of people were still waiting for him. I asked him about the thick skin of Amrutham. 'The lower layer of its skin is bitter. This repels insect pests and pathogens, which is why it has a long shelf life. Its tree, though, is bulky; the tropical cyclones we get in this coastal area have uprooted several Amrutham trees,' he said. It is a highly productive cultivar, all the same: 'We get 800 to 1,000 fruits per tree. The size is small, so it is about six-to-eight mangoes per kg.' He showed trees knocked down or bent by the destructive Cyclone Hudhud in 2014: 'Swagatham trees naturally have big gaps in the canopy; they let through storm winds, protecting the tree. It is better suited to areas vulnerable to storms.' He asked Harita and Venu if they had got me to taste Swagatham mangoes. When I remarked about the fragrance, Ramesh said: 'I called it Swagatham because its aroma calls to you, invites you. And it arrives very early in the season.'

He walked away to look at a tree. When he was out of earshot, Harita said her father did not come to the orchard for two months after the destruction of Cyclone Hudhud. 'He was in mourning, like he had lost a child.' This is when her father called us to a tree that was a cross of Neelum, the most regular bearing variety in India. 'It used to bear fruits so profusely that I began calling it Kamadhenu (the storied cow of wish fulfilment). It was a mistake because it stopped fruiting after the renaming,' he said, pointing to the only fruit in the tree. He showed me a variety that had a strong turpentine smell. Another had three colours: red on top, yellow in the middle, and green at the bottom. 'I will call it Tiranga (tricolour),' he said. It was too early to taste these fruits.

Having met so many people with unsupported claims of novel mango varieties, I was on guard for the slightest hint of hyperbole or bragging. Instead, he replied to each question in detail, with a relaxed and understated tone. I asked him about his education. 'I had to quit school when I was in standard eight to attend to the family farms. That is because my father had been elected the sarpanch (head of village council). I have no regrets about that. If I had completed my formal education, I would have been doing one kind of work. Now, I do several kinds of work,' he replied, spreading his arms wide, encompassing whatever was within reach.

How did he learn the theory and skill of plant breeding? 'The agriculture college at Bapatla was close to our village. Among the scientists who worked there was a rice breeder called Madhusudan Rao and a cotton breeder called Nageshwar Rao. Both of them developed a fondness for me, for my interest in farming. They taught me the basic science of plant breeding. After that, it was about practical learning, about theory to application and vice versa.'

This is how he created his famous chilli variety. 'Chilli fruit droops from the plant. This means dew drops trickle along the fruit, gathering at its lower end, causing moisture to gather and then rot the fruit,' he said. In this coastal region with heavy irrigation, humidity is very high. Ramesh found some unknown varieties locally and began to cross them with the commercial cultivars. One of the crosses had fruits growing upwards, instead of drooping. In this, the moisture from dew ran down the twig, away from the fruit. This variety became popular because chilli farmers could avoid crop losses. But it was not an accident, either. Ramesh has continued to collaborate with agriculture scientists. His name is among the authors of a handful of scientific studies on biopesticides, published in peer-reviewed journals.

We sat down for lunch in front of his small house. A pleasant breeze had taken away the heat, lending an idyllic charm to this laboratory of rural life. There were several small delights in the meal, especially a plantain curry prepared with onions and tamarind. After the meal, we were served buttermilk. It was so refreshing I asked him what he fed his cattle to produce such quality milk. Ramesh pointed towards the dairy he ran near the house. He said he had thirty Holstein-Friesian (H-F) cows there. Why H-F, I asked him, having heard many accounts from dairy and livestock farmers about the vulnerability of the high-maintenance exotic cattle breed to a range of diseases. 'The milk is only a by-product. I do not run a commercial dairy. It is actually a cattle breeding programme. I will turn these exotic animals into desi cows, well adapted to our conditions,' he said.

Over the years, I have met many cattle breeders. Never have I heard anything like this. India's cattle-rearing world is divided between those who push for exotic European breeds with high milk yields and those who insist on the virtues of Indian cattle breeds. 'This is the fourth generation of H-F stock under my care for more than twenty years. I have bred them to acquire heat resistance. They do not pant in the summer heat,' he said. What about the expensive costs of medicines and vaccines that these animals from temperate climes need in our tropical conditions? 'My stock is free of antibiotics, synthetic hormones, and they do not need vaccines. I have reared them on homoeopathic medicines. It is perhaps the only stock of such exotic animals you will find,' he said. We began to walk towards the enclosure of the animals.

How did he see desi cow breeds? 'They tell me punya (merit from selfless deeds) comes from serving desi cows only, not from breeding exotic cows. But these cows are milk factories. I am preparing hardy animals that will support the livelihood of farmers, without making them spend through their nose on maintenance costs,' he said. 'This ideological insistence on desi breeds is useless. Desi is what you make of it, desi is what suits the local physical conditions and the local economy. Ultimately, it is about what you get out of it without making a loss.' By this time, the cows began to moo loudly. Harita said it was time for their green fodder.

When something sounds too good to be true, it usually is! I pressed on with the interrogation. How had the animals avoided diseases altogether? He said they did not avoid them but dealt with them: 'There's a deadly disease called brucellosis, caused by the *Brucella* class of bacteria. There is no treatment for it. Veterinarians recommend culling the entire stock to prevent its spread. Eight of my H-F cows got that infection. I did not cull.' Ramesh isolated the infected animals and treated them for symptoms. He got the infection himself, developing a high fever that gave him the chills. He had to be hospitalized: 'I told the doctors I had brucellosis; they had not seen or heard of the disease. I asked them to test me; the report came positive for a *Brucella* infection. They recommended antibiotics, based on what they could find in medical literature.' He recovered soon. 'I did not lose a single cow. And no other person or animal was infected.'

Whether formally trained or not, plant breeders tend to have an interest in breeding animals also. Often, the shauqeen of fine plants also have the shauq for fine animals. As it is in the forest, so it is in well-run farms and orchards; plants and animals exist in the same ecological space. If Ramesh can breed a mango that survives chilling injury over months and another that can leave a train coach fragrant, if he can breed a chilli that fruits upwards, if he can handle—free of charge—hundreds of sick people, if he can breed a stock of H-F cattle that is desi and hardy, if he can collaborate with formal scholars on scientific studies, if he can explain it all in practical and everyday terms...he has my attention. It is one thing to be sceptical; it is quite another to be blind!

Ramesh's facility with the natural world is extraordinary. But I could not detect a hint of pretence or hubris. He sees himself, first and foremost, as an experimental farmer. His closest partner in his varied experiments is his oldest child, his daughter, whom he lovingly named Harita, a name which means she who is green.

∫

When butterflies collide with fast-moving vehicles, all you hear is a dull splat. Turn that sound into a ceaseless drum roll, and you know what it sounds like to ride a motorcycle on the west-bound highway out of Bengaluru in the month of May. White butterflies mobbed the highway. No matter how slow I went, I could not avoid the guilt of being a serial splatterer of butterflies.

I was covered in butterfly debris when I reached Chittoor in Andhra Pradesh. I turned north and 24 kilometres later arrived at Thalupulapalli village. This region was once famous for its sugarcane and jaggery. The extensive cultivation of the water-intensive crop sucked away the groundwater table—it is down to more than 1,200 feet—labelling this region a 'dark zone'. I had come to meet T. Chandrasekhar Reddy, whose name and number I saw on a document on farmers conserving traditional mango varieties. I found him talking to a neighbour under a blooming tamarind tree. Tiny leaves fell like rain on us, along with the occasional red ant.

Reddy was soft-spoken and reserved, verging on reticent. The way to get an Indian farmer to open up is to ask about the rains. 'The monsoon has disappointed us for over a decade,' he said, as we went past the dry village pond. 'People here took to the mango because it does not need irrigation. We also grow dryland crops like groundnut, millets, and chillies,' he said. He took me home where his wife had readied a lavish breakfast. After, while cutting up a middling Imam Pasand mango, he parsed through the mango's agronomy in Chittoor. One, the mango pulp factories here are a ready market. Two, the incidence of insect pests is low in this dry region, lowering the cost of cultivation. And, three, more than half the total area under mango cultivation grows only one variety: Totapuri.

This humble variety's story is quite the fable. 'Earlier, nobody had Totapuri if they could afford another mango. It was the cheapest mango, consumed by the poorest. But the pulp business changed all that. Its fortunes changed drastically,' Reddy said. Totapuri is the only Indian mango variety, except Neelum, that flowers and fruits each year, regularly. It is naturally robust and does not require much care. Its firm flesh is neither very sweet nor flavourful; but it is remarkably consistent and fibre-free. Commercial mango pulp is 90 per cent Totapuri and 10 per cent Alphonso, for its bouquet. When you gulp down your next mango drink, remind yourself that the meek, in this case Totapuri, do inherit the earth. Reddy's extended family has a total of 120 acres, of which 95 acres is committed to the mango; of that, more than 70 acres is just Totapuri.

I was not there for Totapuri, though. After breakfast, Reddy took me to his orchards. It was too early for varieties other than the Benishan—the mango travel game is a roulette. He first took me to the tree that gives him his best Imam Pasand also known as

Himayat. It barely had any fruits. 'This is the king of all mangoes. But it is a very shy bearer. Each year, we get a few really good mangoes,' Reddy said. 'It has a unique taste, very sweet. The flesh has no fibre and you can cut it like a cake.' The Imam Pasand I had at breakfast, however, tasted ordinary.

We tried some Rumani, a round fruit with an apple-like shape resembling the north Indian variety Kishan Bhog. Rumani commands a sizeable market in Tamil Nadu. What we tasted was unremarkable, but then the best produce had not come yet. Mulgova is a popular variety in south India, especially Tamil Nadu. 'It is a shy bearer, like so many fine varieties,' Reddy said. We tasted one but it was not ripe. There were trees of varieties called Kuddus, Lalbaba, and Kalepadu. A variety called Thorapadu bears large fruits and is used in pickles. 'In Tamil Nadu they call it the "egg omelette". We call it Raja Pasand but we do not like to eat it,' he said. And then we arrived at the seedling trees because of which Reddy has been recognized as a mango conservator. One is called Allikai, it gives large and fibrous mangoes that are used in pickles and chutneys. He has some nameless varieties that scientists have numbered under naati or native.

Reddy's neighbour Sriramulu Achari joined us. He had five seedling-type naati varieties that scientists from the Indian Institute of Horticultural Research (IIHR) in Bengaluru have studied and earmarked for conservation. He showed me his trees that had dried up for lack of water. For all that Reddy had told me, the day had been disappointing. I did not taste anything memorable other than the generous breakfast. This is when I remembered the names of two small varieties: Atimadhuram (extremely sweet) and Shakkar Guthli. He took me to the trees. The fruits were very small and immature. This is when I noticed a cheepi—a small, flat fruit with an underdeveloped seed. (Such fruits ripen on the tree very early. Growers use them to decide when to harvest their crop because they indicate the maturity of the remaining fruits. Across the country, names used for such fruits include cheep or seep or seepi.)

I reached for the Atimadhuram cheepi; it came off readily. Its skin had begun to change colour. Without my pocket knife, I cut it open with my teeth and reached for its warm and fibrous flesh. It was both fragrant and quite sweet. A horticulture professor had told me growers do not sell Atimadhuram, using it for domestic consumption or for gifts to near and dear ones. Both Reddy and Achari confirmed it with a smile.

Shakkar Guthli, meanwhile, is so small it looks like a large grape; it grows in bunches, too. I took off two fruits from a bunch, looking like a couple of ping-pong balls. I bit into them, although they were not mature; there were hints of its strong taste and sweetness. It only seemed natural in this place, once famous for its sugarcane and jaggery. I headed back into the butterflies.

The best maintained orchard I got to see was at Vennangupattu in Kanchipuram district of Tamil Nadu, 115 kilometres south of Chennai. Its main gate opens on the East Coast Road, which is where I met the farm manager B. Venugopal. Inside, the 36-acre orchard was all neatness and order. Every patch testified to love's labour. Each tree was served via drip irrigation. Trees were marked like in IARI research farms.

The Benishan mangoes here were the largest I have seen; the manager said they grow up to 850 grams per fruit. Close by were trees with the largest Mallika mangoes I have encountered. There was the largest Doodh-Pedha mango I have seen, the largest Suvarnarekha, the largest Imam Pasand. Fruits of a variety called Jahangir here looked like small water melons, only uglier. I did not get to taste them because they were immature; the crop was twenty days late. Even if you plan everything, the mango has many ways to surprise you; it elevates as readily as it disappoints.

Tree after tree had more than one variety grafted on to it. When the manager could not identify a mango hanging from the branch, he called the gardener and asked him. In some cases, even the gardener could not identify a variety. The orchard has 3,500 trees and twenty-eight documented mango varieties. It is not just a commercial orchard, either; it is an experimental farm for research. Venugopal took me to rows of trees earmarked for trials. One patch had high-density cropping, another had ultra-high-density cropping. Another patch had old trees undergoing rejuvenation through cutting and pruning. Several horticulturists including M. M. Burondkar of Dapoli have been invited here for consultation.

I tried out the few varieties that were ready. The Neelum there smelled and tasted like sweet-sour seed-grown mangoes. The Jawahar looked like a large Alphonso and had a delightful aroma and aftertaste. I had come here for Imam Pasand, which was not ready. The early produce had been sent to Chennai for ripening. Instead, I was taken to see an 'arboretum' 13 kilometres west of the orchard in Yerrangadu village. Arbor means a tree or a beam made from a tree trunk in Latin. An arboretum is a living collection of a range of tree species; it serves several purposes. A botanical park for visitors, it also serves as a field lab for experiments, especially for students of taxonomy, a branch of biology that badly needs support. Several American cities have their own arboreta but this was the first I saw in India.

The Yerrangadu Arboretum sits along the east coast with its harsh, dry conditions. Yet this 13-acre paradise has 320 plant species—35 species of bamboo, 19 palms, and 266 species of trees. There are at least two plants of each species, labelled, tagged, and marked for visitors. Just as the orchard at Vennangapattu has benefitted from the insight of several horticulturists, the arboretum draws upon the knowledge of a range of ecologists and conservationists, primarily from the Salim Ali Foundation. When I visited, it was overrun with birds, a fitting tribute to the great ornithologist Salim Ali. The arboretum has drawn support from a range of government officials.

It is a novel, public-minded iteration of the pleasure garden. It could be the panchavati or vatika of yore—vat in Sanskrit means the same as arbor in Latin. I was taken to a lovely house in the middle of the arboretum and treated to a six-course lunch, with assorted mangoes. For a day, I felt like a king. What kind of a shauqeen would commit his private property to such a project, I wondered. I had not seen him yet, although I was staying at his charming bungalow in central Chennai. Earlier, when I had cold-called him, he had told me brusquely that I would have to come and stay with him. When I reached there in the morning, a car was ready to take me to his orchard and arboretum.

C. Ramakrishna, CR to his friends, was a reputed lawyer in the Madras High Court. He was from a family of large landholders in Chunampet, near Vennangapattu. He was a bachelor who invested his significant wealth and property into a range of public-spirited projects aligned with his passions: music, philosophy, religion, horticulture, sports, and civic well-being, to name a few. Other than being a regular rasika of Chennai's feted classical music scene, he sat on the board of management of a number of religious shrines. With an old-fashioned faith in building and running institutions, he invested his time and money in these. It was his passion and generosity that brought a number of talented people into projects he supported. Along the periphery of his bungalow was a shop that sold the mangoes from his orchard. This is where I picked up the Imam Pasand grown in his orchard. Finally, I got to taste its delightful flesh; firm and fibreless; sweet and fragrant. A perfect table mango!

Late in the afternoon, I was taken to meet CR. His room had stacked bookshelves and furniture that smelled of seasoned wood. He sat on a reclining chair that resembled those in the first-class waiting rooms of old railway stations. He asked me to sit in a luxurious reclining chair, upholstered in leather. He was pushing ninety, hard of hearing and keeping in poor health. He first wanted to know what I thought of the arboretum. Despite his age and failing health, he spoke with a passion that was as unsettling as his all-round generosity.

After he realized I was from North India, he asked me about the Langda mango. 'It is a great mango, the Langda,' he said, his eyes softening along with his voice. He owned a famed orchard put together through years of labour, investment, and dedication, but he was interested in the mangoes grown in my region. He was very proud of the green paradise he had created, but he had a greater interest in other people and other places. He pumped me for more descriptions of the best Langda fruit I had consumed, where they came from, how they tasted, and the variety's history. He asked if the shape and taste of the best Imam Pasand was comparable with the best Langda. Yes, I said.

He asked if I had tasted mangoes from his orchard. I complimented him on his extraordinary Imam Pasand. He said it is the most difficult variety to grow, given its shy bearing. The rarity adds to its charm, I said. He immediately began to question me about what I had noticed in the orchard. I knew this was my examination: the lawyer in him was cross-examining the witness. Had I seen the experiments with ultra-high-density cropping? Yes. What did I think of it? Essential, yes. Had I seen the size of the Benishan mangoes? Yes, spectacular. How was the Rumani? Great!

When he was satisfied, he asked what I felt when I saw the trees laden with mangoes. I uttered some pleasantry. 'It is fascinating to see a mango tree laden with fruits,' he said, his eyes bulging with rapture. He was tired and it was taking some effort to sit up and talk. He began to drift, saying: 'No sight is as beautiful as a tree full of mangoes...it is abundance...no other sight...the fulsomeness... nothing compares to a tree full of mangoes...' I knew better than to sit there asking questions. As I walked to the room upstairs to pack my bags and move on, I asked myself if CR was the greatest shauqeen I had encountered because of the mango. Definitely up there, I muttered.

As I left the bungalow, as assistant handed me several bags of high-quality mangoes to gift to my friends. He did not let me pay for them because his employer had forbidden it. Those mangoes brought joy to several friends I visited. In June 2021, I read an obituary saying CR had died. He was ninety-two.

⁂

'Naam ke aage degree hona (One must have an educational degree),' Syed Ghani Khan Irfan shouted back from the canopy. He had climbed one of his magic trees to get me a good mango. I had asked him why he went back to college in 1997. After enrolling for a bachelor's degree in commerce in 1994, he had to drop out in the first year because his father suffered his third stroke and became paralysed. He was the eldest son, he had no choice. But in 1997, a regional college offered a course in archaeology and museology. 'I got interested. I have always had an interest in historical heritage. I thought I'd create a museum of historical importance,' he said from up in the tree.

He brought down a small fruit that had ripened on the branch. It did not look like a mango. The skin was greenish-yellow, the pulp fibrous and juicy. It had hints of sweet and sour. It was so refreshing and unusual that I ate it whole; even the skin tasted crisp and fresh, without a hint of bitterness. This was in Khan's village Kirugavalu, 25 kilometres south of the district headquarters at Mandya, which is about 100 kilometres southeast of Bengaluru, on the way to Mysuru.

Khan continued. A kind-hearted professor helped him complete the course while he was farming the family land. Since he had no way out of farming, he decided to do it right. He began boning up on agriculture sciences. He was going to be a progressive farmer. Already, since 1996, he produced hybrid paddy seeds for the Rice Research Station at the University of Agricultural Sciences in Bengaluru. His ambition, though, received a jolt in 2000 when he fainted after accidentally ingesting a synthetic pesticide he was spraying, following a recommendation from scientific literature. I asked him to name at least one of the pesticides he had used. 'I used endosulfan in paddy,' he said. That incident made him re-evaluate his choices.

Something else happened. 'In 1998 my uncle gave me forty grains of a traditional variety of paddy. I slowly multiplied it into two kilograms, then 140 kilograms,' he said. He showed the variety to agriculture scientists he knew; it had not been recorded earlier. They were impressed with its traits and offered to release it in his name as a farmer's variety. Further enquiries revealed it was a paddy variety from northern Karnataka called Ratanchudi. Khan began selling its seeds. A scientist suggested that he should take the story to the press. A relative was a journalist; he wrote an article about Khan's traditional paddy variety in 2003. It got circulated and seed buyers began to contact him.

The fame was transformative. He met people with a wide range of ideas on agriculture. Khan had already gone organic after his accident with the synthetic insecticide; his twenty acres were soon certified as organic. He began collecting traditional paddy varieties, ending up with over 1,300 varieties. He emerged as a poster boy for the champions of small-scale organic farming; for those who see value not just in profit but also in ecological health. He spoke to campaigners working to protect traditional cultivars sidelined by industrial farming. 'This is when I began to notice the value of these mango trees that I'd been taking for granted. The fruit of each tree has a distinct taste,' he said.

Now, he has 116 mango trees registered with the National Bureau of Plant Genetic Resources as indigenous farmer varieties. The village's name owes to the military campaigns of Tipu Sultan, former ruler of Mysore state—Kirugavulu means 'small cantonment'. After the army was disbanded, it is said, the soldiers were given land around here. This region, in the foothills of the Western Ghats, was quite dry and overrun with mango trees. Between 1911 and 1932 the erstwhile Mysore State built the Krishna Raja Sagara dam on the Kaveri River. Irrigated paddy farming took over. Village after village cut down its mango trees for timber or for clearing land. One such tract came to Khan's grandmother. He has inherited about 20 acres from that. He showed me a plot in front of his house: 'That was our kitchen garden. From the age of seven, I grew all the vegetables we needed in that plot. But then my uncle sold it in the face of some financial difficulty.'

His mango trees are older than 100 years, perhaps even 200 years. They were planted more than six generations ago. They represent the mango diversity we have lost already, that we are losing now. It takes an unusual person like Khan to not cut them for profit—and it is difficult for him, too. 'I used to struggle to sell 100 of these mangoes for ₹30. Then, at a meet for organic farmers in Kerala, I met Syed Ateeq. He curates mango festivals in the malls of Gulf countries. The success of his exhibitions depends on showcasing a rich variety, and he needs only organic mangoes. He began paying me ₹35 per kg. That brings me about ₹3 lakh in a season now,' he said. Some other organic outlets began buying his mangoes, as also those who organize mango festivals across India. His fame has finally brought him some income out of conserving agricultural biodiversity, for not taking the shortcut to money.

While most of his work is on traditional paddy varieties, it's his mangoes that have brought him fame. Many a headline has called him the conservator of Tipu Sultan's mangoes. The trees dotted the bund running through his paddy fields and along the periphery. Their branches were thick and moist from the rain. I put a hand on a trunk and a piece of bark broke away into my hand. Underneath, it was crawling with more insects than an entomologist could identify before dusk. It indicated the absence of synthetic pesticides, although Khan has begun using biopesticides in recent years. 'More than forty species of birds get sighted here,' Khan said, as a partridge flew by. While I walked about tasting mangoes, mosquitoes raised on an organic diet made a feast of me.

Khan and a worker were harvesting mangoes to meet an order. He pulled down a small fruit from another tree for me to try. I bit into it and tasted mausambi (sweet lime); even the fibres tasted like fibres of sweet lime. Another had a cumin flavour. Yet another fruit tasted exactly like an apple. Each tree had a unique taste. Some were really sweet with difficult-to-identify aromas, some were sweet-and-sour. After having tasted a few, my mouth began to tingle in confusion. I climbed a tree. Rainwater had gathered in the node of a branch; I blew into it and a few mosquitoes buzzed out, only to bite me through the clothes. The trees were all crawling with critters: insects and birds and fungi. 'The older trees have very few insect pests because they also harbour their natural enemies. The thirty Alphonso trees I planted recently are rife with pests,' Khan said.

How does his village see him? 'My neighbours and other people in the village used to mock me. They laughed at my excessive attention to detail. But that has changed. Now they admire what I do, they make appreciative noises. They like the attention my work brings to the village. One-fifth of the farmers here now grow traditional paddy varieties,' he said. Thousands of farmers have bought paddy seeds from him. Seedlings from his mangoes travel far and wide.

Among the awards he has received is the recognition from a UNEP-GEF-ICAR (United Nations Environment Programme-Global Environment Facility-Indian Council of Agricultural Research) project for being a 'custodian of genetic diversity'. Khan is a new-age star, a darling for journalists looking for a 'positive story' from rural India. He is a multilingual and articulate farmer with a credible voice and biodiverse fields. A self-effacing smile and small stature only add to his charm, his elective vulnerability a foil to his steely resolve. For the urban consumer he is scrutable, with an irresistible story, a biological and historical heritage, and unique cultivars. Not to mention his distinctive mangoes!

The centrepiece of his story, however, is a paddy museum that Khan has set up on the first floor of his house. It has beautiful artwork, hand-drawn images, and carefully arranged patterns in paddy. The kitchen garden that his uncle sold off continues to live in his drawings. The attention to detail, the aesthetic eye, the hard labour...it is all there. Khan did finally set up that museum. His fame outstrips anything that an academic degree could have brought him.

Chapter 16

West
What Money Cannot Buy

'I could earn much more growing cashew, but I love mangoes.'
—NESTOR RANGEL

Several airline flight attendants over many years have told me that one of the most difficult sectors in India for the crew is Delhi–Goa. All too often, they have to deal with boisterous partygoers from Delhi in the mood for a Goan beach getaway. A holiday here is advertised as an escape to carefree beaches and rivers of alcohol. The tiny state is one of India's biggest tourism destinations. Tourism, the biggest part of Goa's economy, is mostly restricted to the western part of the state, close to the beaches. Most tourists barely get a glimpse of what it means to be Goan—that lies outside the package deal. It works both ways. Goans frequently see the worst side of tourists. For all of tourism's economic opportunities, it does not provide a cultural connect. 'Tourism in Goa is divorced from the state's cultural wealth,' said Ana Dias Camara, the state's deputy director of the department of agriculture and a resident of Margaon.

Not that Goa is averse to visitors! From Kerala to Kachchh, the history and culture of India's west coast have always included encounters with travellers. In the sixteenth century, foreign priests invested the latest grafting techniques into the mango. But today, Indian travellers to the state can go away without encountering Goa's fine mangoes! I know, because in my five visits to Goa over the years, both in summer and winter, I did not encounter the mango. Tourism has a way of letting you consume a place without actually getting to know it!

Yogita Mehra came here as a visitor, but she wasn't a tourist. An environmental economist in Mumbai, she came here to study the effects of Goa's second-largest industry: mining. 'My research showed the destructive impact of mining. The desperation of the rural folk left me sickened. I had no way of helping them; all my work just ended up on bookshelves,' she said. 'We wanted out,' said her husband Karan Manral, who used to work for a software magazine. 'We were consumers who wanted clean, healthy food, free of poisonous pesticides. That's how we got interested in farmers. We felt like joining them,' she said.

They ended up leaving Mumbai for Goa, leaving economics and software for ecology. In 2008 they launched Green Essentials, a gardening consultancy. Mehra teaches people how to grow plants with limited resources and in organic conditions. Manral runs the business end. I met them in their beautiful office in Porvorim, about 12 kilometres north of Goa's capital Panaji. The rented bungalow has been converted to a full-fledged gardening school, with plump pineapples growing out of pots. It started with the pots; Mehra began using compost pots to turn kitchen waste to fertilizer. It excited her, and gave her hope that environmental economics did not.

'Goa is ideal for 365 days of agricultural productivity. There's water, there's sunlight, there's fertile soil. Yet, most of the farm produce in Goa comes from outside. Why? Because farming has become so uncool that families here often leave it to the youngest son, of whom they have no expectations,' Mehra said, adding that farmers are the underclass of our times. She has seen the plight of farming up close, the vanishing dignity of farmers, the large tracts of fertile land that remain fallow. If, by some miracle, destructive mining were to stop in Goa, what would people do?

Mehra's kitchen garden became cool, attracting neighbours

and passers-by. Already familiar with farmers and social groups in Goa, she began talking to them, organizing them. She began taking workshops for would-be kitchen gardeners. Hundreds turned up at the first major training in 2011. 'The venue made money. So many people had not come for an event before. There was a hidden demand for what we were doing,' Mehra said. 'People willingly offered us their land. No collateral, no lease!' In their workshops in Mumbai, 90 per cent of the participants were women. In Goa, half of them are men.

In their farmers' markets, they met mango growers, most of whom have only a few trees. They sold their mangoes in bulk, as and when they were ready. 'The agriculture department had no advice on grading, sorting and labelling of fruits to make them more marketable, to obtain better prices,' Mehra said. Green Essentials was a platform for direct sales; soon, it began advising growers on packaging and presentation. Initially, the growers were disappointed with the subtly-designed bags of unbleached cotton. 'They wanted more blingy bags like bright yellow cases of Alphonsos. It took a while to convince them that such tasty mangoes can be presented in a tasteful manner,' she said. Gradually, they adopted cardboard boxes labelled by designers, attractive and elegant.

Several growers now send their produce to the Green Essentials office, from where it is sold. While I was there, a few customers came to pick up their mango boxes. There were multiple accents, ponytails, tattoos, and body piercings. More than anything else, there were big smiles and gratefulness. Just like mango growers, the buyers have also turned into their friends. Goa has more consumers than producers. 'If your farm produce has quality, there's a ready market here,' she said. Green Essentials does not make money from organizing mango sales. The values at stake are far greater than monetary. Mangoes lend a special touch to both their gardening work and their social life.

It started many years ago with the organic farmers' club

they created on Chorao island on the Mandovi River, just north of Old Goa. I took the ferry and went to meet club president Premanand Mhambare, a retired teacher with family farms near St. Bartholomew Church. He speaks little; when he does, he speaks softly. We went walking about his orchard and he remembered the name of the gardener who had grafted some of their trees when he was a young boy. 'It was Pandurang Uskaikar. Each village had skilled graftsmen who learned horticultural skills from their family elders,' he said, adding that trees grafted in the old method of inarching produced better fruits than those grafted with more recent techniques.

He has about eighty mango trees but his regular income is from cashew. 'The mango bears fruit every alternate year while cashew bears regularly. The income from cashew is much higher,' he said. Why does he hang on to the heirloom mango trees? Mhambare smiled and said: 'It's the mango!' Back in the house, he named several customers who come and buy his fruit directly. A few are businessmen who wish to gift something special to their clients.

He got a few ripe mangoes and sliced them on a plate for me. They varied from mildly sweet to sweet-and-sour. There was one outstanding Mankurad, its flesh fragrant, creamy, and caramel-like. 'People haven't heard of this variety outside of Goa because it doesn't travel well, the shelf life is very short,' he said. The Mankurad looks related to the Alphonso; most outsiders won't be able to tell one for the other. Its flesh has more fibre, is sweeter and has a thicker, emulsified taste. 'The Mankurad varies highly from place to place,' said Miguel Braganza, former agriculture officer of Goa. 'The produce of Chorao is different from that of Bardez or Mapusa. And we don't know where the original Mankurad was located.'

There is one outstanding variant of the Mankurad that I did not get to taste. I could speak to Fausto Cardozo of Bardez in North Goa only on the phone. It came to be known in 1992, when the family entered it in a mango competition; it won hands down. It bears regularly and much earlier in the season; sometimes the fruit matures in late February, instead of May. 'The unripe fruit is a green-maroon, turning to red-yellow after ripening. The flesh has less fibre than usual. It is larger in size with an average weight of 300 grams, going up to 500 grams for the biggest. The ripe fruit has a higher pulp-to-refuse ratio. It has better keeping properties, about double the usual,' Cardozo told me.

The tree is not grafted. It was a chance seedling preserved for sentimental reasons. 'It was in 1958 that my cousin from Portugal was visiting. She ate a mango, threw the stone and left. The seed took. It was uprooted twice to be chucked away but my mother replanted it on each occasion out of fondness for her niece,' Cardozo told me. That fondness began to bear fruit in the late 1960s. 'Scientists and other growers have taken up to 1,000 grafts from this tree,' Cardozo said. (In a few years' time, you might hear of this fantastic new variety with a thousand histories!)

'Up to 95 per cent of the Mankurad trees in Goa have grown from seedlings and are 60-100 years old,' said Adavi Rao Desai, senior scientist in-charge of horticulture and fruit science at the Central Coastal Agricultural Research Institute in Old Goa. 'The fruits are not of a uniform quality. The contractors who harvest the fruit often mix them up, so differentiation becomes even more difficult. Still, it has a unique taste that might have to do with the fibre content. Its skin is not bitter,' Desai said.

'Goa's mango story has not been recorded properly,' Ana Dias Camara told me, 'We go crazy over anything Goan but do little to preserve it.' As part of her work for the agriculture department, she has collected 100–200 types from farmers; some she could identify for their varietal character, others were unidentified. Most of the work on mango improvement happened early on, and Goans are eating the fruits of those labours. 'My grandmother knew stone-grafting of mangoes when the seedling is only twenty-one days old. The trees that grow from such grafts do not have the graft line on their trunk. So it is not always possible to tell if an old tree is from a graft or a seedling,' Camara said.

'So many Goans have their own varieties. My son has grown up eating a variety called the Barretto. He likes and eats only that,' she said. Goans are very fond of the Miskut pickle, for which the Musserat mango variety is favoured. 'Many people have between one and ten mango trees on their property. Of these, the fruits of a particular tree are eagerly awaited. The market goes mad for the Mankurad. I had to once buy two mangoes for ₹500 for a friend who was leaving for Canada. But the Fernandin is not seen even. That's because it is consumed or sold before it can hit the market. Likewise, the Hilario is a pleasantly sweet variety but it doesn't have a market,' she added.

The Fernandin makes lovely sweet mango chutney, she explained, although the yield of the variety is poor because it bears late in June, by which time the monsoon sets in and insect pests inflict substantial damage. Which points to a remarkable aspect of Goan mangoes: they are organic, without any aggressive claims of being organic! 'Up to 85–90 per cent of the mango and cashew trees in Goa are not sprayed with any pesticides. Most trees are too old and too tall to be sprayed or manured.'

In 1997, the ICAR Research Complex for Goa released its 'Technical Bulletin No. 1' titled *Mangoes of Goan Origin*, 'because no comprehensive publication on the Goan mango varieties was available so far'. It describes seventy-five-odd mango varieties, although only forty-seven were actually located and collected. The authors say many old varieties might have been lost or 'may be existing in some garden as unknown isolated trees. This is because many of these varieties were protected from indiscriminate multiplication by the owners who originally grafted them. Most of the older generation of Portuguese who left Goa, especially after liberation in 1961, also carried with them information on most of these varieties. Many among the older generation living in Goa also have left the world along with the knowledge they had on these mangoes.' D. G. Dhandar, the former director of the ICAR Research Complex for Goa, wrote: 'The work is, nevertheless, incomplete as several varieties are yet to be located.'

'ICAR agriculture scientists came here only in 1982 for documentation work for an all-India mango conference,' said Braganza. But that did not mean that the Goan mango was not studied or written about. The ICAR booklet draws from a 1979 paper by Fernando do Rego and K. A. Kazi, it had reported seventy-seven named varieties. A former officer of the state agriculture department, Rego wrote a few papers on mangoes. In 2019 he published a bilingual book titled *The Mangoes of Goa*. He died the following year. He reviewed literature, finding more than thirty references from travellers and historians, beginning with Friar Jordanus Catalani in 1328 to the ICAR booklet in 1997. The first register of varieties he found was from 1902. The earliest Goan he found describing the mango was Caetano Xavier Gracias, a physician and a botanist.[1] 'The best mangoes in India come from Goa, where it was grown and the use of grafting was widely spread by the Jesuits,' Rego quoted.

Finding mango growers in Goa can be quite difficult. A mango festival in Furtado was a tepid and underwhelming affair. Most sellers had only one variety in small quantities. Only one stall had multiple varieties and buyers. The man running it was Nestor Rangel, a name I had heard earlier. His orchard is 55 kilometres away in Sattari taluka near the Chorla Ghat tri-state area. The next day I reached his orchard up in the Sahyadri mountains. Rangel was relaxed here and spoke with passion, like a true shauqeen. 'Wherever I go, if I see a nursery, I stop and check what I can take back with me. I could earn much more growing cashew, but I love mangoes,' he said.

The first varieties I noticed there were long ones shaped like gourds—the polyembryonic varieties found along the West Coast!

Other than twenty Goan varieties, he had Kent and Tommy Atkins of Florida, Mallika and Amrapali of Saharanpur, Neelum and Totapuri of Tamil Nadu, and Maya from Israel. His orchard has more than 700 mango trees, a dairy and a free-range goat farm. 'I got eight goats from a man. I returned 100 goats to him,' he said. His goats and cows wander about freely in the orchards: 'They fertilize the soil.' Monkeys, bisons and wild boar do more damage to his farm than the cattle or the goats. 'I don't mind such damage. It's nature taking its due,' he said, adding that he grows his mango without fencing. The crop depredation from monkeys has increased because of mining in adjoining regions; the animals don't have anywhere else to go. Rangel keeps an open mind.

An electronics engineer by training, he grew up and began working in cities like Mumbai, Vadodara. 'I was really impressed by the go-getting jugaad mentality in Delhi. In comparison, Goans are too laid back. They lack enterprise,' he said. He returned to Goa in 2002 and set up an electronics service centre but closed it in 2010. His family had already lost about 300 acres of land, mostly in St. Estevam Island, east of Chorao. He did have a house and a plot of land on which he had planted seventy-odd fruit trees. By 2007, he had bought seven acres of land in Sattari. 'I was told this region is not suitable for mangoes because it gets cold. Which delays flowering and prevents the heat the mango requires in the summer,' he said. He took whatever horticultural advice he could find: 'I used each and every synthetic chemical scientists recommended. But then I realized their poisonous nature. In 2015, I went fully organic, although I don't want organic certification.'

People thought he was mad. 'My wife opposed my shift to agriculture. But I was fed up. I needed to do something else, something that gave satisfaction. This is not just a business to me. It is my passion,' he said. Gradually, his family adjusted to his calling; his wife helped him out in farming. 'You cannot live on passion alone. There's hardly any money in farming. It's taken me years to break even and expand my holding here to 40 acres. I sell mangoes across Goa now. Farming is a kind of social life for me, social service even,' he said.

Mango cultivation has reconnected the deracinated Rangel with his Goan roots. A priest came to him, worried about community lands lying fallow. 'Land mafia and property dealers keep an eye on parcels of lands left uncultivated,' Rangel said. The priest drew Rangel's attention to the Khazan system, a traditional form of estuarine agriculture. It has now become a social campaign to revive Khazan lands.

The mango is becoming a counter to Goa's social flux, as much for Goans Mhambare and Rangel as for migrants such as Mehra and Manral. The fruit is a means to stem the tide of social apathy and land alienation, and to make food production a process of regeneration, social and ecological. That is the mango's power: it is connecting those opposing destruction from mining and tourism. It also introduced me to a Goa that had eluded me previously, and to Goans who are a lot more than caricatures on a beach holiday brochure.

∫

The excitement was at an end. After two weeks in the lively chaos of Konkan's orchards and markets, I was now in Mumbai's departure lounge. Airports are designed after assembly lines; they process people neatly into ordered spaces for safe transit. Brian Eno's *Music for Airports* turns on automatically inside my head while coursing through the foyers, ramps and doors. It is the sound of industrial detachment. 'Ambient music...must be as ignorable as it is interesting,' say the liner notes of the 1978 album.

The departure lounge here, however, was overrun with the sights and smells of attachment. Many passengers carried brightly-coloured cardboard boxes of Alphonso mangoes. In this season, friends and relatives expect a box of Hapus mangoes if you are coming from

Mumbai. Airlines allow one box of mangoes as additional cabin baggage, free of cost. The entire cabin smelled of ripening mangoes. As the aircraft circled over Mumbai, the Konkan coast appeared in the window, scaled down to picture-postcard perfection. I had seen this coastal strip several times. But I was never going to see it with the same eyes again. Looking for the Alphonso's home had changed my perspective. Now, I cannot cut a cheek out of a Hapus without thinking of the fruits of larger slices.

The Konkan coast emerged after Madagascar and India parted ways 88 million years ago, about where the Prince Edward Islands are situated today, between Antarctica and Africa. As the Indian plate moved northwards, the underlying rock got 'weathered' by tropical heat and battering rains. This is why the Western Ghats have porous, permeable laterite rock. The red colour comes from iron. This geography is central to the hallmark Alphonso taste.

Up close, the Konkan region is a strip of land, 45–55 kilometres wide, between the Arabian Sea and the hard basalt of the Sahyadri mountains. It has all the ingredients of a memorable road trip: smooth roads slaloming between the beach and the mountains, spectacular views that can launch social media careers, and a highly developed tourist infrastructure. I rode 340 kilometres southwards of Mumbai to Ganpatipule, a beach resort with a famous temple. Vivek Bhide heads the trust that manages the temple. His family has lived in Ganpatipule for 600 years; his is the nineteenth generation in situ. Since 1880, a member of the family has run a clinic in the town, making Bhide a third-generation doctor who knows everybody here. What makes the Alphonso special here, I asked. It is the soil of Konkan, the air and the water, he replied. 'Get any mango variety here, and you will get the best possible fruit.'

On a March morning, I was waiting for Bhide's son Yadnyesh. A horticulturist trained at the agriculture university in Dapoli, he has his father's calm manner. I had asked him to take me to the trees that produce their best Hapus. He arrived in a pickup truck and we drove to where their workers lived. About twenty workers got into the back of the truck; they are picked up and dropped home, in the manner of BPO (business process outsourcing) workers. We climbed the hills to arrive at a flatland of weathered laterite called 'kaatal' in Konkani. Their best Alphonso comes from trees planted in holes blasted into the laterite here by Yadnyesh's grandfather. The rocks are rich in iron and aluminium. Where the laterite isn't porous, the water stagnates to form aluminium oxide. It turns the soil acidic and is toxic to the plant in excess.

Kaatal is tricky territory. The sun heats up the rocks. The trees endure much greater heat stress here than in the valleys or lowlands. The excessive heat increases the risk of 'spongy tissue', the Alphonso's bane. Since the tableland is higher than the coast, stiff winds from the sea scorch the trees. The yield per tree is much lower here, decreasing precipitously after the tree is 20–22 years old. Besides, the rock underneath does not allow the roots to penetrate, stunting the trees. The abiotic stress on mango trees here is so immense that Yadnyesh began to look stressed just describing it. How do these trees respond to such extreme conditions?

By producing the best fruit! Because stress manipulates the flowering and fruiting of trees, good husbandry turns around the severe disadvantages. Short trees are easier to monitor and manage for diseases and pest attacks. They tend to produce fruit that's round and plump. Fruit from the taller trees in the valley tends to be oblong and lacks the brilliant colour. Being rich in minerals like iron, the laterite imparts the minerals that give the fruit its taste, flavour and that characteristic red hue on the skin. The sea breeze is good for plant growth and pollination. Being more exposed to the light, rain and winds, the trees here flower and fruit much earlier than in the valleys, sometimes by more than a month, ready to cash in on the early-season premium. 'The fruit from here has the best shape, colour, aroma, and taste. It also commands the best price,' said the horticulturist.

All this talk had fired my taste buds. Back home, Yadnyesh found me some small-sized early fruit that was ripe and ready. I had a few in the hotel and packed the rest. They were sweet and fragrant, better than anything I'd tasted in Mumbai. After I left Ganapatipule, I stopped by the beach and had some more. I met other growers on the way. By the time I reached Devgad, 125 kilometres away, the sun was dipping into the Arabian Sea.

My first stop in Devgad was a processing plant run by a cooperative of mango growers. I met its director Ajit Gogate, a well-known political leader here. 'We were sending our Hapus mangoes to Crawford Market in Mumbai by boat in the 1930s,' he said. While the large growers have their own marketing tie-ups, the cooperative provides infrastructure mostly to smaller cultivators, who make up the bulk of its 700-odd members. 'Many farmers want to move out of the Alphonso because the income is very unreliable. But they do not have too many options. This region is rain-fed; without irrigation, what will they grow!'

At their processing facility, I asked the manager to find me the best ripe mango he could identify. He walked about, scrutinizing the crates. 'People go crazy over large fruits. But the best taste comes from medium-sized ones,' he said. He leaned over, pulled out two mangoes from a crate and gave them to me. 'Try these two. You will not be disappointed,' he said. By the time I found a hotel and checked in, it was well past seven. I ordered dinner and before going into the shower, I washed and soaked the two mangoes in cold water. After the shower, I was too impatient to unpack and get out my pocket knife. It was my first real mango of the season, save the small ones Yadnyesh gave me that morning. The red skin was delicate and thin. I tore at it with my teeth. I remember the sensation like it was yesterday.

As the pulp spread into my mouth, an ecstatic cry burst out of me. The flesh was dark orange and sweeter on the outside, getting sour as I worked my way close to the stone. The Alphonso has a sugar-acid blend all its own. My mouth sensed the sweetness, then the mild subacid, then back to the sweet, and so on and so forth. One part of the Hapus charm is its savoury dialectics. I was soon at the end of this delightful mango, my tongue and teeth working the strands of fibre in the stone. The aroma had spread through the room like incense. I felt both relief and desperation. I addressed the second mango with the same gusto but it was slightly overripe and a little disappointing. (Any other time, even the second one would have made the top-drawer of my gustatory memory.)

In the morning, I went looking for Vidyadhar Joshi. I found him inside a compound adjoining his orchard, sorting and packing that morning's harvest. He was wearing a checked shirt and had on a cap. He had a farmer's unshaved, unkempt look, searching eyes and an unusually magnanimous manner. After graduating from high school in Devgad, he went to Sangli for his B.Sc. Then, he enrolled in a master's programme in 'food, drugs and cosmetics analysis' in VJTI College, Mumbai. All others in that batch went into corporate jobs but not Joshi.

'I could have joined a number of pharmaceutical firms. But farming is in my DNA, not pharma,' he said. He returned because he had no tolerance for a boss. His father grew paddy and had a small orchard given away on contract. He was glad to leave it all to his son. 'There's no dignity in farming because the generational relationships are disrupted. When I hold meetings of my farmers' forum, the young men come to drop their fathers and leave. They just aren't interested. But it works both ways: their fathers have not involved them in farming decisions. So they cannot imagine a farming life,' he said. He applied his formal education and training into rethinking farming, in bringing the best of new research to working the land.

At this point, he had to go to his bank for some work. We hopped on his motorcycle and continued the conversation with the wind gushing about our ears. Didn't he miss the urban life, I asked.

He said he loved Mumbai's fast-paced lifestyle and the numerous opportunities it offers. 'I know each and every lane of that city.' He also fell in love with Reshma, a young woman from Dadar. They got married and she moved to Joshi's village with him. How did she take to living in a rural area? Very well, as it turned out. She got involved in a local public library, went on to contest and win an election for its managing director's position. She then entered public life, becoming a member of the Zila Parishad.

Could I meet her? No, she had gone to Delhi to receive an award for the best Zila Parishad in the district. She is the only woman director of a taluka cooperative. 'Now she knows more about village affairs than me. But that is her life. I have no interest in politics and I do not campaign for her. She respects my need for distance from her public life,' he said. She is from the Maratha community and he's from a Brahmin family. 'Like politics, caste and religion do not bother me. I keep busy with my work. In a good year, my profit is above ₹15 lakh. In a bad year, it is a challenge to break even. This is why farmers must take the long view and save up in a good season instead of buying flashy cars.'

We returned to his compound and went for a walk in the orchard. I picked up an early cheepi mango on the floor; it had ripened on the branch and fallen. Joshi realized I had not tasted his mango. On his instruction, a worker sorted through the morning's pickings and sliced up four mangoes. One had spongy tissue but after cutting out the affected part its flesh tasted fine. The second mango hit the spot, it was fragrant, and the flesh was not too sweet, just so.

There is a delicate point between an unripe fruit and an overripe one. (Horticulturists use a mathematical model called the sigmoid curve to explain fruit ripening. The finest ripening occurs at the narrow top of this curve.) The Hapus's reputation rests on several similarly small margins. 'The best time to consume our mangoes is between 20 April and 20 May. The early produce in March, over which the market goes gaga, is not even close to the best.

And do not buy the nonsense the Vashi traders give you about the growers sending in poor produce. It is the traders who push the growers to harvest the fruit early, when it is only 60-70 per cent mature, because they want to capitalize on the early-season prices. My workers know I do not harvest a fruit that is not at least 80 per cent mature,' Joshi said.

What explains the quality of Devgad's mangoes, I asked. 'It is the fruit's chemical composition. That has to do with a unique combination of volatile compounds. The scientist who worked on this was K. H. Pujari at Dapoli. The only Hapus that can compare to ours is from Ratnagiri. Nothing else,' Joshi said. We spoke for another couple of hours; neither of us wanted to stop. I felt sad when we said goodbye. He more than made up for the numerous indifferent growers I had met.

After several unproductive days, salve came at the Fruit Research Station in Vengurle, 50 kilometres north of Goa, from where the Alphonso has spread along the Konkan coast. 'The mother plant was just called tree number 900. It was from 1870 and it died many years ago,' said Ajit Satelkar, an agriculture officer there. I had come here to meet Bharat R. Salvi, an authority on the mango. I found him multitasking in his office, dealing with phone call after phone call, visitor after visitor, colleague after colleague. When I finally got to speak to him, he was very matter-of-fact, saying things I could readily find in published literature. I headed back to Mumbai, where I had to see a mango grower recommended by food writer Vikram Doctor.

I met Noshirwan Mistry and his wife Meher in their apartment in Tata Blocks, Bandra West. An economics graduate, he wanted the life of a producer-farmer. In 2007, he bought 15 acres of orchards in Dapoli. 'My neighbours abuse PBZ indiscriminately because they want to make money quickly,' he said. Mistry had bought land to get out of the 'quick money' mentality. He went fully organic, which is a difficult shift. In 2008, his first year as a grower, he got some mangoes to sell. Then, for three years, there was no crop at all. From 2012, he started getting some fruits regularly. 'The biggest struggle is finding a market. But direct sales are profitable. The difference in income from selling to a trader and selling directly is about six times,' he said. Then comes the tough part. 'There's no regular income and the retail prices remain stagnant. But the expenses just keep piling up with increasing cost of cultivation.'

He knew nothing about farming or horticulture, having grown up in Mumbai. But he kept at it, learning from his neighbours, consulting scientific literature, teaching himself. Gradually, his production increased to about 60–80 mangoes per tree. 'If I were to use PBZ, it will jump to 600–800 fruits per tree,' he said. But that's not why he got into the game; the Mistrys have turned their life into an experiment, doing the opposite of existing practices. Now, his trees flower and fruit even when his neighbours have nothing, despite them using every kind of synthetic chemical. He sells directly through social media and has built a regular clientele in the Bandra area, especially in the Gujarati community that drives the city's high-profile mango business. Consumers like Vikram Doctor have brought attention to his mangoes. There is increasing appreciation for his commitment to ecology and consumer health. 'With a mad husband like him, I have to be prepared for a mad life,' said his wife, Meher.

Before I left, Mistry gave me a taste of the mango wine they prepare from the pulp of the produce that does not sell. It was a heavy and fermented variation of the Hapus taste. Combining mango and alcohol…well! As I headed to the airport, I had no idea the mango-alcohol on my breath will merge with the Alphonso smell in the departure lounge and the aircraft cabin.

The Girnar hill looms to the east of Lal Bagh, an orchard set up in the 1930s under the administration of the nawab of Junagadh. It now belongs to the Junagadh Agriculture University. Most of the trees here are quite young. The oldest tree is about forty years old. It was under this tree that I met horticulturist Viral Chovatia; his PhD is on the Kesar variety that is linked to Lal Bagh.

'Late in the nineteenth century, a man called Salebhai from Mangrol realized the fruit of a particular tree was superior in quality,' said Chovatia. He is said to have brought the fruit to Nawab Mahabat Khanji II. The sweet and flavourful flesh impressed the nawab and he named the mango 'Salebhai Ni Ambdi'. Its grafts were taken to other places in Junagadh. The variety took off during the reign of nawab Mahabat Khanji III (1920–1947). His royal garden superintendent A. S. K. Ayengar was the first to evaluate and describe the qualities of Salebhai Ni Ambdi. The deep saffron colour of the pulp led to the variety getting renamed Kesar in 1934. It was planted in royal orchards and its grafts travelled across the region, establishing the identity.

Was Mangrol the place to go for the best Kesar? No, said Chovatia. The best Kesar mango comes from Talala, a town 70 kilometres south of Junagadh, across the western fringe of the Gir National Park, the last remaining range of the Asiatic lion. Chovatia led a study to evaluate the quality of Kesar mango from nine locations in Saurashtra. The resultant paper said: 'The myth has been proven to be real.... From the conducted experiment over nine different locations, it can be concluded that Talala is more congenial for mango cv. Kesar or it can be truly said that mango orchards located at/near Talala region produce better quality fruits as compared to others.'[2] Among the reasons cited in the study, two bio-chemical factors stand out. One, the soil of Talala is loamy/silty in texture with a unique blend of nutrients. Two, Talala recorded the highest temperature and the lowest humidity of the nine sites studied.

In Talala the mango market is situated at Sat Haveli, Diwan Chowk. At the APMC office, I met Ramnikbhai Sanwaliya, who went through his ledger to report that 360,000 kg of Kesar mango had landed in the market the previous day from neighbouring orchards. It was the cleanest mandi of fresh fruits and vegetables I have seen. It also has the weirdest auction. Kapil Boricha of Bhole Baba Traders, the biggest trader in this market, was standing atop a push-cart. It moved in front of growers lined up with their produce against the wall. A train of traders followed the cart.

After the cart stopped in front of a grower's produce, Boricha assessed the quality and shouted out a baseline rate. The traders quoted their price, tugging at the legs of his white pajamas to place their bids, till the produce was sold. Then the cart was pushed to the next grower's boxes. The market has five other traders like him, authorized to buy the produce in bulk for a commission of 8 per cent. He then sorts and grades the produce, selling it forward according to demand. A trader said housing societies of Gujaratis in Mumbai collectively place direct orders for entire trucks of Kesar mangoes from here, because they seek the best quality.

As luck would have it, I did not get to taste the best mangoes here; the markets deal in unripe fruits. My last chance before getting back was along the way. I had a tip from horticulturist Rajendra Khimani in Ahmedabad. Deeply familiar with the state's agriculture, he had told me the best Kesar is grown along the banks of a small river called Hiran that skirts the Gir National Park and flows into the sea at Veraval. On the way back, I crossed this river at Sasan Gir and stopped near the village of Bhalchhel. A signboard gave directions to Anil Farms and Nursery. In the heat of May, the place had a cool, pleasant air, the nursery stacked with saplings. At the south end of the property is the Hiran River. It's been developed into a tourist resort with twenty-five cottages that can host 100 people. It was overrun by children shrieking in joy, with several overweight adults passionately discussing their next meal.

The chief gardener here was Rafique Baloch. He moved his lean body like an elite athlete, speaking softly with considered pauses. He said the mango crop was halved by a cold wave in February that disrupted flowering. 'The mangoes we have right now are from a subsequent flowering,' he said. He showed me a tree with five varieties grafted onto it. They use only organic compost and do not use any synthetic pesticides. The orchard has 600 trees, almost all of them Kesar; about 100 trees had been cut down recently. 'Neighbours here are cutting down their mango orchards. This orchard survives because of the nursery and tourism,' he said.

Just then, a tourist came to him and began to praise the mangoes, saying they were the best he had tasted anywhere. Baloch asked me if I'd like to try some. I told him I had come all this way just to taste mangoes from the banks of the Hiran River. He took me to the dining hall and instructed a waiter. He cut two ready Kesar mangoes and set them on the table. The first had a peculiar sweet smell that I recognized as another variety. 'We use Rajapuri rootstock for Kesar grafts. So many of our Kesar mangoes smell like Rajapuri,' he said.

The skin of the mango was thin, the stone was small, and the dark saffron flesh had little fibre. When I bit into it, the taste was as sweet and juicy as the best Kesar I had had before. The surprise lay in the heavy fragrance. A portly tourist at the adjoining table sat in front of a mound of discarded mango skin and stones. He guffawed and said to me: 'Aa chhe asal Kesar keri (This is the true Kesar mango).'

Keri, through much of North India, means a green, unripe, juvenile mango fit for pickling. In Gujarat, it is the term for a ripe mango; an unripe one is called, simply, kachi keri and the small fruit is called marva. The words aam/ambo are used for the mango tree. 'The family tree of a Gujarati family is often called kutumb-no ambo (the clan's mango tree),' Prakashbhai Shah, the editor of the journal *Nireekshak*, told me after I returned to Ahmedabad via an overnight bus. As we spoke, a faint smell wafted from my hands. It was the Kesar mango from the banks of the Hiran.

Noorjahan is a mythical mango in Madhya Pradesh. Each mango season, reports from Jhabua call it the largest of all mangoes. You cannot go through a summer in Indore without hearing or reading it. Yet I hadn't met anybody who had actually seen one. The district of Jhabua falls between many cracks. It's between the railway lines and highway networks of western and central India. Its rolling hills are far from the Narmada valley to the south, from the Malwa plateau to the east, and the industrial corridor of Gujarat to the west. Most of the residents hail from the Scheduled Tribes; the district scores low on parameters of literacy and economic development. Having lost a lot of its forests, the hills of Jhabua district look denuded.

Kathiwada, the town that produces the Noorjahan mango, is located in a small enclave inside Gujarat. It is 212 kilometres east of Indore, where lives a cousin of mine who regularly travels to Jhabua for work. One summer morning, I got into his car and we headed west. The town is in a valley surrounded by forested mountains. You can walk through the town in fifteen minutes. In its centre a large haveli built in 1940 called New Palace. It is surrounded by mango trees. Here we met Ishak Mansoori, about seventy, the gardener and caretaker. He is a salaried employee of the owners of the property, a family living in Mumbai who visit once in a few months. He buys the mango crop every year, exactly like his gardener father did. The orchard has some forty mango trees that produce ten kinds of mangoes, mostly the 'Madras Hapus'. All the grafted mangoes sell right here for ₹100 per kg. 'The nearest mandi is at Alirajpur and we never get good rates there,' he said. They did not have a single Noorjahan mango that year because insects ravaged the flowers.

He said he was trained in grafting trees and has visited the agricultural university in Dapoli, Maharashtra, a few years ago. But he got out of the nursery business. 'People buy grafts, especially of the Noorjahan. But when it does not take, they harass me,' he said.

Mangifera indica

'Noorjahan grafts just don't succeed outside.' He knew of only one plant taken from here that had been successful. I took the phone number and called Daya Chandra Patidar. He said he had asked Ishak to plant a Noorjahan seed in a pot three years ago, while he was getting transferred out to Indore. The sapling survived and they are waiting for it to become big enough to fruit. 'Kathiwada has the ideal conditions for horticulture. What grows there doesn't always survive outside,' Patidar said.

The town's famous orchard lies to its northwest near the Raaj Palace, the seat of the erstwhile rulers of Kathiwada. The property has been divided among members of the extended family. The palace is bequeathed to Kunwar Saheb Digvijay Singh, who lives in Mumbai and has converted it into a heritage hotel. The Noorjahan Mango Garden here has come to Shivraj Singh Jadhav. 'My grandfather Sajjan Singh planted this orchard,' he said. 'In those days there was no market for mangoes but he was a real shauqeen. He got grafts from Valsad in Gujarat.' And the name? 'My father named it Noorjahan. He had a great shauq for films and the actress who went by that name.' Today, the orchard has some 350 trees with thirty-five varieties. 'All the produce sells right here. During peak season, there's a traffic jam here on the weekends. People come from far and away to buy mangoes.'

The largest Noorjahan mango recorded here weighed 4.75 kilograms; it was 12 inches long with a diameter of 16 inches. Jadhav pulled out a photo album with pictures of the awards the Noorjahan mango has won. It had a photo of a newborn baby on one end of a weighing scale and a Noorjahan mango on the other; both weighed 3.5 kilograms. Yet the garden has only three trees of Noorjahan. 'Two of them were prepared by my father. The third came from the only successful graft prepared by a friend from the department of horticulture. He actually asked my wife to do the graft in the hope that her hand would bring luck. And it did,' he said. That tree was ten years old, and Jadhav took me to it. Four large mangoes dangled from it. I held the largest in my hand and it weighed about two kilograms; it wasn't mature. A random visitor had come recently, offering him ₹10,000 for the four fruits. Jadhav declined: 'I don't know what I will do with these four. Some lucky person will get them.' Each season, local officials buy the largest of his mangoes, offering him whatever price he quotes: 'Usually, each mango goes for ₹500–1,000.'

Jadhav said no other Noorjahan tree in the region had fruited that year. Several cousins have taken seedlings and grafts from here. None of them produce fruits like the trees in Kathiwada. Horticulturists have told him that this variety lacks enough sap in the grafts for them to be successful. For all its size and fame, it did not look like a pretty mango. Jadhav said the Noorjahan resembles two varieties: the Totapuri of south India and the Rajapuri of Gujarat; both can get large. He said it tastes like Rajapuri. It's a variety popular for pickles and pulp; I have never eaten a Rajapuri worth remembering. Size doesn't matter.

Mangoes come from western Maharashtra. The Vidarbha region in the state's east is known for oranges. But it wasn't always so. Vidarbha is in one of India's seven hotspots of mango diversity but is losing this wealth to commercial orange cultivation. 'The orange has no culture comparable to the mango here,' said Karuna Futane of Rawala village in Amravati district. At her wedding in 1981, the guests were served aam panna at Vinoba Bhave's Pavnar Ashram in Wardha.

She had been raised with the values of a productive rural life, of embracing hard labour. Her husband Vasant was deeply influenced by *The One-Straw Revolution*, an influential book of the 1970s on natural farming by Japanese farmer-philosopher Masanobu Fukuoka. In 1984, they returned to their village Rawala. Instead of continuing to stay in the village, they built a house on their farmland with their hands, using only local materials. It was completed in

2002. It's the house where I spent two days with the Futane family along with their four dogs, two cats, five kittens, and 30-odd cattle rescued from a drought in Gujarat. Their sons Vinay and Chinmay are committed farmers, running the family farms profitably while finding time for socio-environmental projects.

Near the house is an orchard that Vasant Futane has raised into his dreamland. He took me there. The ground was covered with weeds and shrubs and layers of leaf litter. 'This is what Fukuoka meant by zero tillage or do-nothing farming, isn't it!' he said. 'The leaf litter prevents evaporation and keeps the moisture inside the soil that is teeming with microbes. It also prevents soil erosion.' This approach is inconvenient to a lotus eater. A mighty insect had bitten me on the arm. Distracted by the blister, I walked straight into a spider web with silk strong enough to put steel to shame. The clayey black soil had caked the soles of my sandals. I was walking into a mound of leaf litter when he stopped me, asking me to mind my step. That's when I noticed the large red ants crawling over my feet. He laughed and said: 'That's mulch!' He was proud of the patch of forest he had created, turning an orchard into a or an amrai. Land deforested for agriculture was being restored to wilderness, one Fukuoka-inspired step after another.

The grove is full of counter-intuitive ideas, literally. For example, the periphery is lined with bamboo. Sometimes called the farmer's enemy, its roots travel horizontally, taking away nutrition from the topsoil. 'But it's a great windbreaker and prevents damage to flowers and fruits during storms. And, if all else fails, you still have bamboo with its innumerable uses,' he said. He has dug a trench between the mangoes and the bamboo, preventing its roots from spreading into mango territory. The trench doubles up as a rainwater harvesting channel.

The Futane family has been doing the opposite of what commercial mango growers do across India. But they have not let their mangoes go to seed. The trees are numbered. The operations are planned and thought through. Each experiment is observed and recorded in the manner of experimental scientists. 'If the leaf of a young sapling smells bitter like turmeric, we remove it because such trees turn out to be useless,' said Karuna Futane. She has observed changes over time in the taste, size and character of trees. 'One tree gave sour fruit year after year. Then, it changed and began giving delicious, sweet fruit. I was so happy I forgot that I'd told the boys to cut it down. Before I could stop them, they had chopped it up,' she said. 'Vinoba Bhave once told us a tree gives its best fruit in its middle age. It was an allegory for people.'

What's the return from all this investment in ecology and natural farming? Tree No. 179 had emerged from a seed planted about twenty years ago. It was laden with so many fruits as I have never seen on a single tree. The fruit was immature but Vasantbhau spotted a paadh (tree-ripened) fruit on the floor, picked it up and gave it to me. I ripped open the skin and tasted the flesh; it was sweet-sour and fragrant. At the far end of the orchard is Tree No. 177, also seed-grown; its fruit was fragrant but sour. 'Some people bought all its fruit at a cheap rate last year, saying they had to prepare aam ras for a gathering of about 200 people. They added sugar for taste and said it turned out remarkably well,' he said.

'Tree No. 208 is also a chance seedling. It produces our best mangoes that fetch the best price,' he said. Several grafts from this tree have been prepared and distributed. Yet another tree produced sour fruit used for pickling. 'One time, my wife kept its batch in the ripening enclosure, and forgot it for ten days. When she finally took it out, it had turned very sweet. It has a high acid content but it turns sweet if we ripen it for long,' he said.

They have grafted varieties, too, especially Ratna. He took me to his best Ratna tree. A large neem tree overshadowed it, because he was trying out neem-mango combinations. A Ratna tree planted about ten years ago brought new diseases. He accepts those, too. Some trees there had a heavy infestation of powdery mildew. He had left it to the insects and other creatures. He'd rather lose a mango crop than douse his trees with synthetic pesticides.

We went back inside the house for lunch. The highlight was kairi-bhaat, a pulao made with raw mangoes. A little later in the day, I heard shrieks and drums coming from the grove. 'It's the village kids chasing away the monkeys,' said Karuna Futane. The children are paid in instalments for books, stationery, and bicycles, in exchange for crop protection. I went back to the mango trees with her; she had a completely different connection with the orchard as compared to her husband. For him, it was an experimental farm to test his ideas. For her, it was an extension of her kitchen garden, a patch of wilderness that gave her resources and room for creativity.

Her connection with that patch was more sensory, more emotional. She took me to a tree grown from seed, right next to a graft of Chausa and Kali (a black-coloured variety from Gujarat). 'Its fruit is so fragrant I call it Surabhi. It has traits of both the adjoining varieties. Our regular customers demand it most after the Lakshman,' she said. Lakshman is her name for Tree No. 208, the prize of their collection. She gave it that name because its fruit looks like a sukumar (a delicate young man). The children from the village were screaming in joy, running around with drums, and warding away the monkeys. 'I wish somebody had paid me to create a noisy racket when I was a child,' she laughed. The kids also get some mangoes as reward. What was their favourite mango fruit, I asked them. They took me to a Kesar tree.

In the evening, I had aam ras with sewaiyan, along with jackfruit pulao and turmeric-chilli chutney. After dinner, Karuna Futane took me to her ripening enclosure. She spotted a perfectly ripe mango from Tree No. 179. I had that before going to sleep. It was juicy and fibrous, with hints of sweet-and-sour, very close to the Malwi mangoes of my childhood. It made me happy first, then left me wistful.

Chapter 17

North
Memory and Loss

'I'd do anything to taste that mango again.'
—ARUNA RAMAN

It began as a couple's effort to promote a school. Nishat Saiyada and Zahoor Siddiqui had set up the Salma Public School for girls in 1993 in Rataul village of Uttar Pradesh's Baghpat district, 30 kilometres northeast of Delhi. Passionate about quality education for girls in this rural setting, they committed their money, time, and ancestral property to it. Saiyada had retired from teaching in Delhi's government schools; her husband Siddiqui was a well-regarded professor who had taught history at University of Delhi colleges and had once headed its teachers' association. The couple had been involved in a range of socio-cultural efforts all their lives. But their fame today is down to their fine mangoes.

It was their cultural activism that brought them close to Sohail Hashmi many years ago. Zahoor had invited him to see the school; the mangoes from Noor Bagh, which is his ancestral orchard in Rataul, were just a side attraction. Soon, friends began accompanying him. When interest grew, Hashmi turned it into a paid mango trip, offering a glimpse of the mango in its setting. The proceeds from the trip go into the school. It was on such a trip that I first met Siddiqui briefly. Later, I met him at the Nehru Memorial Museum & Library, where he used to go to his research.

He combined his historical research with a public-minded earthiness, explaining complex things in an ordinary idiom.

What was his favourite mango, the one that he eagerly waited for? 'It is called "Haramzada" and its tree came from a chance seedling. It has a fine taste,' he said. Second on his list was a juicy mango called Sharbati Bigrain. Third came the Rataul. He listed another twenty, in the order that they appear in the season, describing their character and their background story. One name caught my attention: Gidwala. He chuckled and then explained: 'Vultures (gidh in Hindi) often rested on that tree and its mangoes were really juicy.' The name was inevitable!

What about Rataul, the variety that's brought the village fame? 'It was a chance seedling noticed in the early twentieth century in Anwar-ul-Haq's orchard Mubarak Bagh in Rataul, named after his grandfather Mubarak Ali,' he said. Siddiqui drew from a paper he had presented on this at the 2015 Indian History Congress in Malda, West Bengal; it was later published as a twenty-page booklet. The variety is special, not just for its fine fragrance but also for how well it was documented. Siddiqui gives the credit for that to a horticulturist called Sheikh Mohammed Afaq, the first to record details of the mother tree; he went on to catalogue 301 grafted mango varieties from that area in Urdu. Afaq took grafts from the tree and planted them in his nursery called Shorah-e-Afaq.

From this nursery, the graft travelled across the region and beyond. Sometime after India's partition in 1947, Anwar-ul-Haq's son Abrar-ul-Haq moved to Multan in Pakistan. He took along Rataul grafts, propagating them there in the name of Anwar Rataul. It became a popular variety. So much so that the government of Pakistan issued a stamp on that mango, and began to use it in diplomacy, gifting its baskets. It was such a basket that the Pakistani President Zia-ul-Haq sent to the Indian prime minister Indira Gandhi in 1981. Newspapers reported the mangoes were much appreciated. This riled up Rataul's mango growers. Afaq's son Jawed took a delegation from Rataul to meet the prime minister with accurate information and mango samples. She became an admirer of the Rataul.

The historian in Siddiqui did not like an inaccurate piece of information in a book four ICAR scientists wrote in 1957. It said the variety is also called Anwar Rataul and that its chance seedling was discovered in Afaq's orchard. The scientists were merely quoting Afaq's original catalogue; after he updated it in 1935, it featured 514 varieties. This revised catalogue did not describe the mother tree in Anwar-ul-Haq's orchard. Although many old orchards have succumbed to real estate development, Siddiqui pointed out the mother tree when I visited him in 2017. It stood there as a scraggy witness to the past.

The first time I tasted the fruits of its grafts was on Hashmi's mango trip. I'd already sampled several kinds of mangoes soaking inside pails of water. 'Where's the Rataul,' I asked Noor Bagh's gardener Nisar Ahmed Abbasi. He peered into the large vessels of water in which the fruits were soaking and pulled out one. It is a small mango with a great fragrance. 'Leave a basket of Rataul mangoes on a table. It will scent up the entire house for a week,' Abbasi said. The Rataul was finer than anything else there. Plump with paper-thin skin, it has very little fibre close to its small stone. In many fine mangoes of North India, the sweetness often overwhelms the subtler elements of taste, leaving it only in the flesh close to the fibres of the stone. Not so with the Rataul! Its sweet flesh retains the more ephemeral notes. They disappear as quickly as the small mango disappears into your mouth.

Where has the Rataul come from? 'It is said its parent is the Doodhia Gola of Amroha and Malihabad,' Siddiqui replied. But even Amroha appreciated the Rataul. In 1928, the founder of a nursery in Amroha, Mumtaz Syed Ali Mutaqqi, described it thus: 'It is wonderfully distinguished mango. No mango can be compared with it...it contains a unique taste (khas latafat) that certainly cannot be described in words...Its exceedingly well bearing quality compensates its small size. Those persons who prefer size over taste cannot appreciate it over the large size mangoes. Rataul can be proud of it.'

Siddiqui had a quiet pride in the mango, as also in his grandfather Hakimuddin Ahmed, an officer of the Bhopal state and the founder of Noor Bagh; he was known for procuring mango varieties for his orchard from places like Amroha and Malihabad. 'He picked up this habit from his father, a revenue official in the British government who was posted in Malihabad and brought several quality mangoes with him,' he said. It is from documents and letters collected from relations and neighbours of Rataul that he has gathered the kind of information often absent on other mango varieties. His paper is rich with the history and details of several orchards. Many are connected to Amroha.

The name Amroha, you will hear, is a combination of aam and rohu fish. The orchards here have produced quality mangoes for long, even if it is not associated with a marquee variety bearing its name. 'The best mango here comes from a particular type of soil,' said Mirza Mohammed Zubair, a Unani hakeem and a poet in Amroha with an eye for detail. 'We call it dumat avval mitti; it's the kind of hard soil that cannot be readily penetrated with the strike of a spade. The fruit is sweeter and more fragrant but the productivity is low.' The town lies close to the Ganga and has large stretches of alluvium. 'Trees in boorh (alluvial soils) give higher yields. The fruits are larger and more colourful. But their taste and fragrance isn't as good as those from the harder soils.'

Some of the more famous orchards here belong to Kaleemuddin Siddiqui's family. 'Our Dashehri tastes better than Malihabad's produce,' he said. 'In the summer of 2005, I took 526 mango varieties to Delhi for the annual mango festival. We bagged several prizes. A painter came and sat next to me and asked me to name the varieties. I could actually identify only twenty or so. The rest I just guessed,' he said. The town has produced several famous poets and writers and there is mention of Urdu manuscripts and unpublished books on the mango. But despite a richer heritage than Rataul, it hasn't made its mark on the popular mango calendar. The town awaits its Zahoor Siddiqui.

I wish each and every mango variety in India gets a historian as scrupulous as Siddiqui, a communicator as generous, too. 'Mangology is a science and it should be treated as such,' wrote the mangologist. Despite his repeated letters to the administration to look after the mother tree of the Rataul mango, the tree died recently. Siddiqui died in 2019. India's mango mania does not translate into caring for its biological and historical heritage or into valuing those who invest their labour, time and money into it. But the Rataul mango lives on. Each summer, if I know somebody travelling from there at the right time, I ask him to bring me some. The fine taste of the Rataul mango reminds me of Zahoor Siddiqui and his genial manner.

⁜

The flight to Lucknow had several connecting passengers from Riyadh. One of them sat next to me; we got talking. He was returning to his village near Lucknow after a few years working as a zardozi craftsman in Saudi Arabia, embroidering expensive fabrics. For years, newspaper reports have documented the exploitative labour racket that takes desperate and unsuspecting workers to the sweatshops of the Gulf countries.[1] As the flight took off, the Himalayas became visible, lording over the Gangetic plains created by its alluvium washed down by rivers. The air-conditioning was broken; the cabin matched the interminably hot June afternoon. As we landed in Lucknow, the pilot said the temperature outside was 47°C.

I headed straight to the Lucknow residence of Sheikh Insram Ali, a political leader who headed an association of mango growers.

There, we got into his car and headed for Malihabad. Along the southern margins of the town lies his ancestral house. Garden chairs were set on the manicured grass of his sprawling garden. 'The mango has lost its charm. When we were kids, nothing else excited us like mangoes,' he said. 'Children today have a wide range of choices—sweets, candies, chocolates, even imported fruits like the kiwi. Those items are very attractive to kids today in a way that the mango isn't.'

His other refrain was the disappointment with UP's regulations for the 'fruit belts' spread across thirty-nine blocks of fifteen districts. (All of them focus on the mango, apart from six blocks in two districts for the guava and two blocks in one district for gooseberry.) The government launched the scheme in 1985 to promote horticulture and protect orchards from polluting industries, like brick kilns, and from changes in land use. Since then, the scheme has undergone several amendments. 'The benefits that were to accrue to horticulture did not come through but land prices have skyrocketed, since this area is so close to Lucknow,' Ali said. This leaves orchard owners neither here nor there; they don't see much of a future in mango cultivation, but they cannot cut down the orchards for real estate development.

Several people came to meet Ali over two days, and he visited many others. Most were large landholders from families that have known each other over generations. Almost all conversations moved between past glory and a hopeless future. Ali kept mentioning ways for them to organize themselves, either to pressure the government to deliver on its promises or to create a better business environment for mango growers. He told me about his other efforts, including meet-and-greet events to bring entrepreneurs together. Nothing seemed to pique the interest of the mango growers here. They were dissatisfied with the state of mango affairs, but not enough to do anything about it.

Ali was a prominent figure. I had hoped that his introduction would give me access to the mango growers here. And yet no matter how many people I visited, no matter what line of conversation I tried, I had little to show for my efforts. The landholders I met did not like questions about their orchards. They wanted to talk about the shauqeen nawabs of yore and the tales of famous poets like Josh Malihabadi. Sometimes, these conversations happened in old havelis, crumbling from a lack of maintenance—havelis that have featured in films like *Junoon* (1978) and *Umrao Jaan* (1981). Each time I asked a landlord to walk into the orchard with me, I was told it was too hot, that we will do it another time.

In one conversation, a landlord spoke about his uncle, the extraordinary talent of his generation who was awarded Lucknow University's gold medal. He retreated to his ancestral lands in Malihabad and gave up on academics. 'Many years later, somebody asked him why he didn't make anything of his academic achievements. He angrily replied the university should've come looking for its gold medallist,' said the nephew. Everybody burst into laughter. Now I am as vulnerable to the devil-may-care charms of glorious underachievers as the next guy. But, at some point after the twelfth anecdote, it becomes difficult to appreciate such stories for the sake of social conformity! Given that many orchards here were established after independence to sidestep land ceiling laws, the atmosphere screamed of a stasis.

'It is a mindset thing,' said Vijay Singh, a major mango grower in Mall, a village 12 kilometres north of Malihabad. He has grown up around here and has friends from several communities. I asked him why Malihabad growers were so stuck in the past. He said it has something to do with a lot of them being Pathan: 'They are a proud lot, glory means so much to them. They also don't like to listen to others. They cannot organize themselves to try something afresh, something new, and something coordinated and collective. They spend so much of their time fighting and feuding among themselves.'

I decided to try the country's most famous mango nursery, situated on the highway going to Delhi, adjoining the railway station. In Abdullah Nursery stands one of India's most famous mango trees. It is about 100 years old and has about 300 mango varieties grafted on to it. Encircled by a paved walkway, visitors walk around and admire the tree, which has labels of varieties that have been grafted. Hardly a mango feature story in North India goes without mentioning this wonder tree and its creator Kaleemullah Khan, about eighty years old.

If the mango is nature's bling, Khan is its living advertisement. He is a skilled nursery man with the gift of graft. That he can tell a compelling story is a godsend for hurried reporters. Headlines often title him 'India's mango man', one that sits well alongside the Padma Shri he was awarded in 2008. At the end of the driveway going into the nursery was an awning above plastic panels with sixteen photographs—all sized 5 inches by 3 inches. They showed Khan with eminent people. His Toyota Etios car was parked right under; above the registration plate was an additional red plate saying 'Padma Shri. Govt. of India'. He is a living example of the places that the mango can get us!

His nursery is idyllic with many kinds of fruit trees—pineapples, guavas, pears, bananas, and jackfruits, among others. Birds gambolled about, their chirps lending melody to the rhythm of a chugging tubewell. People kept coming in, often sitting down to watch a game of chess. Some stayed back and began playing, giving the place an old-world vibe straight out of Munshi Premchand's 1924 short story 'Shatranj Ke Khiladi' that Satyajit Ray made into a feature film in 1977. Each move was accompanied by banter and each game had at least one spectator. In the numerous summer articles on the mango, the chess players make the backdrop to evocative photos of Khan and his mango stories. It bookmarks a nostalgic point or two!

Between games of chess, Khan moved about, handling work, giving instructions and receiving visitors. They streamed in, some known to him and others not so much. They had many questions about mangoes. Khan conducted rapid assessments before allocating time to each visitor, deciding on the level of engagement and 'specialness' to offer. If he thought a person was worth his time, charm appeared on tap. Journalists simply lap up his self-deprecating manner, his account of failing his seventh standard exam, being only marginally literate...yet revolutionizing horticulture with his innovations.

He routinely 'releases' new mango varieties named after celebrities—there are Nargis, Sachin, Aishwarya, Amitabh (Bachhan), Akhilesh (Yadav) and NaMo (Narendra Modi), each breeding new headlines season after season. None of these varieties ever make it anywhere. Both growers in and around Malihabad and scientists at the nearby Central Institute of Subtropical Horticulture are wary of him, some seemed envious. They said his stories and claims are vast exaggerations or that his new varieties are not actually new cultivars. Even among orchard owners, Khan's stories are considered old wives' tales. But his genial conduct and popularity ensure that he gets a wide berth; nobody holds him accountable. Everybody respects his extraordinary nursery skills.

He is a man of many parts who is part tourist attraction, part old-world savant, and part encyclopaedia of gardening practices. During mango season, his nursery is a fulsome course of people watching. Officials and local leaders ask him for saplings and advice about family matters; amateur gardeners from far and away enquire about tricky plant traits; and journalists just stream in. His nursery is a must-stop for visitors or even passers-by. They come to see his wonder tree. He obliges all—a sound bite here, a selfie there, a box of mangoes, some assurance and advice.

When he finally got around to speaking to me, he offered some insights. Down south, he said, they don't like their mangoes very sweet. 'The Lucknowa Safeda is called the Dinga in Delhi and Punjab,' he said. In 1919, Malihabad had 1,300 mango varieties, which were lost over time, he said. He was very disappointed in the CISH scientists, except one or two who had retired. Their knowledge is limited and useless, he said, claiming he can transform mango cultivation across the country if the scientists follow his ideas and listen to him.

I had already read and heard it. I asked him to detail his observations, on managing trees and flowering. Immediately, he decided I was not worth his time. It was the same reluctance I had encountered in others in Malihabad—you cannot ask questions or details of how people learned what they know. I spent a day with the extraordinary 'mango man' and came back with nothing to show for it. The frustration kept mounting. It took several days and serendipity to meet Mujeeb Khan. A formidable, no-nonsense man with his licensed pistol strapped by his side, he had been a successful chartered accountant in London. He and his family had had enough there; they wound up everything and returned to Lucknow.

I met with him at his ancestral haveli in Mohammadpur village. As I stepped into the beautiful haveli ravaged by time, I readied for another gust of the good old days! He handled several matters of property and payments with employees, met neighbours and relatives. After a while, he sat next to me and asked what I wanted to know. He was on point from the get-go. In the manner of engaged growers, he began listing the problems he faces. He was happy to field all kinds of queries, anticipating supplementary questions and sparing no detail. The mango orchard business here is sick, he said, it needs a reboot. The investment needed is not merely that of time and money but in acquiring the best of new knowledge. 'Times have changed. There is no hope in running the orchard business like it was run in the past,' he said. What about heritage, I asked. He answered in rhetoric: 'If our orchards cannot earn their keep, what will we do with our heritage!'

He sounded too good to be true. Could I see his orchards? Of course, he said, he would love to take me around if I was willing to wait for him to attend to some personal matters first. In the half hour that he was away, I walked about the haveli, admiring the workmanship of yore. He then escorted me to his orchards. His serious bearing gave way to child-like enthusiasm. The oppressive afternoon heat did not deter him, neither did the sweat drenching his shirt. He was pleased to be there, proud to show his trees to an interested stranger. He knew the variety growing on most trees; when he did not know, he asked his gardener who was following us. How was he dealing with the large, unwieldy trees typical of this area? 'I am working on rejuvenating old trees by pruning them down. Would you like to see those?' He led me to a patch where all old trees had been cut from the top, leaving no canopy and no high branches. I had seen this on experimental plots of research institutes. But this was in a private orchard of a man proud of his heritage.

He talked of what he had figured out from research material on horticulture. He was frustrated by the limited options for keeping away insect pests. In the entire conversation, he never once spoke of the glory of the good old days. He was occupied with present challenges; with how to improve the future; with how to make his orchard and his life more productive. In a region going to seed, he was a promising tree.

And then I ran into a more interesting way to tackle the past: a project called 'Custodians of Tropical Fruit Tree Genetic Resources in India'. It celebrated farmers who have conserved valuable but unknown mango cultivars grown from seeds. ICAR supported this project funded by the Global Environment Facility (GEF). It created a database and released its report in 2014. Its spearhead was Bhuwon Ratna Sthapit, a Nepalese agriculture scientist with a passion for farmers, their knowledge and the genetic wealth of traditional cultivars. Having trained as a plant breeder specializing in rice, he diversified into ecology and conservation. Bringing together global outfits like UNEP and the Biodiversity International (both project partners), Sthapit contributed to an extensive international network for conservation of farmers' plant varieties. He died in

2017; the obituaries noted his dedication to learning from farmers, to documenting the underrated achievements of ordinary growers.

The project database has several farmers from village Kasmandi Kalan, 8 kilometres east of Malihabad, just north of the CISH campus in Rehmankhedha. I met Nawab Hasan in the village. His father had purchased a plot of land with mango trees in the 1960s; the seller was a wealthy shauqeen family in Lucknow. Today, he has fifty-one grafted farmer's varieties, documented by CISH. His orchard also has trees more than a century old, grown from seeds. Hasan did not cut them down to grow market-friendly varieties. Why? 'This is our heritage. Such rare mangoes fill me with pride,' he said. He hoped they come to command the kind of attention that goes only into the commercial varieties.

The most celebrated mango from this region, the Dashehri, came from one such tree. It still stands at a height of 36 feet in Dashehri village of Kakori block, near a place called Andhe-Ki-Chowki, just beyond Lucknow's western fringe. CISH scientists say the tree is 150–200 years old. All 'pure' Dashehri mangoes come from clonal grafts of this tree. Today, the tree has a barbed wire fence around it and a watchman hangs around during mango season. There are numerous stories about how the tree with its prize fruit was guarded by the landowner, and about how grafts were stolen and propagated.

The mango stronghold here is shifting. 'Mall and its surrounding areas now produce more mangoes than Malihabad and Kakori put together,' said Vijay Singh. 'Our trees are younger. A large number of our trees bear fruit more regularly than the older, tired trees of Malihabad and Kakori,' he said. The erstwhile zamindars of Mall have extensive orchards, now managed by Bhupendra Singh, the eldest of the extant generation of the ruling family. I met him at Kothi Mall, an imposing residence in red and white. 'My great-grandfather Thakur Jangi Singh was the first here to create a commercial orchard in the early 1900s,' he said. 'I cut my teeth on mango cultivation at the age of twenty-two in 1967. I had to return from working in the tea gardens of Assam because my grandfather had died. My father gave me 200 acres of the poorest family land, waiving the revenue. That's how I started.'

'The soil, the weather, the wind, and the drainage along the slope...everything is ideal here for the mango. Everybody here had mangoes during the season, rich or poor. It's just that the aristocrats paid a premium for the best of the fine fruit,' Bhupendra Singh said. 'The Pathan musclemen created the fine mangoes to please the nawabs of Lucknow, who rewarded those who brought excellent mangoes. The refinement comes from the competition to please the nawab.' Now, there's no single person to please but a market of nameless, faceless consumers in faraway places.

In Malihabad, Insram Ali had introduced me to his cousin Hasnain Ali, who manages the family orchards. On a hot June morning, we went for a walk through their orchards. Harvesting of fruit had just begun. We walked past a giant khirni tree, and he spoke about the delightful taste of Amrapali, more expensive here than its parent Dashehri. 'The Unnavi Gola is an old variety with a pale red colour like the phalsa berry. It commands no market but has a remarkable taste,' Hasnain said. He took me to the tree and I could see the small fruit, not yet matured, changing colour. We walked past an irrigation pond. 'Similarly, the Taimur Lang has no market but has a delightful taste. These varieties are kept either for the family or gifted to important people, VIPs,' he said. The Dashehri goes straight to the market but these fine mangoes are kept close. 'Insram Bhai watches the craft varieties quite closely, tolerating no pilferage in their harvest,' he said.

What's his favourite mango, the one he waits for the most? 'It is a small mango called the Gilaas. You can pierce its skin with a needle and suck its fragrant juice,' Hasnain said. Could I see the tree? A fifteen-year-old boy called Ismail was called. He had gashes and bleeding bruises all over his limbs from plucking mangoes. He led us to the 100-year-old tree; it was short and its leaves were so small it looked like a litchi tree. Hasnain said the mother tree of the Gilaas is still standing near the grave of a saint called Doodhia Sharif. Their chief gardener Mohammed Nihaal was with us. I asked him to name his favourite mango. 'Nothing beats the Johari Safeda for me. It's different from the Lucknowi Safeda. If you eat even one, your burps remind you of its flavour for the rest of the day. It is pleasant and a great digestive,' said Nihaal.

We reached a part of the orchard where harvest was in full swing. The contractor who had bought the crop was Kishore Sharma, who runs a mango nursery close by. He was shouting instructions from the ground to five workers up 40 feet in the trees, perched at the nodes of branches, harvesting fruit with a stick into what looked like butterfly nets. The nets emptied into baskets dangling from the branches. When the baskets filled up, they were lowered to the ground, and the fruits were taken to a shelter for sorting and grading. 'A few years ago, a worker called Tata fell down from a high tree and broke both his arms and both legs,' said Kishore, explaining the nets.

Kishore spends two months harvesting and selling mangoes and two months selling mango saplings in his nursery. He explained his economics: 'Last year, I lost ₹2.5 lakh on mangoes because I wasn't able to recover the advance I had paid. This year, I hope to earn ₹4 lakh. But I cannot survive on the mango. I have a zardozi/chikankari operation at home that provides steady income.' I was puzzled because the mango business seems a far cry from embroidering fabrics. Nihaal stepped in: 'What will we do if the orchard doesn't yield? If we depend on the mango only, we will not be able to celebrate a single festival. There will be no new clothes or jewellery for the family.'

Hasnain noticed my surprise. He said the mango business was not the biggest component of Malihabad's economy: 'The total revenue from zardozi/chikankari is three to ten times the revenue from mangoes. As there are no weddings in the summer, it is off-season for the zardozi work, so the workers join the mango operations.' As soon as the mango season ends, work begins on readying fabrics for the post-monsoon wedding season in North India. The saris created here adorn brides in Delhi; the salwar-suits embroidered here embellish wedding dances of Punjab. It is a cottage industry—families handle operations inside the house, with all members contributing.

'Ours was the only pukka house in this locality forty years ago. Now, most houses stand on reinforced cement concrete. All of this is from zardozi income,' Hasnain said. We walked through the locality. House after house had some zardozi/chikankari operations. 'In this locality of 2,000 people, barely ten families have mango orchards. But each family has zardozi operations in the house,' he said. Some male workers go to the sweatshops of Jeddah in Saudi Arabia; the fabric is much in demand across the Arabian Peninsula. This explained the workers I'd met on the flight to Lucknow. The mango is not the mainstay of Malihabad's economy. But it is why the town is famous.

The Langda mango is grown across the northern plains, from West Bengal to Punjab. Several places produce outstanding fruit. Yet mango vendors in North India solicit customers by offering 'Banarasi Langda'. Like Devgad and Ratnagiri for the Alphonso. Other famed varieties are also named after the villages of their origin: Dashehri, Rataul, and Banganapalle. The Langda's reputation, however, stands on a city, not a village. Not just any city, one of the world's oldest living cities!

What makes the Banarasi Langda special? The skin, for one, is very thin. The most cherished variant is called the 'Doodhia Langda' for its whitish skin. 'Stick a knife in and the skin yields effortlessly, getting you into the flesh instantly. There wasn't a trace of fibre even on the stone,' said Shyamsunder Jaiswal, my favourite Banarasi, informally called 'Vidhayakji' because his father was a popular member of the state legislative assembly. He described something he had seen frequently during childhood: 'If a fruit ripened on a branch and fell down, it would burst on the ground, like a water-filled balloon.'

'Once you got past that delicate skin, there wasn't a trace of fibre in the flesh. The stone was so thin and small even an average-sized mango produced a large amount of flesh,' said Aruna Raman, about eighty, former geography lecturer at the Banaras Hindu University, born in a prominent city family and married into another. Sanjay Mehta, about 70, is a doctor and pathologist who grew up in a well-known Banarasi family. He said the skin of the best Doodhia Langda had small white spots, like pretty moles. 'Our family got its mangoes from a shopkeeper called Jawahar, who had his fruits and vegetables shop at Chaukhamba. It wasn't just a business to him, it was his great passion,' said Mehta. 'Jawahar wrapped each mango separately in old newspapers for ripening, checking their progress daily. He used to say mangoes ripened better in the paper from Hindi dailies rather than English newspapers,' he said, bursting into a Banarasi laugh.

When the mangoes came to the Mehta household, they were soaked in water with potassium permanganate to get rid of the germs. The cooled mangoes were eaten by the clan sitting around the buckets. 'As soon as you got past the svelte skin, the flesh was firm, even when the fruit was fully ripe. Then came that overwhelming fragrance,' said Mehta. 'If a batch of quality Langda mangoes arrived in the house in the 1960s, the entire house was filled with its fragrance. Even the neighbours found out.' Seeing as he was a man of science and a man of the senses, I urged Mehta to describe that smell. 'How do you describe a smell in words! I don't know how food writers do it but I cannot. I've never had that sensation from anything else.' Then he switched to physiology, saying olfactory inputs go to the amygdyla, a primitive part of our brain that also handles memory, emotion, and the lower/primal functions of the limbic system, not the intellect. 'I can only say that it was an incomparable, undiluted pleasure.'

None of the fine Langda mangoes I've tasted over the years came close to this description. I asked Vidhayakji to take me to the orchards that grow such mangoes. His family orchard got divided among family members a long time ago. Besides, that area is now part of the urban sprawl of Varanasi, with nary a mango tree spared. So we went to the place that produces some of the finest mangoes in the city now: the campus of the Banaras Hindu University (BHU). Along both sides of the road leading to the Central Office are tall mango trees. People say Madan Mohan Malaviya planted these trees. All things in BHU are attributed to its founder. We saw an average-sized fruit that was days from maturity. Its thin stone wasn't even 2 inches in size with no fibre. Nobody haggles over the price of this mango. The buyers feel lucky even when they know they've overpaid. This mango is one of the most popular gifts sent from Varanasi.

The most expensive mango on this campus comes from the orchards of the university's agriculture department. When Vidhayakji and I reached those orchards, there was nobody to show us around or answer any questions. I saw a Chausa tree laden with fruit and the promise of a good season. I lay down on a bench under it, soaking in the greenery, the chirping of birds, the smell of bountiful vegetation; the cool breeze had put a pleasant, idyllic filter on the setting.

I dragged up myself after a few minutes to go to Chiraigaon, a village north of the city known for horticulture. There, we met Sunil Kant, who has perhaps the most feted family-owned orchard here. He appeared in a white kurta-pajama, with an oversize Labrador retriever by his knees. He knew Vidhayakji since their childhood; they began catching up, as I walked about the orchard. The village has a reputation for its vegetables, guavas and bael fruit (*Aegle marmelos* or Bengal quince). Kant took me to a tree that produces bael fruit with skin so thin it cracks readily. He mentioned it had been brought to the orchard by scientist R. N. Singh. Surprised, I asked: 'The Ram Nath Singh of ICAR?' Kant was pleased to reply: 'Yes. Have you heard of him? He was from around here.' He was a celebrated horticulturist, a key creator of the varieties Amrapali and Mallika. 'There wasn't a big name in horticulture who did not visit our orchards. R. N. Singh used to come here each year to collect plant material and he brought fruit plants he had bred,' said Kant. 'I remember meeting him in the early 1970s when he was working on bael. He planted this variety called ND-49 that is now called by his name,' Kant said.

He took me to a mango tree. 'This variety is unique to here. It is called Gandhraj (king of fragrance). The fruit is large with some fibre, but its smell is unique,' he said, plucking a raw mango from a low branch. The sap squirted from the twig into the air. He asked me to smell it. It was fresh and exotic, like nothing I'd smelled before. This variety is favoured for mango jam and aam panna. Another variety grown here is called Kapooriya because it smells of camphor. Yet another is called Golwa for its round shape; it tastes like the Langda, has a small stone and is rich in flesh.

Do their mangoes draw a premium? 'Not any longer. A sewage treatment plant was built in Dinapur. Its discharge and sludge inevitably end up in the surrounding fields. Now our vegetables are identified by their poor appearance,' Kant said. 'The size of our mango fruit has declined, the fragrance is not the same. We have to face several new diseases, especially fungal ones.' When we left, Vidhayakji took me wandering. 'This area was known for handlooms and skilled weavers in almost every house. Now, it is difficult to find people who remember their traditional skill,' Jaiswal said. The village lies on the 70-kilometre Panchkosi Parikrama route around Varanasi. We stopped at the beautiful tank called Salarpur Kund, adorned with traditional sculpture, but falling to seed.

I still hadn't tasted the marquee variety. If the vendors of North India call customers with the promise of Banarasi Langda, what do vendors promise in Varanasi? 'During mango season, you can hear vendors shout: Langda Moti Jheel ka or Langda State Bank ka. That's the promise of quality,' said Vidhayakji. We reached the main branch of the State Bank of India on the banks of the Varuna River, which combines with the Assi River to give the name Varanasi. This colonial-era building was called Imperial Ganj before it was acquired by the bank. A few mango trees on its premises have survived the transfer. The guard on duty told us they have only three new trees of the Doodhia Langda. None of them had fruited that year.

Our next stop was the Azmatgarh Palace, popularly called the Moti Jheel Mahal. This white-green three-storey structure acquired recognition through the 2018 television series *Mirzapur*. At the palace, in a courtyard of the western wing, was Ashok Gupta, a little over seventy, sitting on a large diwan in a yogic Matsyasana pose. Now, he's the public face of the family that built the palace. The clan has its roots in Hissar, Haryana, but its prominence in the political economy of Bihar and UP goes back to managing Sher Shah Suri's treasury in the mid-sixteenth century. One branch of the family moved to Varanasi in the nineteenth century from Azmatgarh, about 125 kilometres northeast of here.

In 1903, the trader-banker Raja Motichandra commissioned the Azmatgarh Palace. This spot, about 5 kilometres west of the Ganga, was a sparsely populated area called Mahmoorganj. He had a beautiful pond dug up south of the palace, which came to be called Moti Jheel. (Like the Thandi Sadak, this name is common to cities across North India. There's a Moti Jheel each in Kanpur, Lucknow, Gwalior, Vrindavan, Muzaffarpur, Motihari, Murshidabad, and other places.) In the manner of rich families, they set up orchards along the pond with more than forty fruit varieties, including mangoes, citruses, and litchi. Whatever grew here had a cult status. In the lyrics of a popular composition of

Kajri, a light-classical musical form, a woman implores her man to fetch the henna from Moti Jheel!

Motichandra's younger brother Mangla Prasad was Ashok Gupta's grandfather. Vidhayakji's old-world charm eased Gupta and he began to talk. 'Mango here was sold in chhabisi or lots of twenty-six. I remember a chhabisi costing three rupees,' he said. About the Langda of Moti Jheel, he said its skin was so thin it was impossible to massage the flesh without the contents bursting out. 'It could only be eaten after cutting with a knife,' he said. He got misty-eyed remembering a favourite variety of seedless litchi. Where are those trees now? 'When the sewerage was laid in this locality the municipality let out the sewers into Moti Jheel to avoid waterlogging in adjacent areas. We protested but it fell on deaf ears.' The sewage destroyed the fine fruit trees. 'The mango is a sensitive tree; the Banarasi Langda is most sensitive. The trees didn't stand a chance. Over the years, small parcels of land were sold each time the family needed money. Those trees are gone,' he said, hanging his head. He was quiet and reticent. Vidhayakji signalled it was time to go.

Then, he called some people and found a phone number for his older sister, Aruna Raman, who had to quit her geography lecturership at BHU because their father could not stand her drawing a salary from a university for which the first donation had come from their family. 'Did Ashok tell you he used to call himself "mango manager" when we were children?' she asked. 'All the mangoes from our orchards were kept in a large room at the palace for careful ripening. Ashok used to keep the keys to that room. Anybody who wanted a mango had to get his permission first.' Ripe mangoes were gifted across the city to all manner of relations who waited through the year for the gift of Langda from Moti Jheel.

'Our mother forbade us to venture out in the summer heat. But we kids used to steal away in the afternoon, with salt and red chilli powder folded in pieces of paper. We went to the orchards and helped ourselves to raw mangoes. Along with several cousins, we spent entire days climbing trees like monkeys. There was no mischief we did not get up to in those orchards!' she recalled. 'During the mango season, we would go in the morning to bathe in the Ganga. We would return to find ripe Langda mangoes soaking in large vessels of water.' The Langda from the State Bank did not compare to their produce, she said: 'The fruit there was a little more sour than ours. It was larger in size, with a larger stone. Our fruit was unique—the skin so thin, the stone so small, the flesh so sweet and bountiful. I've never had that taste since my childhood! I'd do anything to taste that mango again.'

For about a decade, she has been searching for a tree that has a comparable Langda fruit. If she finds it, she will get a graft or a seed from which she can recreate a fruit of that quality. 'It's not just me. I know several others who are desperate to lay their hands on a graft or a seed of Moti Jheel's Langda. No luck till now,' she said. After talking to her, I stopped my search for the Banarasi Langda. If someone with her rootedness, connections, and history cannot find it...

'In France's Champagne region, they do not allow any changes to the character of the soil or the ecology. That's why their produce is consistently of high quality. That's how you value your biological wealth. We have just squandered ours. How could we let sewage take away such wealth!' said Mehta. 'I remember seeing radishes from Jaunpur in my childhood. They were longer than six feet, the diameter was one foot! Now one cannot find it even in Jaunpur. People there have forgotten about it. The soil's character has changed dramatically.' The Langda consumed in Varanasi now comes from Bihar. 'It tastes okay, it is sweet. But the skin is thick, the stone is large, and the fragrance...well, it is not there!'

The orchards of rich patrons outside the city have been gobbled up by Varanasi's urban sprawl. The only hope is somebody familiar with the taste and quality, like Aruna Raman, will find a tree that survived urban development. Till then, we must make do with the recollections of those who tasted that wonder.

⁂

I do not desire any table-type variety like the Chausa of UP. It is one of the big three commercial varieties of North India. Dashehri appears early to mid-June; Langda follows a fortnight later; mid-July is when Chausa hits the markets. Even though it comes later, the Chausa is more expensive. Fine Chausa is the costliest commercial variety of North India. One reason for that is its low productivity, another is its delicate nature, yet another is that high-quality Chausa is not easy to find in the markets. It does not travel as well as Langda or Dashehri. It transitions from unripe to overripe really quickly. There is barely a half-day window to enjoy it at its ripe peak.

I have asked many a gardener about their favourite mango. For many, it is Chausa. Why? The appearance is the first reason. Unlike the round Langda and the oblong Dashehri, the Chausa is curvaceous, it has attitude. It remains green after ripening, except the produce of a few places; the large-sized Chausa from Pakistan's Punjab that comes to Delhi's markets in August is a bright yellow! The Chausa's skin is not the thinnest but is delicate and gets injured easily. Juice runs out when you cut a cheek. The flesh is a bright yellow and extraordinarily sweet; there is a little fibre in the flesh, especially close to the stone. Its aroma has notes of chalk, it is fragrant. It is a mango I like to eat alone at the table, involved, not distracted by anything.

Naturally, I wanted to find its origin, to eat it where it is at its best. In books on horticulture it is listed as a variety from around Lucknow. But nobody in Lucknow or Malihabad could say exactly where. I found references that connected the variety to a village in Bihar named Chausa. It lies in Buxar district right at the UP-Bihar boundary, at the confluence of the Ganga and the

Karamnasha rivers. This is where Sher Shah Suri's army battled and defeated the Mughal forces in 1539, forcing Humayun to flee to Agra, then Lahore, then Kabul, and then Iran. In Lucknow, I'd heard old people mention that the Chausa mango was irrigated by the blood of the soldiers slain in battle.

So I landed up in Chausa one summer morning. It was not just any day! It was 26 June, the date of the battle in 1539. I visited the mound where archaeological digs have yielded ancient pieces of sculpture. I found the village sarpanch and met several people along with him. Nobody had heard of the Chausa mango! They knew of and consumed Bombay Green and Langda and some others but not the Chausa. I stopped in several villages in that area, right up to the district headquarters in Buxar. Nothing, not a clue! Chausa is actually a generic name in North India, often used for sucking-type mangoes; it draws from choosna aam, Hindi for sucking-type mango. But even the juicy mangoes sucked in this region were not called Chausa. I left disappointed, feeling sympathy for Humayun.

I found a clue a couple of years later in a 2005 book by two CISH scientists. Titled *A Tryst with Mango*, it compiles material published in the popular press on the fruit. (Yes, even a compilation of clippings on the mango finds a publisher!) It reproduced an article from a Hindi paper on 13 June 1981, under a headline borrowed from Mirza Ghalib's comparison of the mango with a sealed glass of honey. It mentioned the legend of a poor widow from a village in Hardoi district northwest of Lucknow: a crow dropped a mango seed in front of her house, from which grew a tree. Its fruits were so good that the Awadh nawab's court gave her ₹100 each year for its fruit, providing the widow subsistence. The mango got its name from the name of the village: Cheensa. Hence 'Chaunsa', with a nasal inflection; not Chausa from choosna aam!

I asked around. Nobody had heard of this village. Ramesh Dixit, a retired university professor in Lucknow, has family in Sandila. I asked him to help. He put me in touch with his young nephew,

Abhishek, a lawyer in Lucknow who lives in Sandila. Abhishek began making enquiries; he knows a lot of people there. It took him weeks to hear about a village with that name. The next opportunity I got, I reached Sandila, had lunch with Abhishek at their ancestral house, and set off for village Cheensa.

It took us more than two hours to cover 50 kilometres on terrible roads. Abhishek stopped along the way to make enquiries. Most people had not heard this name. We reached Victoriaganj, its past name was Chaunsa. Its informal name is Victoriaganj-Chaunsa. We met several old people in the village and nobody was aware of the legend of the Chaunsa mango. In fact, the village does not have any orchards any more. Ashraf Ali, about fifty, said he had seen large Chaunsa trees in his childhood. He had heard that the Chaunsa mango originated in their village. But he could not say anything for sure. Another man recalled people saying the village was settled by warriors coming from the Battle of Chausa!

This village lay on a Mughal-era road connecting Sitapur with Madhoganj, surrounded by dense forests. Chaunsa was one of fourteen villages under the zamindari of Nawab Aijaz Rasool of Sandila. His wife Qudsia Begum was the only Muslim woman elected to the Constituent Assembly in 1946. A renowned parliamentarian, she championed the abolition of zamindari. After she gave up her claim to the fourteen villages, the land was redistributed. The old mango trees were cut down by new cultivators. I'm still waiting for a lead that can take me to the origin of my favourite table-type mango.

Chapter 18

East
Decadent and Digestive

*'This mango does not leave Murshidabad.
You cannot taste it anywhere else in the universe.'*

—RAMAKRISHNA DAS

Two galleries flanked a foyer on the first floor of Kolkata's ITC Royal Bengal hotel. In the centre of the foyer stood a round table with a large mango branch fastened with a wrought-iron frame. Decorated like a Christmas tree, it had a few mango varieties displayed under it like gifts. One gallery, lined with potted mango plants, was approached by a staircase coming up. This had been turned into a makeshift ramp for fashion models, each attired in a dress inspired by a variety of craft mangoes from Murshidabad.

As each model sashayed up the steps, the designer described on the microphone the features of the mango variety that had inspired the dress. The gallery on the other side of the foyer was lined with tables, under which soaked mangoes in buckets of water. Behind them stood women, mostly elderly and traditionally dressed in saris. They were there to perform the fine art of cutting mangoes.

Yet it was the men who caught the eye. Several wore dome-shaped hats straight out of photos of early nineteenth-century Bengal, the brim projecting under it, saucer-like. Only the Sheherwali Jains use such hats ceremonially now. A few wore fine, long-length kurtas over cream-coloured dhotis, tied with intricate pleats in front that bobbed and weaved when they walked. The annual mango festival of the Murshidabad Heritage Development Society (MHDS), held on World Environment Day 2022, had squeezed together varied eras and sensibilities.

The models walked, the special guests made speeches, and the photographers took photos. I stood by my one priority: the tables with mangoes. With wine and appetizers being served, the mango tables were not too crowded. 'It was very difficult to get the mangoes because a severe dry spell destroyed this season's produce,' said Sandip Nowlakha, MHDS vice-president. Only a few varieties were available here. The women behind the counters had begun to peel and dice the mangoes. The first one I tried was the Bimli, apparently named after a favourite servant of an unknown patron; its appearance and taste were close to the Pairi and the Alphonso. Then came Rani Pasand and Sarenga; I could taste a connection with the fine mangoes I'd tasted in Bihar.

Moving to the table that displayed Kohitoor, I watched an elderly woman go through the bucket to get a nice specimen. 'They have not ripened properly,' she apologized. I asked her if I could see one. She held out a large-sized mango that looked a little like the Imam Pasand. Only its skin was dark green and thick, not a sign of refinement. I asked her if I could touch it. She handed it over with an obliging smile. As I moved it around, I felt the skin give way to the gentle pressure of my fingers. She reiterated the story about Kohitoor mangoes being kept in cotton wool.

She picked out the ripest one she could find and expertly removed the skin, rotating the mango with her left hand as the right held the knife. 'Certainly not the best,' she said, putting the diced flesh into my plate. It was a little sour, the taste hinting at Langda mangoes. What we identify usually depends on what we know! Then I tasted a mango called Champa. It had a uniquely refreshing taste, its smell evoking the tree it is named after. As people began to turn towards the mangoes, the Champa table was more crowded than any other.

At each table, I asked for a stone with some flesh on it; the fibres around the stone have their own tastes and smells. As I worked my way through the fibres of a Bimli mango stone, an elderly lady walked up to me. She said she had been watching me for a while, thrilled to see someone enjoying his mango intently. The women cutting and serving the mangoes were also happy to see me return to their counters, selecting the finest specimens at hand, smiling in the manner of indulgent matriarchs. After the formal events, it turned into an all-out mango-themed food party with desserts, such as mango kheer, and mango tiramisu.... Yet for all their experimental attraction, they left me unsatisfied, taking away from the taste of the fine mangoes. I knew I had to go to Murshidabad. But I knew nobody there.

So I first visited a few Sheherwalis in Kolkata. Sanjay Doogar comes from a family of bankers hailing from Kishangarh in Rajasthan. After the Jagat Seths established themselves in Murshidabad, they invited five Oswal Jain clans from Rajasthan to join the booming business. They were: Doogar, Dudhoria, Nowlakha, Nahar and Kothari; other families followed. After the Doogars moved to Murshidabad, they got zamindari in Bihar and Bengal. 'The custom of setting up orchards with fine mangoes came here from Darbhanga. There were elaborate norms of gifting these fine mangoes to other families and important people. Each basket had a label of the variety,' Doogar said. There were elaborate feasts and parties during the mango season, with scores of fine mango varieties served to the guests.

Murshidabad lost its sheen in British times. The Doogars moved to Kolkata in 1902. With their banking business in decline, the family came to rely heavily on revenue from zamindari. The 1947 Partition and the abolition of zamindari took away most of their lands and income. The family's great wealth declined as newer migrants—Agarwals and Maheshwaris, among others—took over business prospects in Kolkata. 'It filled us with great pride because we were like kings there. Here in Kolkata, we were just like everybody else,' Doogar said, remembering going to Murshidabad once a year for a major festival, like Holi. Soon, the family had to send money to Murshidabad for the upkeep of properties there, for the salaries of the caretakers.

'The palatial houses were sold. Two were donated for setting up colleges,' Doogar said. The building donated to set up the Shripat Singh College had separate rooms for ripening mangoes during the season. They had a mango manager responsible for the orchards, especially for carefully ripening the mangoes in controlled conditions. The family still owns one of Murshidabad's biggest tourist attractions, the Kathgola Palace, and its orchards spread over 45 acres. 'It was built to entertain nawabs and British administrators. But our ancestor had a temple built there for his mother. It was not used as a pleasure garden thereafter,' he said.

The move to Kolkata did not take away the mango rituals. The family bungalow at Shambhunath Pandit Street had rooms that were filled with mangoes brought from Murshidabad during the season. He remembers sitting in a circle along with several cousins as their mothers sat in the centre, carefully cutting the mangoes, and giving them pieces on wooden forks; the change in clothes was to prevent the mango juice spoiling their fine garments. 'We had mangoes four times a day during the season. I remember feeling sickened by overindulgence. We've tasted such fine mangoes that it is impossible to eat mangoes outside,' Doogar said, sitting in his apartment on the thirty-seventh floor of a housing complex in southeast Kolkata, overlooking expansive wetlands.

'Zamindari created the surplus income that could support lavish lifestyles,' said Bikram Dugar, a relative of Sanjay, who was dressed in authentic Sheherwali apparel at the mango festival. His ancestors had zamindari in Bihar's Saharsa, Purnea, and Forbesganj. He remembers his grandfather wore itr (oil-based perfume) on his shoes and socks. Rosewater and saffron were used for all kinds of things, and there was a servant whose main job was to fold the intricate pleats of the dhotis worn by his grandfather.

How could the mango escape such a commitment to refinement! 'The truly discerning had only the really good parts of the mango. If it was cut from the top, the bottom was given away to children. Even a rare mango like the Kohitoor!' The fine fibreless mangoes of Murshidabad melt in the mouth, he said: 'If it ripens in the morning, it will turn overripe by the evening. To really enjoy it, you must have the finest mangoes in that brief window,' he said, echoing stories about nawabs in Murshidabad waking up in the middle of the night to have a fine mango at its short-lived peak.

A passionate man and a history buff, he has preserved correspondence from royalty in Britain thanking his ancestors for sending them mangoes. There's an 1898 letter from the Balmoral Castle, a 1901 letter from 10, Downing Street, and a 1912 letter from the Buckingham Palace. But these were the better-travelling Alphonso mangoes from Bombay. Murshidabad's delicate mangoes do not last long enough for such a long voyage. Dugar's family moved out of Murshidabad in 1947, when he was one year old. 'We had all kinds of mangoes here in Kolkata while we were growing up. The Murshidabad mangoes lasted only fifteen to twenty days,' he said. He has noticed a decline in the quality of those mangoes over the years: 'The flavours are disappearing. The Anaras mango I tasted during my childhood is a thing of the past.'

While the Oswal Jains are strict vegetarians from Rajasthan, the Sheherwali food changed dramatically in Murshidabad. 'Our food has Bengali influence, although we have remained strictly vegetarian. This deltaic region has so many vegetables and plant-based foods that are not found in dry Rajasthan,' he said. The Sheherwali interests in things refined is alive in his daughter Sujata, a lifestyle journalist who writes on fine foods and fabrics. I met the father-daughter duo at their heritage bungalow in Kolkata's Ballygunge locality. The Murshidabad mangoes were not ready but she gave me an extraordinarily sweet Himsagar and a sweet-sour Golabkhas. She had some typical Sheherwali dishes ready at home and insisted that I try them; I did not require much persuasion.

The first item was a small kachori, fried so carefully that the jacket was crisp but the filling of shredded cucumber, seasoned with yoghurt and spices, was soft. 'The kheere-ki-kachori is typical of Sheherwali food. You will not find it anywhere else,' she said. The main course was a fine pulao, served to the accompaniment of yoghurt raita, mango-mint chutney, mango launji (syrup of raw mangoes, thickened and sweetened) and Bengali chana (chickpea). Each item had a distinct flavour and taste. The dessert was the real surprise: raw mango kheer! Raw mangoes are shredded and boiled lightly to remove the sourness. After getting strained through a muslin cloth, it is added to milk in the way others add rice. It is an example of how far the Sheherwali go to get their mango kick! A booklet titled *Things You Can Do with Murshidabad Mangoes* by MHDS president Pradip Chopra lists thirty-five mango-based recipes. Some Kolkata hotels offer special Sheherwali cuisines in their top-end restaurants.

A trip to Murshidabad was inevitable. I took a commuter train and alighted at Azimganj on the western bank of the Bhagirathi/Hooghly. A ferry took me to the other side, Jiaganj. There still isn't a bridge here. My first stop was the Kathgola Palace, a neoclassical structure built in 1870 over existing structures. The name connotes an older timberyard stood here once. The palace was under repairs when Sanjay Dey, the manager of the estate, took me around. It has a pond in front; a beautiful stepwell and a zoological enclosure to its east; to its west is a Jain temple that is a tourist attraction. We walked past statues of Sheherwali men on horseback and ruins of a much older guesthouse, made with bricks.

We could not find the contractor who buys the mango crops in the orchards here. Instead we were joined by Tapas Chandra Sarkar, the horticulture field consultant of the Jiaganj local administration.

He was full of the kind of horticultural stats officials have at their tips, including the land area under the mango, major pests, and marketing problems... The kind I did not seek! I asked if any owner had a well-kept orchard. He couldn't think of any. But he did know of a father-son duo managing the same orchard for a long time in Jiaganj's Dubrakhali hamlet.

We left for the orchard immediately. Harvesting operations were in full swing, all the produce landing in an old rest house in the middle of orchard. 'My father has been taking this 65-acre orchard on contract for about fifty years. We visit the orchard 365 days a year,' said Aseem Ghosh, sitting next to his father Gopal Ghosh, who looked remarkably healthy for his advanced age. The land belonged to a Kolkata resident named Baradarshi Prasad Bhattacharya. He died a while ago and his descendants had moved to the US even earlier. 'Some of the trees here were planted more than 100 years ago. Most are about seventy years old,' he said.

For about half an hour, he answered each of my questions about horticulture with specifics, in detail. He was happy that somebody had questions about a subject he knew well. He listed each agrochemical used, from pesticides to fertilizers, and the conditions requiring their use. He could identify seed-grown trees (called aanti in Bengali). The variety with most trees in the orchard was Rani Pasand, followed by Langda, Bombay Green and Himsagar (also called Sadola or Sadakat or Sadakatla or Khirsa Pati, for its sweet, kheer-like taste).

Could I please get to taste a quality Kohitoor or Champa, I asked him. 'You are one day late. We sent the last of those mangoes yesterday to the market,' he said. He must've noticed my disappointment, for he called one of his workers, asking him to search carefully inside a room. A few minutes later, the worker returned with one large, green mango. 'You're lucky,' said Ghosh. I asked his father to assess if this was a quality mango and whether it was ripe. He said it was good but not at its peak. I held it up:

it looked unappetizing, like an ugly, thick-skinned relation of the Imam Pasand. And yet the skin caved in from the slight pressure of my fingers. 'It is delicate. That's why all the stories of keeping it in cotton wool,' said the father. 'One tree gives only fifteen to twenty fruits every alternate year. Only the true shauqeen plant Kohitoor trees, and even they plant only a few,' said the son.

The mango was taken away, washed, sliced, and presented to me on a plate. The bright yellow flesh was firm, with some fibre; the fragrance was enticing. As I bit into it, the flesh disintegrated into pleasure and juice, yellow streaks running down my arms. It was really sweet and reminded me of Chaunsa. The father-son duo began asking me to try other mangoes. 'Which variety is at its peak right now?' I asked. Ghosh asked the worker to fetch Rani Pasand mangoes harvested from a certain part of the orchard.

Several choice specimens of Rani Pasand were cut and put on a large plate in front of me. The father insisted that I try a particular one first. The flesh felt crisp and it was not as sweet as the Kohitoor. The stone was small, with little fibre. Its taste was pleasantly familiar. After the third Rani Pasand, I realized it reminded me of fine Dashehri mangoes, just that the notes here were more delicate, almost fleeting, making me have yet another. One fruit had jelly seed, characteristic of the Dashehri. After that, they served me two Bhavani mangoes. The flesh was even firmer in texture and was as sweet as it was sour. 'Don't judge it harshly. It is not fully ripe yet,' said Ghosh. I left the orchard grateful for that remarkable Kohitoor.

The only old orchard to survive the ravages of time and urban development in Murshidabad is the Raees Mirza Bagh near Topkhana. Little is known about the man who created it. What's known is that his descendants sold off the orchard to the Roman Catholic Diocese of Krishnanagar. The trees are really old, and the floor was overrun with weeds and insects on a hot and humid day. We found the watchman hired by the contractor who had bought the crop. He showed me inside an old and crumbling rest house, where the harvested unripe fruit was stored. Only the Chaunsa and a round-shaped variety called the Laduwa were ripe. He was happy to let me taste one of each. The Chaunsa was bland and unremarkable, the thick-skinned Laduwa was fibrous and sweet-sour.

He had grown up in Murshidabad and knew his mangoes. Without any provocation, he began to describe a variety called the Kalapahar: 'It is sweet-and-sour and very juicy. My father used to have it with sattu (flour of baked chickpea or barley). He said it was a full meal.' It wasn't available to taste. His employer Ananda Sarkar, the biggest trader of fresh fruits and vegetables in Murshidabad, had been buying the crop from the church for fifteen years. Right about that time, an employee of the church showed up and asked me to come back with formal permission from his bosses.

My next stop was the Subhash Chandra Bose Centenary College in Lalbagh. The geography department here is geotagging all the mango orchards of Murshidabad. 'We are looking for a way to link the mango season with the tourism business,' said Debabrata Mandal, assistant professor of geography. 'The tourism season is restricted to the winter just now. If we can extend it to the summer, it will be good for Murshidabad's economy.' Faruque Abdullah, a history lecturer at the college, is conducting interviews and collecting material for a book he is writing on mangoes in Murshidabad's history. I heard of a man called Moharram Sheikh who still has a famous Kohitoor tree. I stopped by that location. The tree stood close to his gate; it had eight gorgeous Kohitoor mangoes.

Before taking my train back to Kolkata, I stopped by the Kathgola Palace. This time I met Ramakrishna Das, who'd been buying the mango crop there for four years. Since the palace is now a museum, a lot of tourists come by; they end up buying mangoes from Das. He described the varieties grown in that orchard, saying it had two Kohitoor trees and four of the Champa. I told him I had tasted Champa in Kolkata. He got annoyed: 'This mango does not leave Murshidabad. You cannot taste it anywhere else in the universe.' He went to a shack where his harvest was stored. He sifted through it and picked out a small mango, washed it with bottled water, handed it to me. 'That is a real Champa. Have you seen anything like it?' It was truly pretty, a glowing green-yellow fruit, shaped not by natural selection but by the reifying hand of a divine sculptor.

Missing my pocket knife, I set my teeth on the task of stripping the skin off the bottom end. Soon, it looked like a little lotus. I took a small bite of the firm, pale-yellow flesh. The sweetness was subtle, fleeting. The fragrance of the Champa flower was refreshing like nothing I had smelled before. I held it in front of me and spent a few moments admiring its appearance and smell. Eventually, I ate it with slow remorse, thinking that people here have this each summer! Regretting that I have never tasted this before! Feeling grateful that I did get to taste it, finally.

∫

Bihar's 'mango man' had to go all the way to Maharashtra in the mid-1980s to realize the value of the mangoes of Bhagalpur. Ashok Kumar Chaudhary had completed a bachelor's degree in law and a batchmate had asked him to join his legal practice in Delhi. But his parents and family needed him to stay in Bhagalpur at that time. He abandoned his legal ambitions. After a while, in 1985, he enrolled on a physical education course in Amravati. 'I did not get a room in the hostel straight away. So I rented a room in the city for ₹600 per month,' he said. The landlord was a ninety-year-old freedom fighter named M. S. Dhawade. It was summertime and Chaudhary had taken a bag of mangoes with him. 'I gifted him two mangoes from my bag. He was shocked by the fine taste. He said he had travelled across the country and had tasted mangoes from all over, but had never tasted anything so fine,' he recalled.

Three months later, he got a room in the hostel. After vacating his rented room, he went to settle the rent of ₹1,800. 'Dhawade refused to accept any money from me. When I tried to pay his son, he forbade him from taking my money. He said the two mangoes I gave him were worth more than ₹18 lakh. I could live in his house forever for free! That he could not imagine dying without tasting those two mangoes,' said Chaudhary. What mangoes were those? 'One Zardalu and one Malda,' he replied. It was his moment of reckoning: 'I realized there's a value in our mangoes that we do not acknowledge. I began reading up on mangoes. I made friends with horticulture scientists.'

Fortunately for him, the Bihar Agricultural University is located at Sabour, east of Bhagalpur city; it is a hallowed campus with a rich history. 'They had some outstanding scientists there. There was a Dr Sengupta and U. S. Jaiswal. They taught me the basics of plant reproduction and the difference between mango varieties,' said Chaudhary. He comes from a caste known for owning land but not for horticulture: 'Nobody in my clan was in the nursery business. Most of the skilful gardeners here are Bengali. But in 1992 I set up a nursery on my family land. Within three years I was selling mango saplings.' His business expanded quietly. Many farmers used his knowledge to grow quality mango trees that fetched good prices. Then a sudden turn of events made him famous.

Bihar's Chief Minister Nitish Kumar had announced in 2007 that he was going to send gifts of the state's best mangoes to dignitaries in Delhi. Bhagalpur's Zardalu was at the top of that list. The district administration and the horticulture department took commitments from mango growers to supply quality fruit. Like so many delicate varieties, quality Zardalu mangoes appear for a fortnight from late May to early June. As it often happens in the bureaucracy, by the time the wheels of procurement set in motion, it was too late. The administrators of the Sabour college were roped in to find quality Zardalu mangoes that the chief minister could gift with pride. Chaudhary heard of the kerfuffle through the Sabour grapevine.

'I saw an opportunity for creating a name, a brand for Bhagalpur's mangoes. I took a risk and committed to the district administration that I'll have the 200 quintals of mangoes they needed in two days,' said Chaudhary. His nursery business had already connected him with numerous farmers across the region. They updated him regularly about the quality and quantity of the produce. He knew which farmer was likely to have good mangoes at that late stage. On the appointed day, the administration brought in forty vehicles. The mango growers were ready with the produce. 'I called the farmers one-by-one, did the maths, ran the show. Since then, I've done this each year, organizing the bulk purchase of mangoes from here. And I send gifts of mangoes to the president and other dignitaries each year,' he said. This has brought him fame: 'I am the mango man. You can look it up on the internet.' He has 'released' two varieties named after Prime Minister Narendra Modi, one each for his electoral victory in 2014 and 2019.

I met him at his nursery in Maheshi-Tilakpur village near Sultanganj, about 30 kilometres west of Bhagalpur. It lies between the highway running along the Ganga's south bank and the railway line that runs further to the south. Chaudhary said this thin band produces the best mangoes. The soil here is not as sandy as that close to the river bank in the north. Neither is it hard and rocky like the land south of the railway line, showing signs of the plateaus and hills that lie southwards.

For all its ideal texture, this strip of land is low-lying. The Ganga has gradually deposited silt along its bank, raising its level. The annual monsoonal flood lingers in the low-lying band south of the river bank, causing waterlogging. For Chaudhary's nursery, this has worsened since a road was built nearby. 'It sends floodwaters in our direction. I've tried a lot to get the officials to change the drainage, but to no avail,' he said. This is a common problem in the eastern floodplains, worsening downstream of the river. During the monsoon, villagers often get into conflicts over embankments and the direction of the floodwaters. Each year, there are instances of violence over embankments and drainage.

'The excess water on the Sultanganj side of Bhagalpur lowers the sweetness of their Zardalu mangoes,' said Mamta Kumari, a scientist at the agriculture college in Sabour. The region to Bhagalpur's west is drier and doesn't get waterlogged. The Zardalu there is now sweeter and more flavourful, she said, mentioning Kahalgaon, in particular. It is about 30 kilometres from Bhagalpur but it took more than two hours to reach by the National Highway 33. It is easily the worst stretch of highway, nay, road, that I've encountered. For the most part, it is boulders and rocks lying on the road in the manner of an obstacle course or an off-piste stretch that rally riders seek.

At Ghogha, halfway to Kahalgaon, a journalist showed me many reports he had written on people killed in accidents caused by the treacherous road. Each year, the highway gets paved. Within weeks, it gets stripped by innumerable trucks and tractors ferrying the Ganga's sand, mined, mostly illegally, to supply brick kilns. A 6-kilometre stretch near Ghogha has sixty-five brick kilns that have combined to turn the region into a noisy, dust-covered dystopia. This strip of land, so close to the Ganga, was once a horticulture hub, producing both vegetables and fruits, including excellent mangoes. The brick kilns that have destroyed the road have also wrecked horticulture. How was I to find villages still producing fine mangoes?

The scientists at Sabour had said such areas lie east of Kahalgaon. So I reached the village of Rampur and met Rajesh Kumar. He told me over 80 per cent of the arable area in the village is under mangoes now. It wasn't always so. Not too long ago, the farmers here were deeply indebted. Agriculture was unviable due to the lack of irrigation. While the Ganga is barely 7 kilometres away, the Kahalgaon region is dry and rocky. The remnants of the hills of south Bihar lie just beneath the soil here, forcing the Ganga on a north-easterly course, before it joins the Kosi river and finally turns south towards the sea. It is not easy at all to harvest the monsoonal bounty here as the rocky substrate instantly converts rainfall into runoff. This also means there's no groundwater to be drawn.

All the while, the family with the largest landholding in the village was making neat profits from the three orchards it had planted in the 1950s. The mango doesn't need irrigation. The indebted farmers began planting orchards. It took a lot of effort and borrowing, given that new orchards take five-to-eight years before yielding fruit. Today, about 80 per cent of the total cropped area in this region is for mangoes. The dry conditions make the mango here sweeter and more fragrant and they also lower the depredations of insect pests. 'Our mango has quickly acquired a reputation for its fine quality. We actually get a premium price in markets all the way to Kolkata,' said Kumar. However, 70 per cent of the trees are just one variety, which is Malda/Landga. This leaves insect pests with a static target!

The Bhagalpur region on the Ganga's south bank has long grown fine mangoes. But the new-found fame of its old mangoes comes from new orchards with younger trees on the Sultanganj side and Kahalgaon. Even as the mango is gaining new ground here, it is losing some hallowed turf. The most high-profile example is Digha, an old settlement about 200 kilometres east of Bhagalpur. This was once the confluence of the rivers Ganga and Sone, which comes all the way from its origin in Amarkantak along the Madhya Pradesh–Chhattisgarh border. The two rivers created an expansive (Dirgha) lake (Hrada), hence the name DirghaHrada in old documents, with mango groves surrounding it. It was a convenient point for crossing the massive, undammed Ganga, so also a major point of trade with north Bihar from antiquity.

Over time, the name DirghaHrada got contracted to Digha. The Sone River's confluence shifted 25 kilometres west, beyond Maner, and Digha acquired a reputation as a centre of horticulture.

Residents of the surrounding areas came here to buy vegetables. Since it is near the state capital Patna, the connoisseurs of fine mangoes came here during the season. The Doodhiya Malda/Langda of Digha is perhaps the only commercially grown variety that can rival the Doodhiya Langda of Varanasi. It has a very small and thin stone with plentiful pulp, its skin is very thin and whitish, and the flesh is famous for its sweetness and fragrance.

Over the past three decades, Patna has sprawled way past Digha, turning it into a messy suburb. It is from here that a bridge called the JP Setu, Patna's second across the Ganga that became operational in 2017, connects Sonepur to north of the river. Urban development has eaten away at the mango groves, with a few stray trees surviving here and there. Barely two or three orchards remain in Digha.

One of them is a new orchard set up at the sprawling grounds of St. Xavier's College of Management And Technology. This happened under the stewardship of Father Robert Athickal, founder of an environmental organisation called Tarumitra. They obtained saplings from old trees of this variety and grafted them on to new rootstock. But the survival rate is quite low. It will take a few years to be able to tell which trees have taken well.

The oldest orchard in Digha with trees of the Doodhiya Malda is inside the Bihar Vidyapith, an important location in India's freedom struggle. A place frozen in time, it belonged to Maulana Mazharul Haque, a leader who has acquired a renewed fame in recent years. Back in 1917, when Mohandas Gandhi arrived in Patna on his way to Champaran in north Bihar, Haque was already a well known leader in the Indian National Congress. During the Champaran Satyagraha, he emerged as one of Bihar's prominent leaders. A successful lawyer with a thriving practice in Patna, he gave up his palatial ancestral house in the tony locality of Fraser Road, in keeping with Gandhi's call for austerity. He bought land from an orchard keeper outside the city in Digha, and established a frugal residence here. He called it the Sadakat Ashram.

Gandhi was so impressed by Haque's commitment that he chose this ashram as one of three sites for establishing universities with an Indian ethos: the Gujarat Vidyapith in Ahmedabad; the Kashi Vidyapith in Varanasi; and the Bihar Vidyapeeth in Digha's Sadakat Ashram. It was created under the guidance of Haque and Rajendra Prasad, another lawyer transformed by the Champaran struggle. The university's list of illustrious alumni includes Jayaprakash Narayan. In 1942, the colonial government seized this campus and took away all its papers, which is why little is known of its early history. After retiring as the first president of independent India, Prasad settled at the Sadakat Ashram and died there. A museum stands at that location.

The museum is surrounded by old mango trees of the Doodhiya Malda variety. I met Pramod Kumar here, who'd been buying the contract for harvesting the mangoes of the ashram campus for about five years. 'The mango grown here is most exquisite. It is the best mango in the world. Nothing can compare with it,' he told me. He took me to a tree that produces particularly fine fruit. 'Last season, just one buyer bought all its produce at the rate of ₹1,300 per kg.'

The mango from here does not go to any market; in fact, it does not leave the gate to get sold. Buyers come and buy it right there, offering to pay advances to book their produce ahead of the harvesting season. Pramod mentioned political leaders and former parliamentarians who come and buy the produce in bulk to gift the precious mango to impress VIPs. The 29-acre campus has 157 mango trees left standing, most of them of the Doodhiya Malda variety. Several of them are old, their branches falling off under the weight of decades. 'Most of the trees here are more than seventy years old,' said Kumar.

The contract he buys is for two years; he has little incentive to protect the trees or conserve the germplasm. Don't be surprised if it goes the way of the Banarasi Doodhiya Langda. While Maulana Haque's fame returned to prominence under the 2017 centenary celebrations of the Champaran Satyagrah, the Malda/Langda trees inside the Sadakat Ashram are withering away.

I visited the ashram in October when there were neither fruits nor flowers on the trees. In Patna, I met two people who in their youth had tasted Digha's Doodhiya Malda and the Banarasi Doodhiya Langda. They said Varanasi's Doodhiya was marginally finer, with thinner skin and more fragrance. But it was touch and go! I haven't yet tasted the Digha Malda at its best time and place! Nor the Zardalu of Bhagalpur. What truly surprised me, however, was Bihar's love for its seedling mangoes, which has disappeared from the rest of the country.

A glimpse of what's been lost came from Chandresh, a retired law professor in Bhagalpur. While he grew up in the city, his father's family came from the village Gogri in Khagaria district, just north of the Ganga. 'My uncle in the village once asked me to taste an unattractive mango called Sadhalwa, meaning rotten. It smelled horrible, I refused. He insisted, washing and cutting the fruit, offering it with much love, telling me I could spit it out if I didn't like it. I reluctantly had it. It was extraordinarily sweet and attractive in taste. But that mango has disappeared from the village. I haven't seen it since,' he said. During his childhood, he looked longingly at a tree that bore a variety called Rasgulwa, named after the sweet rasgolla. 'It was a prolific bearer with thousands of small, round mangoes that fell after ripening on the branch. They were gathered and put in buckets of water. Nobody stopped before having thirty to forty of those sweet mangoes.'

His favourite mangoes, though, came from the other side of his family from Vishambharchak village in Banka district, bordering Jharkhand. 'My maternal grandfather's father was deeply interested in food, he enjoyed feeding other people. Known as Bachhu Babu, he was very well-connected and travelled widely. His lawyer was his friend Rajendra Prasad. Wherever he went, he brought back interesting things to eat and grow.' He also raised a garden and an orchard on the large landholdings the family had. 'When we were children, we spent a lot of time there. Mangoes were the biggest reason we went there.'

Chandresh remembers some outstanding mangoes. 'There was this large fruit weighing more than one kilogram that one person could not eat by himself. It was called Lohajang, meaning rust red, for that was its colour. The fruits were so heavy they weighed down the branches. We loved touching the fruits as they were growing, it was the only mango that came so low.' Another large mango was called Kaduwa after the pumpkin. 'Each fruit was well over half a kilogramme. I've not tasted a mango that was as sour as that; there was no way to eat it. But it made excellent pickle,' he said.

The mango that he longs for more than any other was called Hilsapeti, meaning the stomach of the hilsa fish (*Tenualosa ilisha*). 'My great grandfather had got that mango from Murshidabad, because he had to travel to Bengal for his work. Since the hilsa/ilish is such a popular fish in Bengal, this mango was named after the pink meat that comes out of the middle part of the fish. While the slender mango itself remained green after ripening, it had a pink tinge on top. Its flesh was pink-coloured with a unique fragrance! And it was the sweetest mango I have tasted. Extraordinary! That's the mango we all sought.' Weighing about 250 grams each, the mango was slender and fish-like in shape.

The pulp sticking close to the stone was massaged out and, after processing, left to dry out on muslin cloth. Called amavat in Bihar, this preparation was used during the monsoon months when other foods like vegetables were unavailable. All kinds of mangoes are used for making it. Often, sugar is added. 'No sugar was added to the amavat made with Hilsapeti. And yet when it was served months later, you could see little crystals of sugar. And the pulp dried out so evenly the texture was crisp like no other,' Chandresh said. 'I know that region quite well and travel there often. Nobody had the kind of mangoes my great grandfather had grown in that orchard. The family land was eventually lost in legal disputes. Those trees are all gone. I haven't seen or heard of those mangoes since,' he said.

Most of these varieties were grown from seeds, and were not true-to-type grafts. Ordinary people in Bihar retain a special love for their seedling mangoes. They talk about them with love that I haven't seen elsewhere, even for many choice grafted varieties. This is even more common across the Ganga, in north Bihar.

Mango varieties of north Bihar do not get out of the region. For one, most of them fruit after the monsoon sets in, cutting off transport. This is the playground of some ferocious rivers. Coming down the steep gradient of the Nepal Himalayas, they hit north Bihar with great force. And well they may! It's the rivers that have created this land with the alluvial silt they bring down. The people here knew how to adapt to these seasonally violent rivers, till flawed management in modern times worsened the floods. Today, three-fourths of the population of north Bihar lives in flood-prone areas. Naturally, the Zarda of Champaran has barely been seen or tasted outside the region of the temperamental Gandak River.

One morning in late June, I braved the monsoon and bad roads to drive to the northwestern edge of Bihar. Champaran is named after the aranya (forest) of the Champa tree (*Magnolia champaca*). In the 1970s, it was split into two districts. The eastern half is headquartered in Motihari and the western half in Bettiah. In Motihari, I visited the New Golden Nursery run by Imam-ul-Haq who grew up in the Rizwan caste of gardeners. His forefathers had been invited to settle in Champaran by a local ruler who wanted to create orchards with fine, grafted mangoes. Grafting mangoes is more art than science, he said: 'You can find people who know a lot but that does not mean they will be able to grow desirable plants in difficult conditions.' He said I should go north for the finest Zarda mangoes.

I went north towards Narkatiyaganj, perhaps the last major

town before the thick forests and grasslands of the terai, protected under the Valmiki National Park, adjoining Nepal. It is in these parts that you find the small Zarda mango, a table mango not to be confused with the Zardalu of Bhagalpur. On the way, at the town of Chanpatiya, I turned west to reach village Dhamaura, situated along the Burhi Gandak River. Here, I found the ancestral haveli of Vijay Mohan Mishra who has sprawling mango orchards. Since I had called ahead, Mishra had the best Zarda mangoes available at that time soaked in a bucket of water. They were cut and brought to me on plates, along with a few whole mangoes. It is a pretty fruit, small and plump with some dark spots all over.

Given the high humidity here, the fruit does not retain a good appearance. A number of insects leave their secretions. Besides, the Zarda is a delicate mango. The skin was on the thinner side, though not as thin as the Zardalu. The pulp was light and dainty, showing more juice than firmness. It tasted sweet along with a slight subacid note. There was some fibre on the stone. I picked up a few and massaged them to suck. What stood out was the number of fruits with jelly seed (internal fruit breakdown). Vijay Babu assured me that I can eat as many as I like without fearing indigestion. 'You cannot say that about most fancied table varieties,' he said. (I was to later become familiar with the obsession in Bihar and West Bengal with digestive matters. People here evaluate mango varieties by their effects on the digestive system!)

Downstream of the Burhi Gandak lies Bihar's third largest city: Muzaffarpur. This region is a horticultural hub. Muzaffarpur district is the country's largest producer of quality litchis. To the south is Hajipur, known for its bananas. The most widely consumed mango here is Bambaiya or the Bombay Green. But in ordinary houses, one finds trees—some grafted but mostly seed-grown—of the Malda/Langda and a Bihari favourite called Sukul.

A slender mango with more juice than pulp, Sukul is a late mango that I never got to taste. Across India, I have not heard any other variety described with the kind of love and affection Biharis have for the Sukul. It is easily one of the most beautiful names for the mango. At first glance, it means that which is from an excellent (su-) clan (-kul). But the prefix su- has multiple munificent connotations in Sanskrit: to extract juice, distil, pour out or press out, churn, excite, produce, beget…

Even the mention of Sukul disarms people, making them smile! Tens of people have told me they feel stuffed and lethargic after consuming two Malda/Langda mangoes, but that there is no indigestion or heaviness after having even fifty Sukul mangoes. It is a mild mango yet is has character, carrying off that contradiction with charm. That is the quality of the most outstanding of mango varieties—they manage contradictions, containing contrasts that sit well. Another quality: the Sukul's fame stands on its own, without leaning on the crutches of celebrity. I have not heard one story that associates Sukul mangoes with a king or a nawab or a rich merchant or a celebrated writer or a political leader.

Not that north Bihar lacks in state patronage to the mango. For it is in the Mithila region that the earliest intensely-cropped mango orchard of outsize proportions came about: the Lakhi Bagh or the garden of one lakh trees. But to come to grips with that, it requires a little recap of history and geography. In the medieval times the Mithila region comprised most of what's north Bihar today and several parts that are now in Jharkhand, West Bengal, and Nepal. When Mughal emperor Akbar got control of the Mithila region, he realized it had a tradition of Brahmin kings. He made Mahesh Thakur, a Maithil Brahmin, the governor of Tirhut, one of the larger parts of Mithila.[1] The early capital of Tirhut was in what is Madhubani district today, bordering Nepal.

Lakhi Bagh was a feature of the Mughal promotion of horticulture, particularly mangoes. In the absence of evidence, it is difficult to say for sure what this giant orchard was like. We do know that north Bihar is very vulnerable to floods. Orchards were a way to make productive parcels of land that otherwise did not yield any revenue. What came of the Lakhi Bagh? We will have to wait for a historian to fill in the gaps.

We do know that after the Mughal empire began to lose its strength at the beginning of the eighteenth century, Tirhut became a large state more or less independent of any external control. It had the kind of relations kingdoms have with their neighbours. It fought wars with chieftains like the rulers of Bettiah and Nepal, and often got into conflicts with the nawabs of Bengal. After the British East India Company got control of Bengal in 1757–64, the Tirhut rulers moved their capital to Darbhanga, probably in the early 1770s. The Tirhut state became Raj Darbhanga. In the late nineteenth and early twentieth century, it was the largest and richest state in eastern India.

A major shift came in 1860 when the ruler, Maharaja Maheshwar Singh, died. His heir Lakshmeshwar Singh (1858–1898) was two years old. The estate was placed under the administration of the 'Court of Wards'. The new ruler was educated in both the traditional system and by British-appointed European tutors. He invested in many public works, becoming a patron of the arts and of modernization. (Darbhanga features prominently in the modern history of classical music and arts and even in the creation of India's football association.) In 1881, the ruler wrote to British botanist Joseph Hooker, the director of the Royal Botanic Gardens in Kew, whom we have met earlier; he wrote the encyclopaedic seven-volume *The Flora of British India*. The ruler sought an expert botanist to set up a modern garden around his palace. Hooker recommended Charles Maries, a young botanist just back from explorations in East Asia.

Maries set up the exquisite gardens surrounding the Anand Bagh Palace. The highlights of the gardens were the mango trees. Maries acquired expertise in the mango, studying the fruit in great detail, and writing about the characteristics of each and every variety he encountered. He wrote a book titled *Cultivated Mangoes of India* that was never published. When the botanist George Watt, also recommended by Hooker to the Government of India, produced his six-volume *The Dictionary of Economic Products of India* in the early 1890s, the section on the mango was attributed to Maries. (It mentions the Sukul in the appendices among mangoes grown in Hajipur.)

Like Murshidabad, Darbhanga appears central in the spread of fine mangoes across north India and beyond. Like the former capital of Bengal, it is crying for a talented historian's dedicated attention. The first book written exclusively on the mango in the modern era was published in 1897 under the title *A Treatise on Mango*. Its author, Probodh Chundra De, former superintendent of Murshidabad's royal gardens, drew upon both Maries and Watts. A string of books dedicated to the mango followed, both by botanists and generalists. And just as the remnants of Murshidabad's mango glory can be seen in orchards there, Mithila's mango wealth is surprising, even today.

Bihar does not have the kind of new mango festivals held in Delhi or Lucknow or Bengaluru. A lot of mangoes here are from seed-grown trees. While the rest of the country has turned away from the immeasurable variety of seed-grown mangoes, leaning towards a clutch of commercial grafted varieties, Bihar retains its love of biju mangoes.

I experienced this first hand during the mango season on the campus of the Lalit Narayan Mithila University in Darbhanga. It is the palace and the gardens of the Darbhanga state, given away to set up two universities after independence. I was taken around by Santosh Kumar, a young conservation activist associated with INTACH. He grew up on the campus and had climbed its numerous mango trees. 'If you show me a mango grown here, I can take you to its tree. I can do this for about thirty to forty types,' he said.

We walked around the orchards among trees that were well over sixty to seventy years old. The contractor who had bought the crop from the university came with a guard. He let us taste the mangoes from his harvest, although he could not name the varieties. I tasted more than ten kinds of mangoes. One looked like a small Alphonso—and tasted like it, too. Each mango had a unique taste, their smell was overpowering, and Santosh began to joke at how I was eating them with my nose. 'Smile! You are tasting the fruits that became Charles Maries' obsession,' said Santosh.

Just outside the botany department of the university, a solitary monkey arrived. He was sampling the fruits and dropped a small one that was close to ripeness. Santosh ran out and brought it to me. I was reluctant to have it but tried it after he insisted. It was so fragrant and refreshing that I exclaimed involuntarily. 'Monkeys smell the mangoes before they have it, just like you,' he said. He meant it as a compliment and I took it as one. When it comes to mangoes, primates are primates.

Epilogue
Eat, Discuss, Repeat

'Which mango variety is the best?'
—PARTY QUESTION

Hundreds of books address the mango. Countless titles are about to get written. I have met scores of people with their own book-length plans. Each book can tell only a part of the story—a small part at that. None can hope to do justice to its epic scale. For that, the mango will just have to write its own book. Till then, we must make do with what we have.

This book offers a few slices from the lives of people who grow, sell, and eat mangoes. Or research them! Or just talk about them! They are fleeting reflections of ourselves, mirrored in a fruit that has held us in a bewitching spell for millennia. This compulsion cuts both ways. The outstanding urge to talk about the mango drives memes built on apocrypha or misinformation. Our curiosity in the mango does not take us very far. Why? Because stereotypes work. Because questioning the obvious does not make sense. It is a recipe for failure.

Yet I asked those questions. The risk is worth it because so is the mango. If such a potent conveyor of stories cannot carry inconvenient non-fiction, then what's the point! We need to repeat to ourselves the basic reality of the mango, that it 'is believed to have evolved as a canopy layer or emergent species of the tropical rainforest...' Markets for fresh fruits and vegetables are designed around the products of temperate lands. Apples, potatoes, and apricots have long shelf lives because they evolved to resist the cold. They travel well in refrigerated value chains. They also have relatively few natural enemies, few natural friends.

Shaped by the heat and the ephemeral cornucopia of the tropics, most tropical fruits rot fast. (Discount the coconut and pineapple. Be reminded that the banana in markets around the world is from highly modified varieties grown and marketed in highly controlled conditions that lead to banana republics.) The skin of mango varieties bred for a long shelf life is thick, lacking subtlety of taste and aroma. India and its mangoes lie in the tropics and the subtropics. With the exception of a few varieties, mainly the Alphonso, the Indian mango does not travel well. Consequently, high-profile and influential consumers who buy fruits from the market often have no exposure to the remarkable diversity of Indian mangoes. If you are really passionate about India's most wonderful mangoes, you must lend yourself to inconvenience and risky travels. You have to catch them at the right time on their home turf.

This book has been a long time coming. It has taken me across long distances in this large and complex country. Its limitations are my limitations. I'm a man from Central and North India. In the countryside, it is not very easy for a man to interview women. The book has suffered from that. Not knowing the languages of eastern and southern India is a great handicap, too, limiting my access to their wonderful cultural wealth and their outstanding cultivars. No doubt others will do a better job in those regions. They better hurry up before the remaining mango groves are cut down at the altar of industrial development and mining before traditional varieties and the rich culture around them disappear.

Searching for the origin of prize mangoes has often landed me in front of grandiose ruins. They were bred and refined in the orchards surrounding the palaces of powerful rulers and rich merchants. Their glory has faded, leaving behind buildings in ruins or disrepair. The nawabs of Banganapalle are gone; Kausar Bagh and Pari Bagh are neglected marvels. The Fruit Research Station in Sangareddy, Telangana, is housed in the Hyderabad nizam's erstwhile stables. The Kesar was fostered in the Junagadh nawab's lands. In Malihabad, the grand havelis tell sad stories. Varanasi's Moti Jheel Mahal has lost the Banarasi Langda. Barely one or two old orchards remain standing in Murshidabad. The glorious gardens surrounding the palaces of Raj Darbhanga are hardly maintained by the universities that run out of them now.

Visiting these stories in disrepair carries its own rewards. I've gotten to taste more mangoes than I can remember or describe in my notebook. At times it felt like a burdensome task—but only in moments few and far between. There are sweet mangoes, fragrant ones, those with delicate notes, with firm flesh, with juice that bursts out of the skin with the slightest pressure. There are grape-sized mangoes and those competing with melons for heft. Those that hint at delicate notes and those that overwhelm the senses. Some are prized for sourness. There are fleshy ones and there are juicy ones. Some offer fibre and some creamy flesh. The pulp of some of those slips through like mother's milk, while some attack your throat with irritants to deliver a masochistic joy. There are rare varieties watched over by growers like thoroughbreds. There are trees that produce famous varieties that stand ignored and neglected; a country so immersed in mangoes ought to treat its natural wealth better.

Lending myself to the mango has been rewarding in ways that I have not quite figured out yet. The mango has changed a part of me and lent me new eyes to look at India and its diverse peoples. The mango inside me, however, is unchanged. It is exactly what it was when I was ten years old. It is the small, fibrous, and juicy desi mango of Malwa. The kind my uncle bought from the Indore mandi. The type he massaged and readied for our consumption. A big chunk of that pleasure resides beyond the taste buds. It lies in the memory of how Subhash Chacha's face lit up when he saw me or any of us kids enjoy a mango. The bhoot of our past doesn't leave us, no matter what we taste.

'Which mango variety is the best?' I'm often asked by people expecting well-formed sound bites or objective analysis from vast datasets. I avoid telling my Malwi mango story because it sounds sentimental. I spin some story to match the mood of the gathering and acquit myself. There are so many delightful mango varieties in India and so many stories about each one. See it from any angle; you will find what you seek and also the exact opposite of that.

But the honest truth? That's complicated—and highly subjective. No matter how much you might have spent on an exotic variety, a child somewhere might be running away with the perfect mango they got off a tree or a ripening pile while the caretaker wasn't looking. In that risk lies the purest joy. Desire is sometimes a stone flung at a tree that a bipedal primate struggles to climb. Sometimes, it is a chance mango stone, thrown away by a child who ate the flesh, containing the seed of a sensational variety that nobody has heard of. Yet.

Acknowledgements

It is impossible to put together such a book without the generosity of strangers. I have had more than my share of it. In the course of putting together the material for this book, I dug myself into a hole more often than I should have, without any idea of how to dig myself out of it. On almost every such occasion, I found help from a stranger who had no reason whatsoever to help me.

Orchards and research institutions tend to be located in inaccessible areas. All too often, I followed a hunch into an unfamiliar part of this vast and complex country. After a full day of making enquiries—some fruitful, most not so much—I often found myself stranded in poorly connected areas in the dark. Strangers offered me rides on bicycles, bullock carts, tractors, jugaads, trucks, and suchlike. Complete strangers have handed over their vehicles to me to go chase down a lead. After entire days spent roaming around in oppressive heat, hungry and tired, I've had strangers take me home and offer lavish meals. Hotel managers who offered discounts without being asked; one or two refused to accept any payment from someone who had come to enquire about their heritage.

My first debt of gratitude is to all those strangers who helped me. I cannot hope to name them all. They saw a vulnerable outsider with a quaint aspect and odd questions. While many turned away, they did not. They went out of their way to trust me. Some of this generosity is peculiarly Indian; the rest is universal.

While a majority of mango growers, both contractors and landowners, were indifferent, many were not. Most had not been asked the kind of questions I put to them. They took out time in the middle of the season. The more engaged growers are much busier than the uninterested ones. Yet they are the ones who have the answers. Some growers set aside entire days in peak season to show me around, to explain what they had learned over decades. A few put me up in their house without knowing anything about me. Whether they were engaged or not, almost all growers were generous in letting me taste their mangoes. Many insisted on it. And, given the emotional value of the mango, they watched while I tried their fruits, smiling proudly. They asked if I enjoyed the mangoes. When I expressed satisfaction, they appreciated it. If the fruit was not up to the mark, they apologized. If the fruit delighted me, they just laughed out loud. I thank them and wish them wealth and productivity.

Horticultural research in India is in the public sector. The most knowledgeable scientists work in government research institutions; so do those who are there because they could not find another government job. The former are overworked; the latter have little to offer. Both kinds are not forthcoming because of how the public sector operates in India. My task was to find the ones who know their stuff and are willing to discuss it. I was pleasantly surprised at the number of well-informed scientists who were forthright and open. The mango does have a dilating social effect. I thank the scientists who took out the time to answer questions as if I were their research student, even if the questions were basic enough for a high school graduate. They provided references and access. I wish them engaging puzzles, insightful solutions and dynamic supervisors free of insecurity.

Traders of agricultural commodities are discreet and tactical in all interactions. That's the nature of their business. I'm grateful to the ones who opened up to me, letting me sit close while they negotiated deals, taking me for a ride-along to explain something, offering me names and numbers from their voluminous diaries and their phonebooks. Retailers have an even lower interest in speaking to a reporter. I've met retailers who opened up, explained their operations, and took me walking with them. I acknowledge their kindness and wish them just profits. I hope they generate wealth for mango growers and joy for consumers.

A special vote of gratitude to the scientists at the Birbal Sahni Institute of Palaeosciences in Lucknow, a national treasure. They provided research material, perspective, and explained complicated matters to an untrained ear.

G. V. Ramanjaneyulu of the Centre for Sustainable Agriculture has been a great resource over the years—of research material, of finding the right people, of knowledge and understanding. I'm grateful for all that he has provided, to me and to countless others who wish to understand agriculture and India's farm sector.

A special thanks to Venu Madhav Govindu, Ashish Mehta, and Nisha Susan, friends who reviewed the manuscript and provided detailed comments. Librarians are invaluable to any researcher. My gratitude to librarians for their time and effort, and for using their library networks to help me access rare or out-of-print titles. Thanks to officials in mandi offices and government agencies like APEDA who readily shared information and contacts.

This book has taken a long time to put together, long enough for anybody to doubt its delivery. I thank Aleph Book Company for their patience and trust, for all the small and painstaking tasks that go into bringing out a book.

I must thank Shekhar Tomar and Amit Chaudhary for always keeping my touring motorcycle in sound shape to take on any punishment at the spur of the moment. Thanks also to friends in Royal Enfield, current and former employees, who helped out with traction in tricky terrains.

I am eternally grateful to my masters for what they taught me, for giving me values that hold me in good stead. I lack adequate words to thank my family and friends who have provided love, tolerance, encouragement and a stable base from which to pursue my quixotic enterprises. All that I have is theirs, warts and all.

For all the difficulties of finding research material and good mangoes, there are worse ways to do journalism.

Notes

PROLOGUE

1. Rachana Mishra, 'Aam ka khaas safar', *Sulabh Swachh Bharat*, Vol. 2, No. 24, 2018, pp. 1-5.
2. '3 kids flee home for mangoes', *Times of India*, 21 June 2015.
3. Vladeta Ajdacic-Gross, et al, 'Methods of suicide: international suicide patterns derived from the WHO mortality database', Bulletin of the World Health Organization, Vol. 86, No. 9, 2008, pp. 726-732.
4. 'Right-wing activist Sambhaji Bhide gets bail in case over "infertility-curing mango" claim', *Times of India*, 7 December 2018.

CHAPTER 1. FRUIT

1. Maslow's pyramid or Maslow's hierarchy of needs is a model created by psychologist Abraham Maslow to understand motivations for human behavior, with each level of the pyramid representing a different human need. These include physiological needs, safety, love and belonging, esteem, and self-actualization.
2. '1993 Mumbai Serial Blasts: Timeline of Events from Bomb Explosions to Abu Salem's Life Imprisonment', *Indian Express*, 7 September 2017.

CHAPTER 2. VARIETY

1. M. S. Randhawa, 'The Mango—New Hybrids: Preservation of the Germrplasm of Sucking Mangoes', *A History Of Agriculture In India-Volume IV: 1947-1981*, New Delhi: Indian Council Of Agricultural Research, 1986, p. 475.
2. Damodar Dharmananda Kosambi, *An Introduction to the Study of Indian History*, Mumbai: Popular Prakashan, 1956, p. 334.
3. Damodar Dharmanand Kosambi, *Myth and Reality: Studies in the Formation of Indian Culture*, Mumbai: Popular Prakashail, 1962, p. 160.
4. Parashuram K. Gode, 'History of the Art of Grafting Plants (Between c. 500 B.C. and A.D. 1800', *Studies in Indian Cultural History,*

Vol. I, Visveshvaranand Indological Series-9, Visveshvaranand Institute Publication-189, Hoshiarpur: Visveshvaranand Vedic Research Institute, 1961, pp. 439-451.

5   Parashuram K. Gode, 'References to Grafted Mangoes in India between A.D. 1550 and 1800', *Studies in Indian Cultural History*, Vol. I, Visveshvaranand Indological Series-9, Visveshvaranand Institute Publication-189, Hoshiarpur: Visveshvaranand Vedic Research Institute, 1961, pp. 452-454.

6   Hobson-Jobson entry for mango.

7   Y. L. Nene, 'Mango through Millennia', *Asian Agri-History*, Vol. 5, No. 1, 2001, pp. 39-67.

8   K. T. Achaya, *Indian Food: A Historical Companion*, New Delhi: Oxford University Press, 1994, p. 208.

9   Ibid., p. 162.

10 Nalini Sadhale (Tr.), *Surapala's Vrikshayurveda (The Science of Plat Life by Surpala)*, Agri-History Bulletin No. 1, Secunderabad: Asian Agri-History Foundation, 1996.

11 Kamala Thiagarajan, 'The people resurrecting India's ancient fruit trees', *BBC Future,* 26 September 2022.

12 Paul Craddock, *Spare Parts: A Surprising History of Transplants*, London: Penguin, 2021.

13 Achaya, *Indian Food,* p. 162.

14 Lallanji Gopal, 'Agricultural Technique in Medieval India—Its Central Asian Contacts', A.V. Narasimha Murthy and K.V. Ramesh (eds.), *Giridharasri: Essays on Indology, Dr G.S. Dikshit Felicitation Volume*, New Delhi: Agam Kala Prakashan, 1987, pp. 235-251.

15 M. N. Pearson, 'The New Cambridge History of India I-1', *The Portuguese in India*, Cambridge: Cambridge University Press, 1987, p. 126.

16 'Pioneering the computational linguistics and the largest published work of all time', IBM, 27 March 2012.

17 Charles J. Borges, *The Economics of the Goa Jesuits, 1542-1739*, New Delhi: Concept Publishing Company, 1994, p. 35.

18 M. N. Pearson, 'The New Cambridge History Of India I-1', p. 127.

19 Lizzie Collingham, *Curry: A Tale of Cooks and Conquerers*, United Kingdom: Oxford University Press, 2006, 1987, p. 63

20   Garcia da Orta, Sir Clements Markham (trs.), 'THIRTY FOURTH COLLOQUY: MANGOES', *Colloquies On The Simples And Drugs Of India*, Lisbon: Henry Sotherans and Co, 1895, pp. 284-294.
21   Jonathan Gil Harris, *The First Firangis*, New Delhi: Aleph Book Company, 2014, p. 37.
22   Maria Graham, *Journal Of A Residence In India*, Edinburgh: Archibald Constable And Company, 1813, pp. 6-7.
23   Patricia Ann Alvares 2019, 'The Jesuits and the Mango', *The Times of India*, 15 April 2019.
24   Borges, *The Economics of the Goa Jesuits, 1542-1739*, pp. 28-29.
25   Pearson, 'The New Cambridge History Of India I-1', p. 138.
26   Charmaine O'Brien, *The Penguin Food Guide to India*, New Delhi: Penguin Books India, 2013.
27   Tapan Raychaudhari and Irfan Habib (eds.), *The Cambridge Economic History of India*, Vol. 1, 1982, p. 53.
28   R. N. Singh, *Mango*, ICAR Low-Priced Books Series No.3 (Revised), New Delhi: Indian Council Of Agricultural Research, 1978, p. 34.
29   Manmatha Nath Dutt (Shastri) (trs.), 'Ya'jnawalkya Samhita', *The Dharma Sa'stra or The Hindu Law Codes [English Translation], Vol.1,*
30   Y. L. Nene, 'Mango through Millennia', *Asian Agri-History*, Vol. 5, No. 1, 2001, p. 63.

CHAPTER 3. CULTURE

1   M. S. Randhawa, 'The Mango—New Hybrids: Preservation of the Germrplasm of Sucking Mangoes', *A History Of Agriculture In India- Volume Iv: 1947-1981*, New Delhi: Indian Council Of Agricultural Research, 1986, p. 474.
2   Ibid., p xxxiii.
3   Shankar Vedantam, 'Hidden Brain: Our Better Nature: How The Great Outdoors Can Improve Your Life', *NPR* (National Public Radio), Radio Broadcast, 10 September 2018.
4   Frances E. Kuo and William C. Sullivan, 'Environment and Crime in the Inner City: Does Vegetation Reduce Crime?', *Environment and Behavior*, Vol. 33, No. 3, 2001, pp. 343-367.
5   Frances E. Kuo, 'Coping with Poverty: Impacts of Environment and Attention in the Inner City', *Environment and Behavior*, Vol. 33, No. 1,

2001, pp 5–34.

6. Mark S. Taylor et al, 'Research note: Urban street tree density and antidepressant prescription rates—A cross-sectional study in London, UK', *Landscape and Urban Planning*, Vol 136, 2015, pp 174–179.

7. Frances Kuo, 'How might contact with nature promote human health? Promising mechanisms and a possible central pathway', *Frontiers in Psychology*, Vol. 6, No. 1093, 2015.

8. Ian Johnston, 'Human brain hard-wired for rural tranquillity: Study of brain activity shows it struggling to process complex urban landscapes', *Independent*, London, 2013.

9. George Martine et al, *The State of World Population 2007: Unleashing the Potential of Urban Growth*, United Nations Population Fund, New York, 2007, p.1.

10. Gianluca Crispi and Han Zhang, *Slum Upgrading Legal Assessment Tool*, United Nations Human Settlements Programme (UN-Habitat), Nairobi, 2022, p. 1.

11. Madhav Gadgil, 'Sacred Groves: An Ancient Tradition of Nature Conservation', *Scientific American*, 1 December, 2018.

12. Michael Gross, 'Europe's last wilderness threatened', *Current Biology*, Vol 26, No 14, 2016, pp. R641-R643.

13. Malin Rivers et al, *European Red List of Trees*, IUCN (International Union for Conservation of Nature and Natural Resources), 2019, pp. 9-10.

14. Debal Deb et al, *Forests as Food Producing Habitats*, Living Farms, Bhubaneswar, July 2014

15. Y. L. Nene, 'Mango through Millennia', p. 64.

16. Rabindranath Tagore, 'The Religion of The Forest', *Creative Unity*, London : Macmillan, 1922.

CHAPTER 4. SEASON

1. Leona Anderson, 'The Indian Spring Festival (Vasantotsava): One or Many?', *Annals of the Bhandarkar Oriental Research Institute*, Pune, Vol. 69, No. 1/4, 1988, pp. 63-76.

2. Tridib Nath Ray, 'The Indoor and Outdoor Games in Ancient India', in *Proceedings of the Indian History Congress*, Indian History Congress, Vol 3, 1939, pp. 241-261

3. Kanad Sinha, 'Sporting with Kama: Amusements, Games, Sports and

Festivities in Early Indian Urban Culture', *Journal of the Asiatic Society*, Kolkata, Vol. LV, No. 2, 2013, pp. 73-120.
4   P. Thankappan Nair, *The Mango in Indian Life and Culture*, Dehradun: Bishen Singh Mahendra Pal Singh, 1995, p. 255.
5   Ibid., p. 58.
6   Kusum Budhwar, *Romance of the Mango: The Complete Book of the King of Fruits*, New Delhi: Penguin Books India, 2002, p. 60.
7   Thankappan Nair, *The Mango in Indian Life and Culture*.
8   James McHugh, *An Unholy Brew: Alcohol in Indian History and Religions*, Oxford University Press, 2021.
9   T. N. Mukharji, 'Piuri or "Indian Yellow."', Journal of the Society of Arts, Vol. 22, No. 1(618), 23 November 1883, pp. 16-17.
10 Rebecca Ploeger and Aaron Shugar, 'The story of Indian yellow—excreting a solution', Journal of Cultural Heritage, Vol. 24, March–April 2017, pp. 197-205.
11 Gregory Dale Smith, 'Cow urine, Indian yellow, and art forgeries: An update', *Forensic Science International*, Vol. 276, July 2017, pp. e30-e34.
12 Victoria Finlay, *Colour – Travels Through the Paintbox*, London: Hodder and Stoughton, 2002, pp. 225-240.
13 Emiko Ohnuki-Tierney, Kamikaze, *Cherry Blossoms, and Nationalism: The Militarization of Aesthetics in Japanese History*, Chicago: University of Chicago Press, 2002, p. 12.
14 Ibid., p. 3.

CHAPTER 5. RELIGION
1   Shahid Amin, *Conquest and Community: The Afterlife of Warrior Saint Ghazi Miyan*, New Delhi: Orient Blackswan Private Limited, 2015.
2   Badri Narayan, *Fascinating Hindutva: Saffron Politics and Dalit Mobilisation*, SAGE India, 2008.
3   Christian W. Troll, *Sayyid Ahmad Khan: a Reinterpretation of Muslim Theology*, New Delhi: Vikas Publishing House, 1978, p. 41.
4   Ibid., p. 51.
5   E. B. Cowell (ed.), E. B. Cowell and W. H. D. House (trs.), 'Jataka 539: Mahājanaka-jātaka', *The Jataka*, Vol. 6, Cambridge University Press, 1907.
6   Maurice Bloomfield, *The Life and Stories of the Jaina Savior Parshvanatha*, Baltimore: The Johns Hopkins Press, 1919, pp. 34-35.

7   Ibid., pp. 147-148.
8   M. S. Randhawa, 'The Buddhist Period', *A History Of Agriculture In India- Volume I: Beginning to 12th Century*, New Delhi: Indian Council Of Agricultural Research, 1980, pp. 324-325.

CHAPTER 6. HISTORY

1   Rajindar Sachar, 'My memories of the great Nehrü', *The Sunday Guardian*, New Delhi, Published on the internet on November 19, 2017.
2   Louis Fischer, *A Week With Gandhi*, London: George Allen & Unwin Ltd, 1943, p. 95.
3   Nico Slate, *Gandhi's Search for the Perfect Diet: Eating with the World in Mind*, Seattle: University of Washington Press, 2019, p. 163.
4   Ibid., p. 169.
5   Damodar Dharmanand Kosambi, *An Introduction to the Study of Indian History*, Mumbai: Popular Prakashan Bombay, 1956, pp. 1-2.
6   Lal Behari Singh, *The Mango: Botany, Cultivation, and Utilization*, London: Leonard Hill Books Ltd, and New York: Interscience Publishers, 1960, p. 1.
7   Tapan Raychaudhari and Irfan Habib (eds.), *The Cambridge Economic History of India*, Vol. 1, 1982, p. 462.
8   Lallanji Gopal, 'Agricultural Technique in Medieval India—Its Central Asian Contacts', A.V. Narasimha Murthy and K.V. Ramesh (eds.), *Giridharasri: Essays on Indology*, Dr. G.S. Dikshit Felicitation Volume, New Delhi: Agam Kala Prakashan, 1987, p. 243.
9   Raychaudhari and Habib (eds.), *The Cambridge Economic History of India*, p. 258.
10 Eric Stokes, *The Peasant and the Raj: Studies in Agrarian Society and Peasant Rebellion in Colonial India*, Cambridge: Cambridge University Press, 1980, pp. 73-74.
11 Pallavi V. Das, *Development, and the Environment: Railways and Deforestation in British India*, 1860-1884, Palgrave Macmillan, 2015, p. 1.
12 Richard Mahapatra, 'Waters of Life', *Down To Earth*, New Delhi: Society of Environmental Communications, Published on the internet on March 15, 1999.
13 Pradip Krishen, *Trees of Delhi: A Field Guide*, Dorling Kindersley (India) Pvt Ltd, 2006, p. 36.
14 Mohammad Shaheer, 'Mughal Gardens', *Journal of Landscape Architecture*,

LA 17, Monsoon 2007, pp. 28-30; Ratish Nanda, 'I'm sure Mohammad Shaheer is working on designs for the best garden paradise will ever, *The Indian Express*, New Delhi, Published on November 29, 2015.

15 James L. Wescoat Jr, Rebecca M. Brown and Deborah Hutton (eds.), 'The Changing Cultural Space of Mughal Gardens', *A Companion to Asian Art and Architecture*, London: John Wiley, 2011, pp. 201-229.

16 Abdullah Yousufi, 'Horticulture in Afghanistan: Challenges and Opportunities', *Journal of Developments in Sustainable Agriculture*, Vol. 11, 2016, pp. 36-42.

17 Robert N. Spengler III, *Fruit from the Sands: The Silk Road Origins of the Foods We Eat*, Oakland: University of California Press, 2019, p. 374.

18 Anku Bharadwaj, 'Feasts and Food Symbolism in the Court Culture of the Early Mughals (1504-1605 CE)', *Research Journal of Humanities and Social Sciences*, Vol. 6, No. 4, 2015, pp. 255-268.

19 Lizzie Collingham, *Curry: A Tale of Cooks and Conquerors*, Oxford University Press, 2006, p. 35.

20 Thankappan Nair, *The Mango in Indian Life and Culture*, 1995, p. 182.

21 KT Achaya, *Indian Food: A Historical Companion*, Oxford: Oxford University Press, 1994, p. 161.

22 Maria Graham, *Journal Of A Residence In India*, Edinburgh: Archibald Constable And Company; London: Longman, Hurst, Rees, Orme, and Brown, 1813, Second Edition, pp. 6-7.

23 Francois Bernier, as quoted in: Raychaudhari and Habib (eds.), *The Cambridge Economic History of India*, Vol. 1, 1982, p. 331.

24 Audrey Truschke, 'Early Years', *Aurangzeb: The Life and Legacy of India's Most Controversial King*, Stanford: Stanford University Press.

25 Colleen Taylor Sen, *Feasts and Fasts: A History of Food in India*, London: Reaktion Books Ltd, 2015, pp. 189-190.

26 Truschke, 'Later Years', *Aurangzeb: The Life and Legacy of India's Most Controversial King*.

27 Truschke, 'Aurangzeb's legacy', *Aurangzeb: The Life and Legacy of India's Most Controversial King*.

28 Raychaudhari and Habib (eds.), *The Cambridge Economic History of India*, p. 182.

29 Yousufi, 'Horticulture in Afghanistan: Challenges and Opportunities', pp. 36-42.

30 Ajit Prasad Jain in SR Gangolly et al, *The Mango*, New Delhi: Indian Council of Agricultural Research, 1957, p. iii.
31 Vikram Doctor, 'How India's mango diplomacy has been winning friends and foes over the years', *Economic Times*, Mumbai, 10 June 2017.
32 Om Prakash and RM Khan, *A Tryst With Mango*, New Delhi: APH Publishing Corporation, 2005, pp. 29-31.

CHAPTER 7. ORIGIN

1 Pratik Chakrabarti, 'Gondwana and the Politics of Deep Past', *Past & Present*, Vol. 242, No. 1, 2019, pp. 119–153.
2 Oliver Burkeman, 'Jared Diamond: 'Humans, 150,000 years ago, wouldn't figure on a list of the five most interesting species on Earth'', The Guardian, 24 October 2014.
3 Arunima Kashyap and Steve Weber, 'Harappan plant use revealed by starch grains from Farmana, India', *Antiquity*, Vol. 84, No. 326, December 2010.
4 R. C. Mehrotra, D. L. Dilcher, and N Awasthi, 'A Palaeocene Mangifera-Like Leaf Fossil From India', *Phytomorphology*, Vol. 48, No. 1, 1998, pp. 91-100.
5 Douwe J. J. van Hinsbergen et al, 'Greater India Basin hypothesis and a two-stage Cenozoic collision between India and Asia', *PNAS*, Vol. 109, No. 20, 2012, pp. 7659–7664.
6 Prakart Sawangchote, Paul J Grote, and David L Dilcher, 'Tertiary Leaf Fossils Of Mangifera (Anacardiaceae) From Li Basin, Thailand As Examples Of The Utility Of Leaf Marginal Venation Characters', *American Journal of Botany*, Vol. 96, No. 11, 2009, pp. 2048–2061.
7 Anon, Christopher Cumo (ed.), *Encyclopedia of Cultivated Plants: From Acacia To Zinnia*, ABC-CLIO, 2013, p. 619.
8 M. Ducousso et al, 'The last common ancestor of Sarcolaenaceae and Asian dipterocarp trees was ectomycorrhizal before the India–Madagascar separation, about 88 million years ago', *Molecular Ecology*, Vol. 13, 2004, pp. 231–236.
9 The Angiosperm Phylogeny Group, 'An update of the Angiosperm Phylogeny Group classification for the orders and families of flowering plants: APG III', *Botanical Journal of the Linnean Society*, Vol. 161, 2009, pp. 105–121.

10. Suryendu Dutta et al, 'Eocene out-of-India dispersal of Asian dipterocarps', *Review of Palaeobotany and Palynology*, Vol. 166, No. 1, 2011, pp. 63-68.
11. Jes Rust et al, 'Biogeographic and evolutionary implications of a diverse paleobiota in amber from the early Eocene of India', *PNAS*, Vol. 107, No. 43, 2010, pp. 18360–18365.
12. Anumeha Shukla, R. C. Mehrotra, and J. S. Guleria, 'Emergence and extinction of Dipterocarpaceae in western India with reference to climate change: Fossil wood evidences', *Journal of Earth System Sciences (Indian Academy of Sciences)*, Vol. 122, No. 5, 2013, pp. 1373–1386.
13. Ben H. Warrena et al, 'Why does the biota of the Madagascar region have such a strong Asiatic flavour?', *Cladistics*, Vol. 26, No. 5, 2010, pp. 526-538.
14. André JGH Kostermans and Jean-Marie Bompard, *The Mangoes: Their Botany Nomenclature Horticulture and Utilization*, London: Academic Press, Harcourt Brace & Company, 1993, p. 104.
15. Jacob E. Bronowski 1973, *The Ascent of Man*, BBC Books, London
16. Richard Feynman, interview with BBC television programme Horizon, excerpted from *The Pleasure of Finding Things Out: The Best Short Works of Richard P Feynman*, Helix Books, July 1999, p. 146.
17. V. Prasad, A. Farooqui, S.K.M. Tripathi, R. Garg, and B. Thakur, 'Evidence of Late Palaeocene-Early Eocene equatorial rain forest refugia in southern Western Ghats, India', *Journal of Biosciences (Indian Academy of Sciences)*, Vol. 34, No. 5, 2009, pp. 777–797.
18. R. Dietmar Müller, 'An Indian cheetah', *Nature*, Vol. 449, No. 7164, 2007, pp. 795-797.
19. Prakash Kumar, Xiaohui Yuan, M. Ravi Kumar, Rainer Kind, Xueqing Li, and R.K. Chadha, 'The rapid drift of the Indian tectonic plate', *Nature*, Vol. 449, No. 7164, 2007, pp. 894-897.
20. P. D. Gingerich, 'Mammalian responses to climate change at the Paleocene-Eocene boundary: Polecat Bench record in the northern Bighorn Basin, Wyoming', *Causes and Consequences of Globally Warm Climates in the Early Paleogene*, Geological Society of America Special Paper 369, Colorado, 2003, pp. 463–478.
21. Ross Secord et al, 'Evolution of the Earliest Horses Driven by Climate

Change in the Paleocene-Eocene Thermal Maximum', *Science*, Vol. 335, No. 6071, 2012, pp. 959-962.
22. Vandana Prasad et al, 'Apectodinium Acme And Palynofacies Characteristics In The Latest Palaeocene-Earliest Eocene Of Northeastern India: Biotic Response To The Palaeocene-Eocene Thermal Maxima (PETM) In Low Latitude', *Journal of The Palaeontological Society of India*, Vol. 51, No. 1, 2006, pp. 75-91.
23. Gaurav Srivastava and Rakesh C. Mehrotra, 'First Fossil Record of Alphonsea Hk f & T (Annonaceae) from the Late Oligocene Sediments of Assam, India and Comments on Its Phytogeography', *PLOS ONE*, Volume 8, Issue 1, Art No e53177, January 2013.

CHAPTER 8. ANGIOSPERMS

1. Barbara Ehrenreich, 'What a cute Universe you have!', *TIME* magazine, August 25, 1997.
2. S. K. Mukherjee and R. E. Litz, 'Introduction: Botany and Importance', in The Mango: Botany, Production and Uses (2nd Edition), Richard E. Litz (Ed.), Oxfordshire, UK: CAB International, 2009, p. 2.
3. Jonathan Silvertown, *An Orchard Invisible: A Natural History of Seeds*, Chicago and London: The University of Chicago Press, 2009, p. 126.
4. Peter Crane, Ginkgo: *The Tree Time Forgot*, Yale University Press, 2013, p. 22.
5. Silvertown, *An Orchard Invisible*, p. 126.
6. Bruce H. Tiffney, 'Vertebrate Dispersal Of Seed Plants Through Time', *Annual Review of Ecology, Evolution, and Systematics*, Vol. 35, 2004, pp. 1–29.
7. Dong Ren et al, 'A Probable Pollination Mode Before Angiosperms: Eurasian, Long-Proboscid Scorpionflies', *Science*, Vol. 326, No. 5954, 2009, pp. 840-847.
8. Silvertown, *An Orchard Invisible*, p. 126.
9. David Dilcher, 'Toward a new synthesis: Major evolutionary trends in the angiosperm fossil record', *PNAS*, June 20, Vol. 97, No. 13, 2000, pp. 7030–7036.
10. Michael D. Crisp and Lyn G. Cook, 'Cenozoic extinctions account for the low diversity of extant gymnosperms compared with angiosperms', *New Phytologist*, Vol. 192, 2011, pp. 997–1009.

11 Charles Darwin, in *More Letters of Charles Darwin*, F. Darwin and A.C. Seward (eds.), London: John Murray, 1879, Vol. 2.

12 Ruth A. Stockey, Sean W. Graham, and Peter R. Crane (eds.), 'Darwin Bicentennial: The 'Abominable Mystery'', *American Journal of Botany*, Vol. 96, No. 1, 2009.

13 T. Jonathan Davies et al, 'Darwin's abominable mystery: Insights from a supertree of the angiosperms', *PNAS*, February 17, Vol. 101, No. 7, 2004, pp. 1904-1909.

14 Richard J.A. Buggs, 'The origin of Darwin's 'abominable mystery'', *American Journal of Botany*, Vol. 108, No. 1, 2021, pp. 22-36.

15 Luca Comai, 'The advantages and disadvantages of being polyploid', *Nature Review Genetics*, Vol. 6, No. 11, 2005, pp. 836-846.

16 Sharda Khandelwal, 'Chromosome evolution in the genus Ophioglossum L.', *Botanical Journal of the Linnean Society*, Vol. 102, No. 3, 1990, pp. 205–217.

17 Yuannian Jiao et al, 'Ancestral polyploidy in seed plants and angiosperms', Nature, Macmillan Publishers Limited, May 5, Vol. 473, 2011, pp. 97-102.

18 Jeffrey A. Fawcett, Steven Maere, and Yves Van de Peer, 'Plants with double genomes might have had a better chance to survive the Cretaceous–Tertiary extinction event', *PNAS*, April 7, Vol. 106, No. 14, 2009, pp. 5737–5742.

19 Patrick S. Herendeen et al, 'Palaeobotanical redux: revisiting the age of the angiosperms', Nature Plants, Vol. 3, No. 3, 2017, Article No 17015.

20 Hong-Tao Li et al, 'Origin of angiosperms and the puzzle of the Jurassic gap', *Nature Plants*, Vol. 5, 2019, pp. 461–470.

21 EM Friis, P R. Crane, and K. R. Pedersen, *Early Flowers and Angiosperm Evolution*, Cambridge University Press, 2011, p. 16.

22 Peter R. Crane, Else Marie Friis, and Kaj Raunsgaard Pedersen, 'The origin and early diversification of angiosperms', *Nature*, Vol. 374, No. 6517, 1995, pp. 27-33.

23 Friis, Crane and Pedersen 2011, *Early Flowers and Angiosperm Evolution*, pp. 50-54.

24 Ayelet Salman-Minkov, Niv Sabath, and Itay Mayrose, 'Whole-genome duplication as a key factor in crop domestication', *Nature Plants*, Vol. 2, 2016, Article No. 16115.

25 Sarah P. Otto, 'Sexual Reproduction and the Evolution of Sex', *Scitable by*

Nature Education, 2008, Article No. 1(1):182.

26 Peter Schulte et al, 'The Chicxulub Asteroid Impact and Mass Extinction at the Cretaceous-Paleogene Boundary', *Science*, Vol. 327, No. 5970, 2010, pp. 1214-1218.

27 Claire M. Lorts, Trevor Briggeman, and Tao Sang, 'Evolution of fruit types and seed dispersal: A phylogenetic and ecological snapshot', *Journal of Systematics and Evolution*, Vol. 46, No. 3, 2008, pp. 396–404.

28 Kjell Bolmgren and Ove Eriksson, 'Fleshy fruits—origins, niche shifts, and diversification', *Oikos*, Vol. 109, 2005, pp. 255-272.

29 D. W. Snow, 'Evolutionary Aspects Of Fruit-Eating By Birds', *Ibis*, Vol. 113, No. 2, 1971, pp. 194-202.

30 John N. Thompson and Mary F. Willson, 'Evolution Of Temperate Fruit/Bird Interactions: Phenological Strategies', *Evolution*, Vol. 33, No. 3, 1979, pp. 973-982.

31 Omer Nevo et al, 'Chemical recognition of fruit ripeness in spider monkeys (*Ateles geoffroyi*)', *Nature: Scientific Reports*, Vol. 5, 2015, Art No. 14895.

32 Omer Nevo et al, 'Fruit Odor as A Ripeness Signal for Seed-Dispersing Primates? A Case Study on Four Neotropical Plant Species', *Journal of Chemical Ecology*, Vol. 42, No. 4, 2016, pp. 323–328.

33 J. K. Brecht and E. M. Yahia, 'Postharvest Physiology', in *The Mango: Botany, Production and Uses* (2nd Edition), Richard E. Litz (Ed.), Oxfordshire, UK: CAB International, 2009, p. 505.

34 Charles H. Janson, 'Adaptation of Fruit Morphology to Dispersal Agents in a Neotropical Forest', *Science*, Vol. 219, No. 4581, 1983, pp. 187-189.

35 Rachel L. Jacobs et al, 'Less is more: lemurs (*Eulemur spp.*) may benefit from loss of trichromatic vision', *Behavioral Ecology and Sociobiology*, Vol. 73, No. 2, 2019, Article No. 22.

36 Yoav Gilad et al, 'Loss of Olfactory Receptor Genes Coincides with the Acquisition of Full Trichromatic Vision in Primates', *PLOS Biology*, Vol. 2, No. 1, 2004, pp. 120-125.

CHAPTER 9. PRIMATES

1 Gerry Everding, 'Obituary: Robert Sussman, professor of anthropology, 74', *The Source*, Washington University in St. Louis, USA, June 9, 2016.

2 Erika Lorraine Milam, *Creatures of Cain: The Hunt for Human Nature in*

Cold War America, Princeton University Press, 2019.

3   Donna Hart and Robert Sussman, *Man the Hunted: Primates, Predators, and Human Evolution* (Expanded 2009 Edition), Westview Press, Boulder, Colorado, 2005, p. 19.

4   'In Memoriam: Robert Sussman', Department of Anthropology, Washington University in St. Louis, 9 June 2016, <https://anthropology.wustl.edu/articles/1683>.

5   D. T. Rasmussen, 'Primate origins: Lessons from a neotropical marsupial', *American Journal of Primatology*, Vol. 22, No. 4, 1990, pp. 263–277.

6   F. S. Szalay, 'The beginnings of primates', *Evolution*, Vol. 22, No. 1, March 1968, pp. 19-36.

7   M. Cartmill, 'Arboreal adaptations and the origin of the order Primates', in *The Functional And Evolutionary Biology Of Primates*, R Tuttle (Ed), Chicago, Aldine, 1972, pp. 97-122.

8   M. Cartmill, 'Rethinking primate origins', *Science*, Vol. 184, No. 4135, 1974, pp. 436-43.

9   Robert W. Sussman, 'Primate Origins and the Evolution of Angiosperms', *American Journal of Primatology*, Vol. 23, 1991, pp. 209-223.

10 Matt Cartmill, 'New Views on Primate Origins', in *Evolutionary Anthropology*, Volume 1, No. 3, 1992, pp. 105-111.

11 Robert W. Sussman, 'How Primates Invented the Rainforest and Vice Versa', in *Creatures of the Dark: The Nocturnal Prosimians* (Eds. L. Alterman, Gerald A. Doyle and M. Kay Izard), 'Proceedings of an International Conference on Creatures of the Dark: the Nocturnal Prosimians' held June 9-12, 1993, in Durham, North Carolina, Springer Science+Business Media, New York, 1995, pp. 1-10.

12 Jonathan I. Bloch and Doug M. Boyer, 'Grasping Primate Origins', *Science*, Vol. 298, November 22, 2002, pp. 1606-1610.

13 Robert W. Sussman, D. Tab Rasmussen and Peter H. Raven, 'Rethinking Primate Origins Again', *American Journal of Primatology*, Vol. 75, 2013, pp. 95–106.

14 Rasmussen, 'Primate origins: Lessons from a neotropical marsupial', pp. 263–277.

15 Robert W. Sussman, D. Tab Rasmussen and Peter H. Raven, 'Rethinking

Primate Origins Again', *American Journal of Primatology*, Vol. 75, 2013, pp. 95–106.

16 Jonathan I. Bloch et al, 'New Paleocene skeletons and the relationship of plesiadapiforms to crown-clade primates', *PNAS*, Vol. 104, No. 4, 2007, pp. 1159–1164.

17 Connor J. Burgin et al, 'How many species of mammals are there?', *Journal of Mammalogy*, Vol. 99, No. 1, February 1, 2018, Pages 1–14.

18 J. D. Mollon, "'Tho' she kneel'd in that place where they grew...' The uses and origins of primate colour vision', *Journal of Experimental Biology*, Vol. 146, No. 1, 1989, pp. 21–38.

19 Mariano Bond et al, 'Eocene primates of South America and the African origins of New World monkeys', *Nature*, Vol. 520, No. 7548, 2015, pp. 538–541.

20 Josh Gabbatiss, 'The monkeys that sailed across the Atlantic to South America', *BBC.com*, January 26, 2016.

21 Grant Allen, *The Colour-Sense: Its Origin and Development. An Essay In Comparative Psychology*, Houghton, Osgood & Company, Boston, USA, 1879, p VI.

22 Gerald H. Jacobs, 'Evolution of colour vision in mammals', *Philosophical Transactions of Royal Society B*, Vol. 364, No. 1531, 2009, pp. 2957-2967.

23 'About Colour Blindness', *Colour Blind Awareness*, <http://www.colourblindawareness.org/colour-blindness/>

24 Katherine Mancuso, 'Gene therapy for red-green colour blindness in adult primates', *Nature*, Vol. 461, No. 7265, 2009, pp. 784–787.

25 J. Neitz and M. Neitz, 'Evolution of the circuitry for conscious color vision in primates', *Eye*, Vol. 31, No. 2, 2017, pp. 286–300.

26 Alison K Surridge, Daniel Osorio and Nicholas I Mundy, 'Evolution and selection of trichromatic vision in primates', *Trends in Ecology and Evolution*, Vol. 18, No. 4, 2003, pp. 198-205.

27 Mark A. Changizi, Qiong Zhang and Shinsuke Shimojo, 'Bare skin, blood and the evolution of primate colour vision', *Biology Letters*, Vol. 2, No. 2, 2006, pp. 217–221.

28 Nathaniel J. Dominy and Peter W Lucas, 'Ecological importance of trichromatic vision to primates', *Nature*, Vol. 410, No. 6826, 2001, pp. 363-366.

29 Peter W. Lucas et al, 'Evolution And Function Of Routine Trichromatic Vision In Primates', *Evolution*, Vol. 57, No. 11, 2003, pp. 2636-2643.
30 B. C. Regan et al, 'Fruits, foliage and the evolution of primate colour vision', *Philosophical Transactions of the Royal Society B*, Vol. 356, No. 1407, 2001, pp. 229-283.
31 Mary F. Willson and Christopher J. Whelan, 'The Evolution of Fruit Color in Fleshy-Fruited Plants', *The American Naturalist*, Vol. 136, No. 6, 1990, pp. 790-809.
32 Claire M. Lorts et al, 'Evolution of fruit types and seed dispersal: A phylogenetic and ecological snapshot', *Journal of Systematics and Evolution*, Vol. 46, No. 3, 2008, pp. 396-404.
33 Renske E. Onstein et al, 'Palm fruit colours are linked to the broad-scale distribution and diversification of primate colour vision systems', *Proceedings of the Royal Society B*, Vol. 287, Art 20192731, 2020.
34 Pedro Jordano, 'Fruits and Frugivory', in *Seeds: The Ecology of Regeneration in Plant Communities*, 3rd Edition, Robert S. Gallagher (Ed), CAB International, Oxfordshire, UK, 2013, pp. 19-20.
35 R. D. Martin, 'Vertebrate phylogeny: Are fruit bats primates?', *Nature*, Vol. 320, 1986, pp. 482-483.
36 A. R. DeCasien, S.A. Williams and J.P. Higham, 'Primate brain size is predicted by diet but not sociality', *Nature Ecology & Evolution*, Vol. 1, No. 5, 2017..
37 Robert Dudley, 'Evolutionary Origins Of Human Alcoholism In Primate Frugivory', *The Quarterly Review of Biology*, Vol. 75, No. 1, 2000, pp. 3-15.
38 Robert Dudley, *The Drunken Monkey: Why We Drink and Abuse Alcohol*, University of California Press, 2014.
39 Ryan Wallace, 'What Drives Monkeys to Drink—The Fruit-Filled Tale Of Why We Imbibe', *The Science Times*, 2 December 2014.
40 Nathaniel J. Dominy, 'Ferment in the family tree', *PNAS*, Vol. 112, No. 2, 2015, pp. 308-309.
41 Matthew A. Carrigan et al, 'Hominids adapted to metabolize ethanol long before human-directed fermentation', *PNAS*, Vol. 112, No. 2, 2015, pp. 458-463.
42 K. J. Hockings et al, 'Tools to tipple: ethanol ingestion by wild chimpanzees using leaf-sponges', *Royal Society Open Science*, Vol. 2, Art No 150150, 2015.

43 Edward Slingerland, *Drunk: How We Sipped, Danced, and Stumbled Our Way to Civilization*, Little, Brown Spark, 2021.
44 Dudley, *The Drunken Monkey*, p. xii.
45 Andrew F. Smith, *Sugar: A Global History*, 'The Edible Series', Reaktion Books Ltd, 2015, pp. 128.
46 James Walvin, *Sugar: The World Corrupted From Slavery To Obesity*, Robinson, 2017, pp. 325.
47 Ibid., p. 286.
48 Frederick Kaufman, 'How Fruit Became So Sugary', *NPR*, in an interview published on the Internet on 7 October 2018.
49 Katherine Ellen Foley, 'Humans have bred fruits to be so sweet, a zoo had to stop feeding them to some animals', *Quartz*, 1 October 2018.
50 '"This made my Eid", says Delhi mango seller overwhelmed by donations from people after loot', *ANI*, 24 May 2020

CHAPTER 10. GENETICS

1 Michael E. Ruse, 'Are there Laws in Biology?', in *Australasian Journal of Philosophy*, Vol. 48, No. 2, 1970, pp. 234-246.
2 Lindley Darden (ed.), 'Proceedings of the 1996 Biennial Meetings of the Philosophy Of Science Association Part II: Symposia Papers', in *Philosophy of Science*, Volume 64, Number 4, Supplement, 1997.
3 Pawan K. Dhar and Alessandro Giuliani, 'Laws of biology: why so few?', in *Systems and Synthetic Biology*, Vol. 4, No. 1, 2010, pp. 7-13.
4 John S. Wilkins, *Species: A History of the Idea*, University of California Press, 2009, p. 222.
5 F. J. Ayala, 'Darwin's greatest discovery: Design without designer', *Proceedings of the National Academy of Sciences*, 104 (Supplement 1), 2007, pp. 8567–8573.
6 M. Sankaran et al, 'Botany of Mango', in *The Mango Genome*, Chittaranjan Kole (ed.), Springer Nature Switzerland, 2021, p. 25.
7 D. K. Sharma and R. N. Singh, 'Self incompatibility in mango (*Mangifera indica* L.)', in *Horticultural Research*, vol. 10, 1970, pp. 108–115.
8 S. K. Mukherjee and R. E. Litz, 'Introduction: Botany and Importance', in *The Mango: Botany, Production and Uses* (2nd Edition), Richard E Litz (ed.), CAB International, Oxfordshire, UK, 2009, p. 5.

9   R. N. Singh, *Mango*, ICAR Low-Priced Books Series, No. 3 (Revised in 1990), Indian Council of Agricultural Research, Delhi, 1978, p. 23.
10 T. L. Davenport, 'Reproductive Physiology', in *The Mango: Botany, Production and Uses* (2nd Edition), Richard E Litz (Ed), CAB International, Oxfordshire, UK, 2009, p. 119.
11 C. P. A. Iyer and R. J. Schnell, 'Breeding and Genetics', in *The Mango: Botany, Production and Uses* (2nd Edition), Richard E Litz (Ed), CAB International, Oxfordshire, UK, 2009, p. 76.
12 Davenport, 'Reproductive Physiology', p. 115.
13 Ibid., p. 98.
14 Ibid., p. 131
15 William R. Chaney, 'Growth Retardants: A Promising Tool for Managing Urban Trees', *FNR-252-W Purdue Extension*, Department of Forestry and Natural Resources, Purdue University, West Lafayette, Indiana 47907, USA, 2018.
16 Shailendra Rajan et al, 'Genetic Resources in Mango', in *The Mango Genome*, Chittaranjan Kole (ed.), Springer Nature Switzerland, 2021, p. 65.
17 Singh, *Mango*, p. 31.
18 S. K. Mukherjee, 'Mango: Its Allopolyploid Nature', in *Nature*, Vol. 166, 1950, pp. 196-197.
19 Nagendra K. Singh et al, 'The Genome Sequence and Transcriptome Studies in Mango (*Mangifera indica* L.)', in *The Mango Genome*, Chittaranjan Kole (ed.), Springer Nature Switzerland, 2021, p. 175.
20 Ibid.
21 K. V. Ravishankar et al, 'Assessment of Genetic Diversity of Mango (*Mangifera indica* L.) Cultivars from Indian Peninsula Using Sequence Tagged Microsatellite Site (STMS) Markers', in *Proceedings of the IXth International Mango Symposium*, (Ed. Ping Lu), in *Acta Hortic*, vol 992, 2013, p. 269.
22 S. K. Mukherjee and RE Litz, 'Introduction: Botany and Importance', in *The Mango: Botany, Production and Uses* (2nd Edition), Richard E. Litz (Ed), CAB International, Oxfordshire, UK, 2009, p. 9.
23 Victor Galan Sauco et al, 'Mango Propagation', in *The Mango Genome*, Chittaranjan Kole (ed.), Springer Nature Switzerland, 2021, p. 33.

24 Y. Aron et al, 'Polyembryony in mango (*Mangifera indica* L.) is controlled by a single dominant gene', in *HortScience*, Vol. 33, 1998, pp. 1241–1242.
25 D. K. Sharma and P. K. Majumder, 'Further studies on inheritance in mango', in *Acta Horticulturae*, Vol. 231, 1988, pp. 106–111.
26 Singh, *Mango*, p. 23.
27 Ram S. Kulkarni et al, 'Geographic variation in the flavour volatiles of Alphonso mango', in *Food Chemistry*, Vol. 130, 2011, pp. 58–66.
28 Iyer and Schnell, 'Breeding and Genetics', p. 89.
29 Shailendra Rajan et al, 'Genetic Resources in Mango', in *The Mango Genome*, Chittaranjan Kole (ed.), Springer Nature Switzerland, 2021, p. 60.
30 James C. Scott, *Against the Grain: A Deep History of the Earliest States*, Yale University Press, 2017, pp. 89-90.
31 Ram Chandra Jena and Pradeep Kumar Chand, 'Multiple DNA marker-assisted diversity analysis of Indian mango (*Mangifera indica* L.) populations', in *Scientific Reports*, Vol. 11, Article No 10345, 2021.
32 I. S. Yadav and S. Rajan, 'Genetic resources of mango', in KL Chadha and OP Pareek (eds.), *Advances in Horticulture*, Vol. 1, Malhotra Publishing, New Delhi, 1993, pp. 77–93.

CHAPTER 11. ECOLOGY

1 Cary Fowler and Pat Mooney, *The Threatened Gene: Food, Politics and the Loss of Genetic Diversity*, The University of Arizona Press, 1990, p. xi.
2 J. E. Pena et al, 'Pests', in *The Mango: Botany, Production and Uses* (2nd Edition), Ed. Richard E Litz, CAB International, Oxfordshire, UK, 2009, pp. 317-318 [Of these, 127 (39%) are foliage feeders, 87 (27%) are fruit feeders, 36 (12%) feed on the inflorescence, 33 (10%) inhabit buds and 39 (12%) feed on branches, the trunk and roots.].
3 R. P. Srivastava, 'Pests of mango', in *Mango Cultivation*, RP Srivastava (ed), International Book Distributing, Charbagh, Lucknow, India, 1998, pp. 175–299 [Hoppers, stem borers, shoot borers, fruit flies, stone weevil, leaf miners, termites, scale insects, thrips, aphids and ants are 11 of the major pests.].
4 D. Prusky et al, 'Fruit Diseases', in *The Mango: Botany, Production and Uses* (2nd Edition), Ed. Richard E Litz, CAB International, Oxfordshire, UK, 2009, p. 214.

5. R. C. Ploetz and S Freeman, 'Foliar, Floral and Soilborne Diseases', in *The Mango: Botany, Production and Uses* (2nd Edition), Ed. Richard E Litz, CAB International, Oxfordshire, UK, 2009, p. 232.
6. J. E. Pena et al, 'Pests', in *The Mango: Botany, Production and Uses* (2nd Edition), Ed. Richard E Litz, CAB International, Oxfordshire, UK, 2009, p. 318..
7. NAIP sub project on Mass Media Mobilization, DIPA and IIHR, Benguluru. As viewed on <https://icar.org.in/node/2027>.
8. Hemanth K N Vasanthaiah et al, 'Influence of Temperature on Spongy Tissue Formation in 'Alphonso' Mango', *International Journal of Fruit Science*, Vol. 8, No. 3, 2008, pp. 226-234.
9. Galan Sauco, 'Physiological Disorders', *The Mango: Botany, Production and Uses* (2nd Edition), Ed. Richard E Litz, CAB International, Oxfordshire, UK, 2009, p. 307.
10. V. Ravindra and S. Shivashankar, 'Spongy tissue in Alphonso mango. II. A key evidence for the causative role of seed', *Current Science*, Vol. 91, No. 12, 2006, pp. 1712-1714.
11. S. Shivashankar, 'Physiological Disorders of Mango Fruit', *Horticultural Reviews*, Vol. 42, 2014, pp. 313-348; Seshadri Shivashankar et al, 'Premature seed germination induced by very-long-chain fatty acids causes jelly seed disorder in the mango (Mangifera indica L.) cultivar 'Amrapali' in India', in The Journal Of Horticultural Science And Biotechnology, Vol. 91, No. 2, 2016, pp. 138–147.
12. S. Shivashankar et al, 'Do seed VLCFAs trigger spongy tissue formation in Alphonso mango by inducing germination?', *Journal of Biosciences*, Vol. 40, No. 2, 2015, pp. 375–387.
13. S. Prabhudesai, 'Special Quality Production of Mango in Coastal Zone of Maharashtra, India', *Advanced Agricultural Research & Technology Journal*, IMC-2018, Special, Vol. III, No. 1, January 2019, pp. 67-73.
14. J. S. Katrodia, 'Spongy tissue in mango-causes and control measures', *Acta Hortic*, Vol. 231, 1989, pp. 814-826.
15. Prabhudesai, 'Special Quality Production of Mango in Coastal Zone of Maharashtra, India'.
16. Galan Sauco, Richard E Litz (ed.), 'Physiological Disorders', The Mango: Botany, Production and Uses (2nd Edition), CAB International, Oxfordshire, UK, 2009, p. 307.

17 Y. R. Chanana, J.S. Josan, and P.K. Arora, Malik et al (eds.), 'Evaluation Of Some Mango Cultivars Under North Indian Conditions', *Proceedings of the International Conference on Mango and Date Palm: Culture and Export*, 2005, University of Agriculture, Faisalabad, Pakistan, 2005.
18 Muzaffar A. Talpur and Rab Dino Khuhro, 'Relative Population of Mango Hopper Species on Different Mango Varieties', *Journal of Asia-Pacific Entomology*, Vol. 6, No. 2, 2003, pp. 183-186.
19 Iyer and Schnell, 'Breeding and Genetics', p. 78.
20 R. J. Jr. Knight, et al, Richard E Litz (ed.), 'Important Mango Cultivars and their Descriptors', *The Mango: Botany, Production and Uses* (2nd Edition), CAB International, Oxfordshire, UK, 2009, p. 60.
21 P. D. Kamala Jayanthi and Abraham Verghese, 'Studies on differential susceptibility of selected polyembryonic varieties of mango to oriental fruit fly, Bactrocera dorsalis (Hendel)', in *Pest Management in Horticultural E.*
22 Iyer and Schnell, 'Breeding and Genetics', p.78.

CHAPTER 12. CULTIVATION

1 Shamani Joshi, 'A Couple Accidentally Grew the World's Most Expensive Mango', Vice.com, 18 June 2021.
2 Sunita Dubey and Kushal Pal Singh Yadav, 'US fines Dow Chemicals for bribery by Indian subsidiary', *Down To Earth*, 14 March 2007.
3 US Securities and Exchange Commission, *Accounting and Auditing Enforcement Release No. 2554 / February 13, 2007, Litigation Release No. 20000 / February 13, 2007, Securities and Exchange Commission v The Dow Chemical Company, Civil Action No. 07CV00336 (DDC)*, as viewed on the internet on April 24, 2020, <https://www.sec.gov/litigation/litreleases/2007/lr20000.htm>.
4 Bhartesh Singh Thakur, 'Court discharges Dow as CBI fails to attach proof', in *The Hindustan Times*, Chandigarh, dated 11 May 2014.
5 Major Uses Of Pesticides (registered under the Insecticides Act, 1968) up to 31.10.2019, Protection, Quarantine & Storage, Central Insecticide Board & Registration Committee NH-IV, Faridabad, Haryana, under Department of Agriculture, Cooperation & Farmers Welfare, Ministry of Agriculture & Farmers Welfare, Government of India, 2019.

6   Ritwik Mukherjee, 'Insecticide banned in West widely used here', *The Asian Age*, Kolkata, 11 December 2019.
7   Probodh Chundra De, *A Treatise on Mango*, Raghunath Press Calcutta, 1897, p. 28.
8   Shalendra D. Sharma, *Development and Democracy in India*, Lynne Reiner Publishers, Colorado, USA, 1999, p. 116.
9   B. B. Jadhav et al (eds.), *Hard Rock Mango Plantation*, Dr. Balasaheb Swant Konkan Krishi Vidyapeeth, Dapoli, 2009, p. 4.
10 Ibid., pp. 1-2.

CHAPTER 13. MARKETS
1   Mike Davis, *Late Victorian Holocausts: El Nino Famines and the Making of the Third World*, New York: Verso Books, 2001.

CHAPTER 14. EXPORTS
1   Myles Karp, 'We Were Promised the World's Most Delicious Mangoes. They Never Came', *VICE*, New York City, published online on 7 February 2018
2   'Last Mango in Paris', *Forbes* (magazine), New Jersey, 8 June 2007.
3   Sidhartha, 'India's farm products face US import hurdles', *Times of India*, New Delhi, published on 12 July 2013.
4   Tushar Pawar, 'First batch of mangoes sent to US from Nashik', *Times of India*, 14 April 2022.
5   'India registers 19% growth in mango export to USA in 2023-24 over last year', *Press Information Bureau*, 27 October 2023, https://pib.gov.in/PressReleaseIframePage.aspx?PRID=1971931.
6   Miguel Angel Miranda, 'Forecast of Mexican Mango Exports to the U.S.', *PRODUCEPAY*, 28 February 2022.
7   Miguel Angel Miranda 2023, 'Mango Market Situation in the United States and Latin America', *PRODUCEPAY*, 14 March 2023.
8   'EU bans Indian Alphonso mangoes, 4 vegetables from May 1', *The Hindu*, 28 April 2014.

CHAPTER 16. WEST
1   Klaus Karttunen, 'GRACIAS, Caetano', *Who Was Who of Indian Studies*, https://whowaswho-indology.info/2368/gracias-caetano-francisco-xavier/.

2   V. M. Chovatiya *et al*, 'Bio-Chemical Evaluation Of Mango (*Mangifera Indica* L.) cv. Kesar At Different Locations In Saurashtra Region (Gujarat)', *Journal of Horticulture* (OMICS International), Vol, 2, Issue 4, 2016.

CHAPTER 17. NORTH
1   Shaju Philip, 'Feeling the Gulf: Meet the men who returned from Saudi Arabia in wake of its 'nitaqat' law', *Indian Express*, 22 December 2013.

CHAPTER 18. EAST
1   Stephen Henningham, *A Great Estate and Its Landlords in Colonial India: Darbhanga 1860-1942*, New Delhi: Oxford University Press, 1990, pp. 17-20.

Index

aam balaai (maana ka salva), 22
aamer maachhi (mango fly), 214
aam panna (a drink made of raw mango pulp), 21–2, 68, 308, 329
Aam Phir Baura Gaye (Mango Trees Flower Again, essay by Hazari Prasad Dwivedi), 68
aam ras, xi, 310–11
Abdullah, Faruque, 105, 344
Abdullah Nursery, 319
abiotic stress, 185, 201-202, 205, 297
absentee landlords, 223–4
acetylene, use of, 241
Achaya, K. T., 27
ADH4 (alcohol dehydrogenase class IV), 155
Adivasi communities, living inside forests, 53
Adventures of Princes Amarasena and Varasena, The, 83
aerosol spraying, of synthetic insecticides, 198
Afaq, Sheikh Mohammed, 313
Afghan Civil War, 108
African great apes, 155
Aga Khan Trust for Culture (AKTC), 108
Agarwal, Anil, 159
agri-business, 211
Agricultural and Processed Food Products Export Development Authority (APEDA), 249, 255
Agricultural Produce Marketing Committee (APMC), 228, 231, 241, 303
agricultural techniques in medieval India, 28
agri-markets, 227
Agrosurg Irradiators, 248
Aishwarya mango, 320
Akbar, Emperor, 34-35, 102-103, 210, 356
Akhand Gondwana, 113
Akhilesh (Yadav) mango, 320
alcohol, xi, 63, 97, 287, 302
    addiction, 157
    consumption, 155-156, 157-158
    fruits containing, 154
Aligarh Muslim University (AMU), 78-79,
Allahabadi, Akbar
    mango-related anecdotes of, 68
all-India mango conference, 293
Alphonso (Hapus) mangoes, 3–7, 10–11, 33, 168, 180, 201, 211, 250, 279, 290, 295, 359, 361
    of Devgad, 233, 242
    grown by Konkan's growers, 217
    mango wine made of, 157
    productivity of, 224
    risk of 'spongy tissue' in, 297
#alphonso-vs-dashehri, 72
alternate bearers, 168, 173, 181, 218
Ambani, Dhirubhai, 8, 208
Ambani, Mukesh, 8

Ambekhana (House of the Mango), 105
Ambewale, Desai Bandhu, 212
Ambika (Jain yakshi/goddess), 84
Amin, Shahid, 78
Amitabh (Bachhan) mango, 320
amrais (traditional mango groves), 21, 36, 56, 210,
  Hijdon Ki, 51
Amrapali (mango woman), 61-62, 83, 175
Amrapali mango, 174-176, 192, 200, 270, 294, 324, 329,
Anacardiaceae, 62, 119, 121-122
An Agricultural Testament (1940), 52
Anandbagh Botanical Gardens (Darbhanga), 105, 357
Anand Kanan (forest of joy), 41
ancient cult, of the mango, 48
Andheria Bagh (Delhi), 40
Andijan nashpati, 98
angiosperms, 136, 140, 149, 162
Anil Farms and Nursery, 304
Animal and Plant Health Inspection Service (APHIS) (US Department of Agriculture), 249
'anna prashan' ceremony, 23
Anpadh (film), 95
anthracnose (fungal blight), 189–90, 201, 229
anthrax, 189, 191, 229
antsy mango growers, 213
Anwar Rataul mango, 110, 313–14
aperitif effect, 155
arboreal mammals, 134
archaea, 132
Aristotle, 95
arsenic poisoning, risk of, 241
Arthashastra (Kautilya), 77
artificial ripening of fruits, 240

Ashoka Chakra, 81
Ashtadhyayi (Panini), 77
Asian Agri-History Foundation (AAHF), Secunderabad, 27–8
Aspergillus, 192
Atif, Khan Mohammed, 223
Atimadhuram (extremely sweet), 277–8
Auliya, Nizamuddin, 64
Aurangzeb, Emperor, 104
Aurat (1940), 230
autotrophs, 131
Ayengar, A. S. K., 303
Azmatgarh Palace (Moti Jheel Mahal), 330

Babur, Emperor, 100
  Baburnama, 98
  battle with Ibrahim Lodhi, 98
  control of Delhi, 101
  fondness for mangoes, 98, 101, 103
  gardens in Kabul, 108
  plantmania, 99
  yearning for melons, 101
bacteria, 155
  Age of Bacteria, 132
  Brucella, class of, 274
  green-coloured, 131
  pathogenic, 186
  production of oxygen, 131–2
  Xanthomonas campestris, 201
Bada Bagh, 42
Badami mango, of Karnataka, 12, 250, 264
Badam mango, of Maharashtra, 12
Badayuni, Shakeel, 43
bael (*Aegle marmelos*) leaves, 60
Bajoria, Lata, 16
Bajpai, Himanshu
  mango-related anecdotes of, 68

Bajrang mango, 9, 210–11
Baloch, Rafique, 304
Bambaiya (Bombay Green), of Bihar, 334, 355
Banaras Hindu University, 161, 327–8
Bang, Ashok, 172
Banginapalle mango, of Andhra Pradesh, 12, 175, 236, 265
banyan tree, 77
BARC Krushak Kendra, 247
'bare-skin theory' of colour vision, 154
Battuta, Ibn, 92, 97
beeju mangoes, 36
Begum, Qudsia, 335
Begum, Shamshad, 42
Bemaraha woolly lemur (*Avahi cleesei*), 144
Bend It Like Beckham (2002), 62
Benishan12
, 264, 266, 278, 281
Bernier, Francois, 103
beverage, made of fruit pulp, 21
Bhabha Atomic Research Centre (BARC), 247
Bhagwadgita, 59
Bhandarkar Oriental Research Institute, Pune, 26
Bhatambrekar, Prakash, xvii, 48
Bhide, Vivek, 229, 231, 257
Bhopal gas disaster (1984), 213
biennial bearers, 168
Bihar Agricultural University, 346
Biju mangoes, of Bihar, 358
biodiversity hotspots, in Southeast Asia, 119
Biodiversity International, 322
bioinformatics, 175
biological turnover, between the Cretaceous and Paleogene, 142
biotechnology, 57

'biotic' stress, 185, 196, 202, 205
Birbal Sahni Institute of Palaeosciences (BSIP), Lucknow, 114–15, 122
black mangoes, 98
black marketeering, 230
black mould, 192
blue flies, 214
Book of Genesis, 99
Boricha, Kapil, 304
Braganza, Miguel, 290
breast milk, 158
Brihadaranyaka, 76
Brihatsamhita (Varahamihira), 26–7, 50
British East India Company, 35, 41, 96, 104, 357
British forest laws, 85
British Raj, 229
zamindari system of, 221
Bronowski, Jacob, 121
brucellosis, 274
Buddhist literature, references of mangoes in, 81–3
Budhwar, Kusum, 64
Burondkar, Murad, 174
Busa, Roberto, 30
Bush, George W., 246
Bushido: The Soul of Japan (Nitobe Inazo), 71
business process outsourcing (BPO), 297

calcium carbide, use of, 241–2
Cambrian Explosion, 132
Carabao mangoes, of the Philippines, 256
Carboniferous, 133
Carnatic music, 63
Cartmill, Matt, 148
cash-crops, 93

Cashew tree, 26
castor beans, 134
cattle breeders, 273
Cautley, Proby, 26
Central Asian fruits, 103, 105–6
Central Bureau of Investigation (CBI), 213
Central Coastal Agricultural Research Institute, in Old Goa, 20
Central Institute for Subtropical Horticulture (CISH), Lucknow, 187, 216, 320
Chaman (town on Afghanistan-Pakistan border), 107
Champa mango, 337, 345
Champaran Satyagraha, 350–1
Champa tree (*Magnolia champaca*), 354
Chandragupta, King, 164
Charakasamhita, 63
Chaudhary, Ashok Kumar, 345
Chaunsa mango, 335, 344
Chausa, Battle of (1539), 95–6
Chausa mango, 12, 96, 311, 333–5
chauvinist militarism, symbol of, 71
cheetah plate (tectonic Ferrari), 123
cherry blossom festival, in Japan, 70–1
cherry saplings, state gifts of, 71
chhavani, 49
ChinaGAP certification system, 258
Chiroptera, 143
Chittoor, mango's agronomy in, 276
chlorpyrifos, 212–13
Chopra, Deepak, 232
Chowdhrani, Sarladevi, 90
chowk poorna, 75
Christianity, spread of, 30, 47
chromosomes, 138
Chundra De, Probodh, 358
chutneys, 39

Cities and Canopies: Trees in Indian Cities (2019), 48
civil rights movement, 146
'climacteric' fruit, 240
climate change, 140
Clive, Robert, 26
Collingham, Lizzie, 101
Colloquies on The Simples and Drugs of India (1563), 33
colonialism, ecological impacts of, 94
colour-coded ripe fruit, 153
The Colour-Sense: Its Origin and Development (1879), 150
Columbian Exchange, The (1972), 28–9, 124, 150
Columbus, Christopher, 29
commission agents, 226, 239
community forests, 94
Congress Socialist Party, 88
consumption of mangoes
  factors influencing, 13
  in Indus Valley Civilization, 114, 130
  in Mughal Empire, 98, 101, 103
contractors, 221
Cooper, Joseph, 129
Covid-19 pandemic, 70, 161, 249
cow dung compost and urine, 258
Craddock, Paul, 27
Crawford Market (Mumbai), 6, 12–13, 233
  Muslim fruit traders at, 235
Cretaceous-Paleogene extinction event, 139
crop plants, 186
Crosby, Alfred, 28
Crow, Sylvia, 99
cultivation of mango, 15, 114
cultural activism, 312
custodian of genetic diversity, 285

Cyclone Hudhud (2014), 271–2

Dagadusheth Halwai Ganapati temple (Pune), 8
da Gama, Vasco, 29
Dancing in the Streets: A History of Collective Joy (2006), 57
da Orta, Garcia, 33
Darbhanga (Bihar), 19
Darogah-e-Bagh (inspector of gardens), 103
Dar-Sifat-e-Amba (In the Praise of the Mango), 67
Darwin, Charles, 136, 141
  On the Origin of Species (1859), 136, 165
Dashehri mango, of Uttar Pradesh, 10, 12, 20, 175, 201, 236, 315, 323, 333
Das, Rai Krishna, 60
Dastan-e-Aam, 68
day-time predators, 148
de Albuquerque, Afonso, 33
Deccan Plateau, 140
deforestation, in India, 94
Delhi Press Club, 10
Development and Democracy in India (1999), 222
Devgad's mangoes, 301
Dharmachakra (the wheel of Dharma), 81
Dhirubhai Ambani Lakhibag Amrai, 210
Diamond, Jared, 114
dicotyledons (dicots), 137
diffuse coevolution, notion of, 149
Dinga mango, 320
Dipterocarpaceae, 119, 121
diseases, affecting mangoes
  black mould, 192
  black root rot, 193
  black spots, 189–90, 192
  calcium deficiency, 200
  chilling injury, 271
  fungal infestation, 191
  internal fruit breakdown (IFB), 200
  jelly seed, 200
  mango malformation, 191
  powdery mildew, 192
  stem-end rot, 192
Dixit, Ramesh, 223
DNA, 137–9
  Junk DNA, 140
  of primate species, 155
Doodhia Gola of Amroha and Malihabad, 314
Doodhiya/Dhulia Malda mango, 15, 350, 352
Doodhiya Langda of Varanasi, 326, 350–2
Doodh-Pedha mango, 278
dosimeters, 248
Dow Chemical Company, 212–13
Dr. Balasaheb Sawant Konkan Krishi Vidyapeeth, in Dapoli, Maharashtra, 174
dried fruits, 107
drunken monkey hypothesis, 154, 157
dry-nosed (*haplorhini*) simians, 144
Dudley, Robert, 154
Dursban insecticide, 212
Dwivedi, Hazari Prasad, 68

Ehrenreich, Barbara, 57, 156–7
Elaichi mango, 201
Electra Glide motorcycle, 247
electrified fence, for protection of mango crops from monkeys, 196–7
Elkunchwar, Mahesh, 51
endosulfan pesticide, 283

endosymbiosis, 132
English Civil War, 34
Eomangiferophyllum damalgiriensis, 116
erotic art and pornography, 152
ethanol, 156
    caloric value of, 155
Ethephon, 242
ethylene gas, used for ripening of fruits, 240–1, 255
Etruscan Places, 47
eucalyptus wood, 116
EurepGAP certification system, 257
European 'Age of Discovery', 28
European Union (EU), ban on the import of Indian mangoes, 256
even-toed ungulates (*Artiodactyla*), 124
evolutionary biology, 130, 136
export of mangoes, 255

Faiz, Faiz Ahmed, 68
'false ashok' (*Polyalthia longifolia* or *Monoon longifolium*), 69–70
farmers' training programme, 263
farmer-trader relationship, 231
farm houses, 40
Far West Technology in Santa Barbara, California, 248
fast food industry, 159
fast-moving consumer goods (FMCG), 225
Feroze Shah Kotla (Delhi), 92
fertility, mango as symbol of, 64
fertilizers
    inorganic, 205
    organic, 205
Feynman, Richard, 122
Fischer, Louis, 89
flavour in mango, factors influencing, 12

flower-feeding animal, 151
flower-fruit schema, 135–6
flowering plants, 136
    age of polyploidy of, 139
    diversity of, 136–7, 140
    survival in the extinction event, 141
    whole genome duplications in, 139
fly traps, 199, 206
foliovory hypothesis, 152
Food and Agriculture Organization (FAO), UN, 118, 257
food items, made of raw mangoes, 21
food party, mango-themed, 338
food poisoning, 86
food safety, 248
    European concerns over, 257
    Food Safety and Standards Authority of India (FSSAI), 241
    US norms on, 249
food security schemes, 86
food trails, 10
forest-dwellers, of ancient India, 53
forest dwelling communities, 85, 94, 121
forest laws, violation of, 85
fossils, of marine animals, 132
fragrance and aroma of ripeness, of mangoes, 158
Frampton, Ian, 45
French East India Company, 96
frenemy fungi, 186
fructose, 24
fruit-bearing plants, 106
fruit-bearing trees, 43, 223
fruit belts, in UP, 317
fruit borer, 188, 214
fruit-eating animal, 151
fruit-eating arboreal mammals, 148
fruit-eating bats, 143, 153
fruit fly (*Drosophila melanogaster*),

187, 197–8, 245, 256
  ways to control, 198–9
fruit juice, 107
'Fruit King' brand, 218
fruit of immortality, 83
fruit-primate relationship, 146
Fruit Research Station in Sangareddy, Telangana, 170, 361
fruits, gift of, 101
Fukuoka, Masanobu, 308
fungal infestation, 191–2, 205
fungi and pathogens, 134
Fusarium moniliforme, 191
Futane, Karuna, 310–11

Gadaba tribe, of Odisha, 62
Gadgil, Madhav, 47
gametes (sex cells), 138
gamma rays, use for restricting germination of onions in transit, 247
Gandharva, Kumar, 59, 65
Gandhi, M. K., 89
  relationship to mangoes, 90
gardening practices, of the British Raj, 70
Garden of Eden, 46
Garden-of-eight-paradises, 101
Gardens of Mughal India (1972), 99
Geet Vasanta, 59
genetic inheritance, 164, 179
genetic mutation, 155–6
genome mapping, 177, 179
Ghalib, Mirza
  comparison of the mango with a sealed glass of honey, 334
  fascination with mangoes, 67
Ghildiyal/Gildiyal, Rushina Munshaw, 6, 238
Ghorapade, Sujay, 235

giant mealybug, 193
'Gidwala' mango, 313
gift of mangoes, 5–6, 8
  as bribe, 21
  influence on geopolitics, 89
  during Jamai Shashthi festival, 18
  from Nehru to Lohia, 88
  during Saraswati Puja, 18
Gilaas mango, 324–5
Girnar hill, 302
Gir National Park, 303–4
Global Environment Facility (GEF), 322
GlobalGAP (Global Good Agricultural Practices) certification system, 257–8
global health problem, 160
global South, 186
global warming, 123
Goa Inquisition, 32
Goan mangoes, 14, 20
  Mankurad mango, 20
Gode, Parashuram Krishna, 26
Gond tribe, of central India, 86, 113
Gondwanaland (Gondwana), 113, 124
Gopal, Lallanji, 92
Gould, Stephen Jay, 132
grafting, technique of, 26–7, 354
  and agricultural techniques in medieval India, 28
  of Jesuits, 35
  as mode of improving plants, 28
  modern practices of mango grafting, 28
  under Mughal administration, 27–8
  origins in ancient India, 27
  stone-grafting of mangoes, 291
graft mangoes, developed from local varieties, 26

Graham, Maria, 34
grain-producing grasses, 162
Grand American Touring line, 246
great famine (1876–78), 229
Green Essentials, 288–9
Green Revolution, 231
gudhmi, 205
Gunjate, R. T., 174, 209
Gupta, Hanuman Prasad, 239
Guru, Ghanshyam, 51
gymnosperm conifers, 186
gymnosperms, 133, 135–6, 141, 186

habitat selection theory, 44
Hali, Altaf Husain, 67
hanami (international tourism event), 70–1
Hande, Ashok Seth, 233
Haneef, Mohammed, 52
Haque, Maulana Mazharul, 350
"Haramzada" mango, 313
Harimela (an island in the middle of the ocean), 83
haritaki (*Terminalia chebula*), 77
Harita mango, 271
Harley-Davidson (H-D) motorcycles, 244–5, 249
  'mango-for-H-D' deal, 246
  US protection of, 245
harvesting of mangoes, 198, 219, 221, 284, 324–5, 342, 351
Hashmi, Sohail, 10, 39–40
  mango picnic, 43
hattha (digital counter), 232
hemibiotrophy, 190
herbivorous animals, 134
hermaphrodite flowers, 169
Hero Motocorp, 250
higher plants, 136–7
high-yielding seeds, 231

hijras (transgender persons), 51–2
hill glory bower (*Clerodendrum infortunatum*), 195
hilsa fish (*Tenualosa ilisha*), 353
Hilsapeti mango, 353
Himalayan range, 124
Himayat mango, 270
Himsagar mango, 12, 18–19
Hind Swaraj, 89
Hiran River, 304–5
Historia Plantarum (Enquiry into Plants), 95
Hiuen Tsang (Chinese monk), 81
Hobson-Jobson, 26
Holstein-Friesian (H-F) cows, 273–5
Homo sapiens, 46, 166
honey, 23
honey mushroom, 190
Hooker, Joseph, 64, 105, 119, 136, 357–8
  Dictionary of Economic Products of India, The, 358
  Flora of British India, 357
hormone ethylene, released while ripening by fruits, 190, 240
horticulture, xvii, 13, 15, 93, 105, 131, 203, 223–4, 333, 342
  commodities, 107
  conditions for growing mangoes, 208
  in DirghaHrada region, 349
  in Kathiwada, 307
  Mughal promotion of, 356
  production, 107–8
  promotion of, 317
Howard, Albert, 52
howler monkeys (*Alouatta palliata*), 152
human behaviour, evolutionary study of, 146

Mangifera indica

human resource management, 163
human spring festivities, 56
Humayun, Emperor, 96, 99–100, 102, 108, 333–4
Hussain, Javed, 251
Hussain, Tanveer, 12, 251
hydrological cycle, 94

Ibrahim, Dawood, 7
ice caps, melting of, 123
Ignatius of Loyola, Saint, 30
Imam Pasand mango, of Andhra Pradesh and Tamil Nadu, 14, 175, 270, 276–8, 281, 337
Imperial Economic Botanist, 52
independent genome duplications, 139
Indian agricultural practices, 52
Indian Agricultural Research Institute (IARI), 264
Indian classical music, ragas of, 65
Indian Council for Agricultural Research (ICAR), 24, 168, 322
Research Complex for Goa, 292
Indian Food: A Historical Companion (1994), 27
Indian Institute of Horticultural Research (IIHR) Bengaluru, 277
Indian life and culture, role of mango in, 73–5, 79
Indian National Congress, 350
Indian obsession, with mangoes, 204
Indian Yellow pigment, 64
indigenous communities, 85
Indo–US trade deal, for export of mangoes, 248
'mango-for-H-D' deal, 246
industrial development and mining, 47, 361
industrial farming, 283

Indus Valley Civilization, 114, 130
Inigo of Loyola, 30
insect-hunting bats, 153
insecticides
    chlorpyrifos, 212–13
    Dursban, 212
    endosulfan, 283
    Ethephon, 242
    impact of indiscriminate use of, 214
    organic, 283
    paclobutrazol (PBZ), 173, 217–18, 242
    spurious, 213
    synthetic, 283
integrated pest management (IPM), 188
inter-cropping of herbs, 257
internal fruit breakdown (IFB), 200
International Crops Research Institute for the Semi-Arid Tropics (ICRISAT), Hyderabad, 188
International Treaty on Plant Genetic Resources for Food and Agriculture (2001), 118
Interstellar (2014), 129
irradiation plant, for irradiation of onions, 247
irradiation, process of
    BARC Krushak Kendra, 247
    for extending the shelf life of foods, 248
    for food safety, 248
    killing of insects and microbes by, 248
    for mangoes, 248
    to prevent the germination of onions, 247
Islamic law, associated with eating of mangoes, 78

jackfruit pulao, 311
Jadhav, Shivraj Singh, 307–8
Jadoo (Hindi film), 42
Jafar, Mir, 96
Jagat Seth, 15
Jahangir mango, 278
Jainism, 83
Jain tirthankars (founders and teachers), 83
Jaiswal, Shyam Sunder, 13, 41
Jamaican sugar plantation, 159
Jamai Shashthi, 18
Jambudwipa, 113
James, LeBron, 160
jamun (Indian blackberry), 113
jelly seed, 200
Jesuits' Bark, 31
Jha, Binodanand, 91
Jonko, Kunwar Singh, 84–5
Joshi, Vidyadhar, 234
Judaism, 33
juicy mangoes, 36
Junagadh Agriculture University, 302
Jungle Andolan, 85
Junoon (1978), 318
Justicia adhatoda, 240

Kabuliwala (film), 106
Kaduwa mango, 353
Kaggalahalli mango market, 251
kairi-bhaat (a pulao made with raw mangoes), 311
Kala Amb Park, 98
Kala Ashram, 56
Kalidasa (Sanskrit poet), 66
Kali mango, from Gujarat, 311
Kalpavriksha, 76, 84
Kamadeva (god of love), xii, 58–9, 63, 66, 74, 168
Kamasutra (Vatsyayana's ancient Sanskrit treatise), 27, 62–3
Kapooriya mango, 15, 329
Karimbhai, Faiyyumbhai, 10
Kathiwada, 306–8
Kaufman, Frederick, 160
Kausar Bagh, 266–7, 361
Keh-Mukarni, 65
Kesar mango, 9–10, 209–10, 250, 303–5, 361, *see also* Reliance Kesar
Khajuraho, 64
Khanji III, Mahabat, 303
Khan, Shabbir Hasan, 68
Khan, Syed Ahmad, 78
kheere-ki-kachori, 341
Khusro, Amir, 65
Kishan Bhog mango, 277
Kohitoor mangoes, 17, 337, 340, 344
Kol revolt (1831), 85
Kond tribe, of Odisha, 62
Kosambi, Damodar D, 25, 90
Kostermans, Andre, 121
K-Pg boundary, 139–42
Krishen, Pradip, 94
Krishi Vigyan Kendra, 265
Krishna, Lord, 59
Krishna Raja Sagara dam, 283
Kumar, Arun, 106
Kumar, Deepika, xv
Kumar, Manoj, 230
Kumar, Nitish, 346
kutumb-no ambo (the clan's mango tree), 305

Laduwa mango, 344
Laingda mango, in eastern Uttar Pradesh, 15
Lakhi Bagh (Darbhanga), 103, 105, 210–11, 356
Lalit Narayan Mithila University, Darbhanga, 358

land ceiling laws, 222
land deforested, for agriculture, 309
land mafia, 295
Langda mango, 19, 42, 161, 177, 281, 332, 334, 337
   of Bihar, 14
   of Uttar Pradesh, 10, 12, 14, 326
Lewis, Martin, 125
Liddle, Swapna, 40
Living Farms, in Odisha, 49
Lodhi, Daulat Khan, 100
Lodhi, Ibrahim, 98, 100
Lohajang mango, 353
Lohia, Ram Manohar, 58, 88
London Vegetarian Society, 89

'M13-1' mango, from Israel, 210
Madana Panchami, 60
Madanotsava (Vasantotsava), 59, 69, 71
Mahabharata, references of mango in, 77
Maharajah of Darbhanga, 26
Maharashtra State Agricultural Marketing Board (MSAMB), 249
Mahavira, Lord, 83
Mahidpur, Battle of (1817), 41
Malaviya, Madan Mohan, 328
Malda mango, 15, 346, 351, 355
Male Annihilation Technique (MAT), 199
Malihabad (Uttar Pradesh), 20
Malihabadi, Josh, 68, 318
Malihabadi mango, 315, 325
Mallika mangoes, 174, 201, 278
Malwi mangoes, 37, 310–11, 363
mammals, 142, 145
M. andamanesis, 201
Mandi Parishad, in Uttar Pradesh, 227
mandi tax, 227

Mandovi River, 32
Mane, Abhimanyu, 258–9
Mangifera indica trees, 28, 118, 122, 171, 176
Mangifera laurina, 181
mango
   agronomy, xvii, 276
   Babur's fondness for, 98, 101, 103
   bagh, 52
   as best fruit of Hindustan, 103
   as 'climacteric' fruit, 240
   common word for, 95
   cultivation, 15, 114
   as divine gift, 86
   eating contests, xvi
   flowering, 68
   flowers, 61
   as gesture of detente in statecraft, 89, 92, 108–10
   gift of, 88–9
   grafting, 25–6
   Indian obsession with, 204
   juice, xii
   kernel cakes, 86
   kitchen, 50
   linking with
      Alexander, 95
      Buddha, 83
      Mahavira, 83
   markets, 9
   medicinal values, 77
   megafossils of, 142
   Mughal obsession with, 98, 101, 103, 104–5
   origin of, 113, 122, 186
   passion of Awadh's nawabs and Pathan warriors, 91
   pests of, 187
   pickles, 22, 39
   as political spectacle in China's

National Day Parade of 1968, 109
pollen fossils of, 142
as qaumi or awaami fruit, 91, 94
in Ramayana and Mahabharata, 77
recipes, 7
references in Buddhist literature, 81–3
season, 9
sold in chhabisi (lots of twenty-six), 331
species of, 118
types of, 24, 81
wholesalers, 6
mango badlands, 221
mango breeders, 170, 179, 201
mango-breeding experiments, 174
mango crops, 304, 341
mango cultivation, 223, 295
mango culture, of Bengal, 17
mango diplomacy, 89, 92, 108–10
   between India and China, 109
   between India and Pakistan, 110–11, 313
   between India and US, 109
   'mango-for-H-D' deal, 246
   between Pakistan and China, 109
   political and non-political, 111
mango ecosystem, 195
Mangoes of Goan Origin (1997), 292
mango traders, 240
mango festivals, xvii, 19, 284, 358
   in Lucknow, 21
   of Murshidabad Heritage Development Society (MHDS), 337
'mango-for-H-D' deal, 246
mango groves, 44, 51, 89, 361
   battles fought among, 94, 96
   Buddha's stay at, 84

Chunda's grove, 84
culture of, 48
Kala-e-Aam, 98
Laksha Bagh, 96
mango growers
   in Chorao Island, 34
   in Uttar Pradesh, 168
mango hoppers, 187, 201
mango husbandry, spread of, 107
mango inflorescence, 56, 70–1
Mango in Indian Life and Culture, The (1995), 73
mango leaf webber (*Orthaga euadrusalis*), 189
mango leaves
   impact of cool temperatures on, 170
   role in Indian life and culture, 74–5
   used for decoration, 79
mango malformation, 191, 201
mango-marauding monkeys, 196–7
mango markets
   black marketeering, 230
   in China, 256
   Crawford Market (Mumbai), 6, 12–13, 233, 235
   emergence of, 223
   Kaggalahalli mango market, 251
   makeshift, 225
   mandi tax, 227
   mango marketing chain, 240
   non-tariff barriers in the US, 245
   retail sales of mangoes, 236
   return on investment, 226
   in US, 256
   Vashi market (Navi Mumbai), 229, 238, 247
mango orchards, 4, 43, 79, 204
   of Awadh in Uttar Pradesh, 35
   cutting down for real estate

    development, 317
    in Digha, 350
    as feasting grounds of insect pests, 224
    in Goa, 21
    of grafted trees, 25
    jungle paalna, 207
    at Lakhi Bagh (Darbhanga), 103, 105
    of Malihabad (Uttar Pradesh), 20
    Noor Bagh, 312, 315
    planting of, 222
    protection from polluting industries, 317
    of Rataul, 43
    Reliance Industries Ltd (RIL), 8
    of Vennangupattu in Kanchipuram district of Tamil Nadu, 278
    wedding between, 79
mango seed
    act of planting, 168
    high-yielding seeds, 231
    as a sign of rebirth, 76
mango trade, 255–6
    access to European markets, 257
mango trees, xv, xix, 17, 54, 77–8, 83, 306
    Alphonso-bearing, 34
    beeju/tukhmi tree, 25
    biennial bearing, 221
    chemical defence to tackle their natural enemies, 206
    grown from grafting, 25
    as India's favourite festive tree, 78
    as indigenous farmer varieties, 283
    kalami tree, 25
    Kalepadu, 277
    Kuddus, 277
    Lalbaba, 277
    marriage of, 79
    planting of, 92
        along the Grand Trunk Road, 92
        along the railroad, 94
    registered with the National Bureau of Plant Genetic Resources, 283
    revelry and penance performed under, 64
    sacredness of, 80
    Thorapadu, 277
    wild mango trees, 122
mango varieties
    of North Bihar, 354
    of North India, 160
#mangowars, 36, 72
mango wine, made of Alphonso, 157
mango wood
    fire of, 74
    use for cremation, 73–4
mango writing
    gold standard of, 66
    by Kalidasa, 66
    by Mirza Asadullah Beg Khan 'Ghalib', 66–7
Manikchand, Seth, 15
Mankurad mango, 34, 175
    of Goa, 20
Mankurad trees, in Goa, 291
'Man the Hunter' culture, 147–8, 154
Mao's Golden Mangoes and the Cultural Revolution (2013), 109
Maries, Charles, 26, 105, 357, 359
    Cultivated Mangoes of India, 357
marsupials, 150
marva fruit, 305
Maslow's pyramid, 6
Matrapaksha, 163
matrimonial issues, role of mango in, 79
Mauryan Empire, 164
mausambi (sweet lime), 285

M. captosperma, 201
McHugh, James, 63
medicinal plants, 33, 77, 240
medicinal values, of mangoes, 77
megafossils, 120–2, 142
Meghalaya, 130
Mehrotra, Rakesh C., 115–16, 121
Meiji revival, 70
Mendel, Gregor, 137
Menezes, Vivek, 32
Midsummer Night's Dream, A (William Shakespeare), 167
Mirzapur television series (2018), 330
Mirza, Saeed Akhtar, 11
Mishra, Anupam, 50
Miskut pickle, 292
Miyan, Ghazi, 78
Miyazaki mangoes, of Japan, 207, 256
Miyazaki prefecture, 207–8
Mochi Bagh, *see* Moti Bagh (Delhi)
Modi, Narendra, 110, 347
Mojave Desert, 209
moneylenders, 230
monoculture, practice of, 257
monotremes, 150
mood-related disorders, 45
Morde, Ram, 237
Morris, Desmond, 232
Mother India (1957), 230
mother's milk, nutrition of, 158
Moti Bagh (Delhi), 40
Motichandra, Raja, 330
Moti Jheel, 330–1
Mughal Empire, 357
Mughal garden, 35, 99
Mughal obsession, with the mangoes, 98, 101, 103, 104–5
Mukharji, T.N., 64
Mulgoa mango, 277
Mundoli, Seema, 48

Muneer, R. M., 251
Murck, Alfreda, 109
Murshidabad's mangoes, 15–17, 19, 105, 341
history of, 105
muskmelon, 103
Mussolini, Benito, 47
The Myth of Continents: A Critique of Metageography (1997), 125

Nagendra, Harini, 48
naghz tareen mewa-e-hindustan (the most supreme fruit of Hindustan), 65
Naidu, Sarojini, 90
NaMo mango, 320
Narayana sage, 62
Narayan, Badri, 78
Nargis mango, 320
Narmada River, 51
Nathwani, Parimal, 208
National Institute of Standards and Technology (NIST), US, 248–9
Native American tribes, 85
natural farming, 51, 308, 310
natural selection, theory of, 166
nature's farming, 53
Nava Rasas (nine essences of ancient aesthetics), 61
Navlakha, 41
Neelum mango, 174–5, 201–2, 279
Nehru, Jawaharlal, 233
gift of mangoes to Ram Manohar Lohia, 88
Neminatha (Jain tirthankara), 84
Nene, Yeshwant L, 27
New World monkeys, 150
'New World' of the Americas, 29
Niranjan mango, 9, 210
Nireekshak journal, 305

nocturnal predators, 148
non-tariff barriers (NTBs), 245, 256
Noorjahan mango, 306–8
Noorjahan Mango Garden, 307
North–South mango debate, 14
Novitiate of Chorao, 32

obesity, global epidemic of, 160
odd-toed ungulates (*Perissodactyla*), 124
Ohnuki-Tierney, Emiko, 70
oil-based perfume, 339
oil palm, 188
Old Testament, 46
'Old World' of Africa and Asia, 29
One-Straw Revolution, The, 308
onions, export of, 247
optichromic dosimeter, 248–9
orange cultivation, in Vidarbha region, 21, 308
organic certification, 294
organic farmers, 53
organic farmers' club, 290
organic farming, 187, 283
organic farming movement, 52
organic insecticide, 283
organic mangoes, 284
organochlorides, 199
organophosphates, 198
Oriental fruit fly, 198
Oxygen: The Molecule that Made the World (2002), 132

paclobutrazol (PBZ), 173, 217–18, 242
Palace, Azmatgarh, 330
palaeobotany, 121
Palaeocene-Eocene Thermal Maxima (PETM), 123–4, 142–3, 152
Palashi, Battle of (1757), *see* Plassey, Battle of (1757)
pancha indriya (five senses), 77
pancha kosh, 77
panch amrita, 77
pancha mukhi, 77
panchavati, 77
panch mahabhoot (five elements), 77
Pandey, Ghanshyam, 187
'Pan-Gondwana' movement, 113
Panipat, battle of
  first battle (1526), 98
  third battle (1761), 98
Parashara, 92
Pari Bagh, 361
Patel, Vallabhbhai, 88
Patidar, Daya Chandra, 307
peepal tree (*Ficus religiosa*), 76–7, 80
Permanent Settlement, 93
Persian melons, 102
pesticides, 205–6, 212–14, 216, 221, 226, 231, 242, 266, 292, 342
  biopesticides, 273, 284
  chlorine-containing, 194
  endosulfan, 283
  poisonous, 288
  residues in fruits, 199, 245, 256
  synthetic, 211, 282, 284, 305, 311
pests, of the mango, 187–8
  aamer maachhi (mango fly), 214
  anthracnose, 207
  blue flies, 214
  fruit borer, 188, 214
  fruit fly, 187, 197–9, 202–3, 206, 207, 245, 256–7
  mango hoppers, 187
  methods to control
    'biocontrol' measures, 199, 206
    vapour heat treatment, 207, 258
  seed weevils, 187
  stone weevil, 245

tree borers, 187
war against, 214
pheniconazole (fungicide), 191
photosynthesis, 131, 140, 172
Pitrapaksha, 163
placentals, 150
planetary biotic exchange, 124
plant–animal relationship, 134
plant breeding, 170, 269, 272
plant fossils, 148
plant growth retardants, 173
plant lineages, 139
plant systematics, 120
plant taxonomy, 120–1
Plassey, Battle of (1757), 15–16, 96
Plato, 95
platyrrhini, 150
pleasure garden, 279
plesiadapiforms, 142
poison ivy, 134
political utility, of mangoes, 108–10
pollination, among mango trees, 169, 192
polyploids, 138–9, 141
  in domesticated plants, 140
polyspermy, 138
Portuguese empire in India, establishment of, 29
Portuguese Estado de India, 35
powdery mildew disease, 192
Prakash, Gyan, 106
Prasad, Jaishankar, 60–1
Prasad Ki Yaad, 60
Prasad, Rajan, 11
Prasad, Vandana, 122, 124
pregnancy yearning, 59, 62
Premchand, Munshi, 319
  Godaan (1936, *The Gift of a Cow*), 68
primate order, 124, 143, 148, 150

primates
  bare skin theory, 151
  behaviour and ecology of, 146
  body-to-brain size ratio of, 153
  colour-blindness, 151–2
  colour perception, 150
  diurnal fruit-eating primates, 155
  DNA of, 155
  evolution of, 147–8
  and fruit-bearing plants, 147
  howler monkeys (*Alouatta palliata*), 152
  New World monkeys, 150
  owl monkey, 150
  spread of obesity among, 160
  sugar addiction, 160
Puck (aka Robin Goodfellow), 167
Puggalapannatti, 81
pulp-to-refuse ratio, of ripe fruits, 291
purnakumbha (purnakalash), 75, 79

R
Raga Bahar, 65
rainwater harvesting, 309
Raival, 37
Rajapuri mango, of Gujarat, 308
Raj Darbhanga, 19, 105, 357, 362
Ramayana, references of mango in, 77
Ramdas, 40
Rana, Munawwar, 38
Randhawa, M. S., 24, 42, 84
Ranga Rao, G. V., 214
Rangbhari Ekadashi, 60
Rani Pasand mango, 337
Rasgulwa mango, 352
Rasna Vilas (luxury of the tongue), 104
Rasool, Aijaz (Nawab of Sandila), 335
'Rataul' mango, 313–16
Ratna mango, 175, 201

Ratna tree, 310
raw mangoes, 62, 341
    kairi-bhaat (a pulao made with raw mangoes), 311
Ray, Satyajit, 319
Reagan, Ronald, 245
Rebbapragada, Ravi, 51
red-banded mango caterpillar (RBMC), 188
Reddy, B. M. Chandrasekhara, 179
Reddy, T. Chandrasekhar, 275–7
Reliance Fresh, 209
Reliance Industries Ltd (RIL), 208, 210
    mango orchards in Saurashtra, Gujarat, 8
    Reliance Kesar mango, 9
Reliance Kesar, 210
Report on Arboriculture (1892), 93
Report on the Land Revenue Settlement, Nagpur District (1899), 93
reproduction, in microorganisms, 132
reproductive physiology, 171
Reunion Islands, 120
RGB (*Red, Green, and Blue*) colour vision, 153
rhinarium, 144
Rhizopus, 192
ricin, 134
Rihla, 92
Rishi Panchami, 49
Rituraj (king of ritus), 59
Rituraj Mehfil, 59
RMM, 251–2, 254
Roe, Thomas, 102
Romance of the Mango (2002), 63–4
rootstock, 180
Royal Botanical Gardens, Kew, 119, 136, 357
royal gardens, administration of, 103

rufous treepie, 195

saasandari (tombstones), 84
Sabarmati Ashram, Ahmedabad, 90
sabre-toothed smilodons, 134
Sachar, Rajinder, 88
Sachin mango, 320
Sadakat Ashram, 350
Sadhalwa mango, 352
Safeda mango, of North India, 12, 264, 320, 325
Sagai (film), 43
Sahayadris (Western Ghats), 196
sakura, 70
sakura-dan (the cherry blossom bomb), 71
Salim Ali Foundation, 279
sal tree (*Shorea robusta*), 80, 119
Salvi, Bharat R., 12
Samar Behisht mangoes, 96
Samudra Manthan, story of, 76
Sanskrit Var Yatra, 50
Saraswati Puja, 18
Sarcolaenaceae, 119
Sarenga mango, 337
Sarnath, 81
Scheduled Tribes, 53, 85–6, 306
School of Oriental and African Studies, in London, 79
scorpion stings, mango flower as antidote to, 69
Scott, James C., 182
seasonal affective disorder (SAD), 22
season, for growing mangoes, 206
seed-bearing plants (spermatophytes), 133, 135
seed dispersal, by birds, 142–4
seed-grown mangoes
    in Bihar, 358
    versus clonal graft mangoes, 36

notion of wholesomeness associated with, 37
use in religious ceremonies, 37
seed-grown tree
growth of, 180
wood of, 74
seed weevils, 187
self-pollination, 170
Semnan pomegranate, 99
Settlement Report of Budaon District (1873), 52
Sewagram Ashram, in Wardha, 89
Shah II, Akbar, 40
Shah Jehan, Emperor, 34, 164
Shah, Prakashbhai, 305
Shakkar Guthli mango, 267, 277–8
Shakuntala (Kalidasa), 66
'Sharbati Bigrain' mango, 313
Sharif, Nawaz, 110
Sharma, Ravindra, 56, 75, 79
Sharma, Shalendar D., 222
'*Shatranj Ke Khiladi*' short story (1924), 319
shelf life, of mangoes, 13,
of the Mankurad mango, 290
Shivaratri, 60
Shorah-e-Afaq, 313
Siddhartha (Gautam Buddha), story of, 80–4
Silence of the Lambs, The (1988), 193
Silk Road, 100
Silvertown, Jonathan, 135
Sindhu mango, 175
Singh, Lakshmeshwar, 357
Singh, Lal Behari, 91
Singh, Ram Nath, 175, 180, 329
Singh, Vijay, 22
Slate, Nico, 89
Slingerland, Edward, 156
Smith, Andrew F., 158

social brain, 154
social networking, 154
Society of Jesus, 25–6, 30
Spanish Inquisition, 33
spring festival, in Japan, 70
Sthapit, Bhuwon Ratna, 322
Stokes, Eric, 93
stone weevil, 245
sucking-type juicy mangoes, 19, 24, 38
sucrose, 24
Sudha Ras (nectar), 104
sugar addiction, 159
sugar tax, 160
Sugar: The World Corrupted From Slavery To Obesity (2017), 159
sugary carbonated drinks, 160
Sukul mango, of Bihar, 355–6, 358
Sukuta mountain, 83
Sulabh International, xv
Sulabh Swachh Bharat, xv
Suri, Sher Shah, 92, 96, 330
battle against Humayun, 333
Sussman, Robert, 146
Suvarnarekha mangoes, 278
Swagatham mangoes, 271, 272
sweet mango chutney, 292
synthetic insecticides, 187, 198, 213, 214, 283

table-type mangoes, 24, 38, 335
Taft, William Howard, 71
Tagore, Rabindranath, 53, 55
'Taiyo no Tamago' mango, 207
Talala mandi (Saurashtra), 22
Tarash, Aam, 105
Tarikh-e-Salatin-e-Afghaniyah (Ahmad Yadgar), 100
TASA Foods, 251–2
tea gardens, of Assam, 324

Tethys Sea, 113
Thandi Sadak, 42–3, 72
thanwla, 205
Thar Desert, 42
Theophrastus, 95
Theravada, 81
Third World, 186
Tipu Sultan, 283
tokkotai, 71
toran, 79
Totapuri (Bangalora/Collector) mango, of south India, 236, 251–3, 276, 308
tourism, in Goa, 287
transit rot, 192
Treatise on Mango, A (1897), 358
tree borers, 187
tribal customs, 86
tribal people, diets of, 86
Tripathi, Suryakant mango-related anecdotes of, 68
Troll, Christian W., 78
tropical rainforests, 119, 123, 130, 142, 144, 147, 152, 155, 186, 360
Truschke, Audrey, 104
Tryst with Mango, A (2005), 334
Tulsi flowering, 55
turmeric-chilli chutney, 311
types of mangoes, 24, 81

ud-Daulah, Siraj, 96–7, 105
Umrao Jaan (1981), 318
UNEP-GEF-ICAR (United Nations Environment Programme-Global Environment Facility-Indian Council of Agricultural Research) project, 285
Union Carbide Corporation, 213
Union Ministry of Health and Family Welfare, 241

Upkar (1967), 230
Upper Palaeocene, 130
urban greening, 44, 46
urushiol, 134
Urvashi apsara (celestial nymph), 62
US Securities and Exchange Commission, 212

Vaishali, republic of, 83
Valmiki National Park, 355
Vantalamammidi, 50
Vapour Heat Treatment (VHT), 207, 258
Varahamihira, 92
Varanasi, 81
Vasanta Panchami, 60, 68
Vashi market (Navi Mumbai), 229, 238, 247
vatika, 77
Vat Poornima, 242
Vedic ceremonies, use of mango twigs in, 74
Vedic schools, 84
Vegetable Kingdom, 136
venture capital, 61
Venus flytrap, 134
vihara, 81
volatile compounds, in mangoes, 144
Vrikshayurveda, 27
Vrikshayurveda, 27

Wahhabism, 79
Wallace, Alfred Russel, 166
Walvin, James, 159
wasteland, cultivation of, 93
Watt, George, 357
wedding ceremonies, use of mango twigs in, 74
wet-nosed (*strepsirhini*) prosimians, 144

whole genome doubling (WGD),
139, *see also* independent genome
duplications
whole genome duplication (WGD)
events, 176
Wigen, Karen, 125
wild mangoes, 118, 122, 181–3, 202
Wilson, Edward O., 44
World Environment Day (2022), 337

Xavier Centre of Historical Research
(XCHR), Goa, 31
Xavier, Francis, 30
X chromosome, 151
Xuanzang (Chinese monk), 81

Yadgar-e-Ghalib (1897), 67
Yajnavalkya (Vedic rishi), 37
Yajnavalkya Samhita, 37
Yitzhak, Avraham ben, 33

Zafar II, Bahadur Shah, 40, 67
zamindari, abolition of, 221, 335
zamindari system, 93
Zardalu mango, of Bhagalpur, 15,
346, 348, 352, 354
Zarda mango, of Bihar, 19, 355
zari (gold-plated silk thread), 41
Zia-ul-Haq, General, 110
mango diplomacy, 313
Zubair, Mirza Mohammed, 315